Hyperspectral Image Processing

Liguo Wang · Chunhui Zhao

Hyperspectral Image Processing

国防工业出版社
National Defense Industry Press

Springer

Liguo Wang
Harbin Engineering University
Harbin
China

Chunhui Zhao
Harbin Engineering University
Harbin
China

ISBN 978-3-662-47455-6 ISBN 978-3-662-47456-3 (eBook)
DOI 10.1007/978-3-662-47456-3

Jointly published with National Defense Industry Press, Beijing
ISBN: 978-7-118-08646-1 National Defense Industry Press, Beijing

Library of Congress Control Number: 2015941114

Springer Heidelberg New York Dordrecht London

Printed on acid-free paper

Springer-Verlag GmbH Berlin Heidelberg is part of Springer Science+Business Media
(www.springer.com)

Preface

With the rapid development of the modern science and technology, the hyperspectral remote sensing science, as a comprehensive high and new technology, has gained the extensive as well as considerable development in the theory, technology, and application. The hyperspectral remote sensing will acquire the land object information by utilizing the space image and spectral features, to achieve the unity of image and spectrum and provide the powerful technical support for the humans to learn the world and change the world. At present, the hyperspectral remote sensing has been widely applied to the fields such as military affairs, agriculture, forest, meadow, ocean, geology, and ecology. In recent years, the quantity of the remote sensing SCI periodicals and the impact factors have been increased internationally, and the international academic conference specially aiming at the hyperspectrum is gradually developed and continues to mature.

The development degree of the hyperspectral image processing technique directly determines the information acquisition and utilization degree. Our country attaches importance to the development of this technique. The supporting strength of the fields involving 863, 973, and Natural Science Foundation of China is bigger and bigger, and the human powers input by various national colleges and universities and the scientific research institutions are more and more, with more and more abundant achievements. However, this current situation still cannot meet the demands of the scholars on the breadth and depth of the technique. In combination with the research of the author engaging in the related technique for many years, this book carries out the systematic classification and elaborate interpretation for the main processing techniques of hyperspectral remote sensing, i.e., classification, endmember extraction, spectral unmixing, sub-pixel mapping, super-resolution reconstruction, anomaly detection, dimensionality reduction, and the newer research results, for the purpose of contributing the small gains for the readers to understand, learn, and research the hyperspectral image processing technique.

The whole book is composed of nine chapters. Chapters 2–6 are mainly written by the first author Liguo Wang, involving classification, endmember selection, spectral unmixing, sub-pixel mapping, super-resolution reconstruction, and other hyperspectral image processing techniques, which are the research results gained by this author for many years, in the hope that these contents can provide some references or enlightenments for the readers in the thought or method. Chapters 1 and 7–9 are mainly written by the second author Chunhui Zhao. Chapters 7 and 8 are the innovative achievements gained by this author in recent years. In order to facilitate the reader to have a comprehensive understandings on the principles, current situation and main processing technique of the hyperspectral image, this book especially add the contents of Chaps. 1 and 9. Chapter 1 firstly makes the simple introduction to the basic theory of the hyperspectral remote sensing and mainly refers to the related works of Qingxi Tong, Jiabing Sun, and other persons. Then, it makes the systematic introduction to various typical and mainstream processing techniques of hyperspectrum and mainly refers to a lot of domestic and overseas academic literatures. Chapter 9 makes brief introductions to the application of the hyperspectral remote sensing technique. These contents mainly refer to the published articles of the domestic scholars in this field. It is important to note that the main body of this book, namely Chaps. 2–8, regards the creative contents of the author as the textual basis, so the current situation and review of the relevant technique are mainly introduced in Chap. 1. On the other hand, the sequencing determination of various knowledge points takes consideration into their inclusion relations front and back, rather than designing in accordance with the sequence relations from the perspective of the image processing.

A part of job for the contents in Chaps. 2–4 is finished by the first author during studying as a PhD candidate in Harbin Institute of Technology under the guidance of Professor Ye Zhang; therefore, this author especially thanks to his own old school and tutor. At the same time, the contents of Chap. 4 are finished in cooperation with Doctor Xiuping Jia in University of New South Wales. Here, we express the sincere gratitude and respect for her industrious work and rigorous academic spirit. The contents of Chap. 5 are finished by the first author together with his graduate students. The contents of Chaps. 7 and 8 are finished by the second author together with his graduate students. Many graduate students of the authors participate in the research or classification work of some contents, mainly involving Danfeng Liu, Qunming Wang (gainer of outstanding master's thesis of Heilongjiang Province), Chunhong Liu, Feng Mei, Jia You, Lingyan Zhang, Jing Zhang (gainer of outstanding master's thesis of Heilongjiang Province), Yaxin Ji, Chunmei Hu, Luqun Deng (gainer of outstanding master's thesis of Harbin Engineering University), Yan Zhao, Lijuan Sun, Wensheng Zhang, Guofeng Wu, Fangjie Wei, Qian Xiao, Yao Shi, Zhengyan Wang, Jian Tan, Qiong Wang, Liang Zhao, Jinghui Yang, Siyuan Hao, Fanwang Meng, and Jie Sun. Here, we express the gratitude to them. Here, we express the sincere thanks to the authors of references by this book, and especially thank to all authors with the key references in this book. The first author includes Qingxi Tong, Ruiliang Pu, Jiabing Sun, Ersen Li, Chunhong Liu, Wu Ren, Feng Ling, Jianwei Wan, Anguo Teng, Bingxiang Tan, Lei Zhou, Quansheng Lou, Chengkai Pei, Yanjie Yang, Chaoyang Zhang, and numerous experts and scholars in the related fields internationally. Their outstanding achievements provide the favorable materials for the compilation of the summary and introduction of this book. For the above-referenced literatures, due to huge numbers and limited space, this book does not make the further comments. Here, we express the apology to the relevant authors.

On account of the limited ability of the author, the large time span of the research contents, big differences of the programming software and hardware conditions, many involved researchers and other practical problems, the main technical contents of Chaps. 2–8 inevitably have the problems in various levels. The contents of Chaps. 1 and 9 are difficult to make the most reasonable and scientific arrangement for the mainstream technique and typical application, and the comments made can only stand for the statements of a school. We sincerely hope that the readers criticize, correct, and grant instructions. We will further perfect in the subsequent work.

January 2015

Liguo Wang
Chunhui Zhao

Contents

Symbols and Abbreviations

Symbols

ND	The number of features/dimensionalities
Nd	The number of reduced features/dimensionalities
Np, Ntr and Nte	The number of pixels, trainning pixels, and testing pixels
p and *P*	Pixel and the matrix formed by pixels
Nc and Ne	The number of classes and endmembers
e, *E*	Endmember and the matrix formed by endmembers
v, *V*	Endmember difference vector, and the matrix formed by endmember difference vectors
x or s_0	Sample
F	Fractional abundance
Vol	Volume
Dist	Distance

Abbreviations

ASVDD	Adaptive kernel parameter estimation-based support vector data description
FCLS	Fully constrained least squares
HSI	Hyperspectral imagery
LS	Least squares
LSMA	Linear spectral mixing analysis
LSMM	Linear spectral mixing modeling
LSSVM	Least squares support vector machine
LSVM	Linear support vector machine
MAP	Maximum a posterior probability
MPS	Modified pixel swapping algorithm
MRF	Markov random field

MSAM	Mixed spatial attraction model
MSPSAM	Sub-pixel/pixel spatial attraction model
POCS	Projection onto convex sets
PSA	Pixel swapping algorithm
SPSAM	Modified sub-pixel/pixel spatial attraction model
SSRSI	Sub-pixel shifted remote sensing images
SVDD	Support vector data description
SVM	Support vector machine

Abstract

The whole book is composed of nine chapters. To cater to different readers with different requirements, in the first and last chapters, basic theory on hyperspectral remote sensing, development trend of some important techniques for hyperspectral remote sensing, and application of hyperspectral remote sensing are briefly introduced. Chapters 2–8 mainly consists of the authors' research achievements on hyperspectral imagery in recent years, including classification, endmember extraction, spectral unmixing, sub-pixel mapping, super-resolution reconstruction, anomaly detection, and dimensionality reduction. All these techniques are introduced in detail to offer readers complete framework of the mentioned novel knowledge on hyperspectral imagery.

This book can be used for undergraduates and graduates in different domains, including remote sensing, surveying and mapping and geoscience and information system. It can also provide some reference information for researchers at different levels.

Chapter 1
Basic Theory and Main Processing Techniques of Hyperspectral Remote Sensing

In order to enable the readers to have a better understanding of the main body contents of this book, this chapter firstly introduces the basic theory of the hyperspectral remote-sensing technique, involving the electromagnetic wave, solar radiation, imaging spectrometer and spectral imaging modes etc. Then, it introduces the development status of the main technique, involving classification, endmember extraction, spectral unmixing, sub-pixel mapping, super-resolution processing, anomaly detection, dimensionality reduction and compression, etc.

1.1 Basic Theory of Hyperspectral Remote Sensing

Basic theory of hyperspectral remote sensing mainly includes the theoretical basis of electromagnetic wave, interaction of solar radiation and materials, imaging spectrometer and spectral imaging modes, and imaging characteristics of HIS etc. (Tong et al. 2006).

1.1.1 Theory of Remote Electromagnetic Wave

Remote sensing means a technique of the long-range detection and perception for the target or the natural phenomenon without the direct contact (Sun 2003). Remote sensing can judge the land object target and the natural phenomenon in accordance with the collected electromagnetic wave, because all objects possess entirely different electromagnetic wave reflection or emitted radiation features due to different categories, features, and environmental conditions. The remote-sensing technique is mainly established on the principle of the object reflection or electromagnetic wave emission.

The varying electric field can cause the varying magnetic field all round, and the varying magnetic field causes the new varying electric field in further region, and causes the new varying magnetic field in the farther region. The varying electric field and magnetic field are alternated, and the process of transmission in the space

L. Wang and C. Zhao, *Hyperspectral Image Processing*,
DOI 10.1007/978-3-662-47456-3_1

Table 1.1 Electromagnetic wave spectrum

Wave band		Wave length	
Long wave Medium wave and short wave Ultrashort wave		More than 3000 m 10–3000 m 1–10 m	
Microwave		1 mm–1 m	
Infrared band	Ultrafar infrared Far infrared Mid-infrared Near-infrared	0.76 | 1000 μm	15–1000 μm 6–15 μm 3–6 μm 0.76–3 μm
Visible light	Red Orange Yellow Green Black Blue Purple	0.38 | 0.76 μm	0.62–0.76 μm 0.59–0.62 μm 0.56–0.59 μm 0.50–0.56 μm 0.47–0.50 μm 0.43–0.47 μm 0.38–0.43 μm
Ultraviolet rays		$10^{-3} \sim 3.8 \times 10^{-1}$ μm	
X ray		$10^{-6} \sim 10^{-3}$ μm	
γ ray		Less than 10^{-6} μm	

from the near to the distant at the limited speed is called as the electromagnetic wave. The increasing or decreasing arrangement in accordance with the wave length or frequency of the electromagnetic wave in the vacuum shall constitute the electromagnetic wave spectrum divided as per the section (see also Table 1.1). The electromagnetic wave spectrum is arranged from high frequency to low frequency, and can be divided into γ rat, X ray, ultraviolet rays, visible light, infrared ray, and radio wave. In the hyperspectral remote sensing, it mostly uses the visible light and the near-infrared band. The visible light band has very narrow spectrum section, but it is very important to the hyperspectral remote-sensing technique.

1.1.2 *Interaction of Solar Radiation and Materials*

The interaction mechanism of electromagnetic wave and materials is scattered, reflected, emitted, absorbed and transmitted. The electromagnetic radiation on the earth is mainly from the sun. When the solar radiation reaches the ground through the atmospheric layer, some is reflected to the outer space by the atmospheric layer (about 30 %), and the other is absorbed by the atmosphere (about 17 %). Also some is scattered to be diffuse radiation and reaches the earth's surface (about 22 %). Only about 31 % of the solar radiation as the direct solar radiation reaches the earth's surface. This part of the electromagnetic radiation is partly reflected by the earth's surface, and the remaining part is absorbed by the land objects. After the earth's surface absorbs the solar radiation, it will generate self-radiation, known as the earth's radiation. The earth's radiation can be divided into the short wave radiation (0.3–2.5 μm) and long wave radiation (more than 6 μm). The short wave radiation focuses on the reflection of the earth's surface for the sun, and the long wave radiation can only consider the thermal radiation of the object on the earth's surface. When the sun passes through the atmosphere, it will be influenced by the atmosphere on the scattered, reflected, and absorbed effect. When the electromagnetic radiation reaches the earth's surface, it will make three basic interactions of reflection, transmission and absorption with the earth's surface.

1.1.3 Imaging Spectrometer and Spectral Imaging Modes

The imaging spectrum technique can be divided into the prism/raster dispersion type, interference type, optical filter type, computed tomography type, binary optical element type, and the three-dimensional imaging type in theory. According to different imaging means, the imaging spectrometer can also be divided into four types, involving the linear detector and optical mechanical scanning type, area array detector and space push broom type, spectrum scanning type, and the spectrum and space cross-scanning type.

Generally, the multiband scanner divides the visible light and infrared band into several or more than 10 bands. As for the remote sensing, with the range of a certain wave length, the more the divided bands, i.e., the more the sampling points of the spectrum, and all the more it approaches the continuous spectrum curve. Therefore, it can make the scanner obtaining the spectrum composition of the land object while gaining the targeted land object image. This technique of "integration of spectrum and image" which forms the image and also the obtains targeted spectrum curve is called the imaging spectrum technique. The scanner made in accordance with the principle is called as imaging spectrometer.

The hyperspectral imager is the new technique in the remote-sensing progress. The image is composed of very narrow continuous spectral band with hundreds of bands. The spectral band covers all spectral bands of the visible light, near-infrared, mid-infrared and thermal infrared areas. While imaging of the spectrometer, it mostly adopts the scanning type or push broom type, and can collect the data of the band with 200 or above. It enables each pixel in the image to obtain the continuous reflectivity curve, rather than having the interval between the bands like other general traditional imaging spectrometer.

Two main imaging modes of the hyperspectral imager are, respectively, the optical/mechanical scanning mode and push broom-type area array scanning mode. The first array imaging spectrometer should generate above 200 continuous spectral bands. After color separation of the optical dispersion device, different bands of the radiations are exposed to each element of CCD linear array. Therefore, the radiation intensity from the ground instantaneous field of view is recorded by the color separation, and the number of the spectral channels is the same as the number of the linear array elements. During the process of the pixel scanning one by one progressively, it generates the continuous spectrum images composed of hundreds of narrow bands. Such scan-based hyperspectral imager is mainly applied to the aerial remote-sensing detection, and the slower flight speed makes the enhancement of the spatial resolution possible. The later imaging mode is the two-dimensional area array. One dimension is the linear array, and the other dimension is regarded as the spectrometer. The image will record the data line by line, and will not move the elements. The imaging device will measure the radiation intensity of all bands of each pixel in a line horizontally, and the quantity of the detecting elements depends on the quantity of the bands. During working, the radiation energy from the ground is transmitted to the register for recording the data through the shutter exposure. The photoelectric detector adopts CCD or mercury—cadmium—tellurium/CCD hybrid device. The space scanning is finished by the solid state scanning of the device. Due to long camera time of the pixel, the enhancement of the sensitivity and space resolution ratio of the system can be achieved.

In 1983, the first AIS-1 (aero imaging spectrometer-1) was prepared successfully in the American Jet Propulsion Laboratory, and was successfull in the application of mineral mapping, vegetation, chemistry and other aspect, showing a huge potential for imaging spectrometer imaging. Thereafter, it successively prepared the airborne visible infrared imaging spectrometer (AVIRIS) of America relevant to aero-imaging spectrometer, the fluorescence line imaging spectrometer (FLI) of Canada, the small airborne imaging spectrometer (AIS) developed on this basis, MIVIS of Deadaulus Company of America, and 79-band AIS (DAIS-7951) of GER Company, the airborne multipurpose imaging spectrometer (DAISA) of Finland, the reflection-type imaging spectrometer (ROSIS-10H and ROSIS-22) of Germany, and the hyperspectral digital imaging capture the experiment instrument (HYDICE) of the Naval Research Laboratory (NRL). AVIRIS has the largest influence, which is a revolutionary imaging spectrometer and greatly promotes the development of the hyperspectral remote-sensing technique and application. In recent years, some qualified countries the world over compete for putting into the research and application of spectrometer imaging.

Our country has always tracked the development frontier of the international hyperspectral imaging technique, and developed our own hyperspectral imaging system at the middle and later phases of 1980s. Under the support of the major projects such as “the Seventh Five-year Plan” to “the Ninth Five-year Plan” scientific research projects and “863” high-tech development research plan of China, the development of imaging spectrometer in our country experienced the development process from the multi-band scanning to imaging spectrometer scanning, from the optical mechanical scanning to solid-state scanning of area array CCD detector.

During the period of “the Seventh Five-year Plan,” the Chinese Academy of Sciences (CAS) supported the national key project program of the aerial airborne remote sensing use system, and Shanghai Institute of Technical Physics of CAS developed multiple relevant special scanner. These jobs laid the solid foundation for preparing and developing the high performance hyperspectral imaging spectrometer in China. During the period of “the Eighth Five-year Plan,” the successful preparation of new type modular aero imaging spectrometer (MAIS) marked the major breakthrough of the aero imaging spectrometer technique and application in China. Thereafter, the push broom imaging spectral imager (PHI) and utility-type modular imaging spectrometer system (OMIS) voluntarily prepared in our country had occupied the important place in the big family of the world aviation imaging spectrometer, on behalf of the technological level of imaging spectrometer in Asia, and had repeatedly participated in the cooperation with foreign countries and executed the flight missions abroad. See Tables 1.2 and 1.3 for the main technical parameters of PHI and OMIS.

Shenzhou-III unmanned spacecraft launched by China in March 2002 embarked the China moderate resolution imaging spectroradiometer (CMODIS). CMODIS had 34 bands, and the wave length was ranged between 0.4 and 12.5 μm. In the environmental and disaster reduction minisatellite constellation to be launched in China, it also included the hyperspectral remote sensor of 128 bands. In addition, in "Fengyun-3" meteorological satellite and lunar exploration program in the plan of our country, it also included the preparation and development of aero imaging spectrometer. CMODIS embarked by "Fengyun-3" meteorological satellite possessed 20 bands, and the imaging range includes the visible light, near-infrared and thermal infrared. The satellite of "Chang'e—I" will carry the interference imager voluntarily prepared by our country to detect the lunar surface substance.

Xi'an Institute of Optics and Precision Mechanics of CAS is the unit which took the lead in researching interference spectrum imaging theory and techniques systematically and deeply in China and prepared the first airborne engineering prototype of our country successfully. In 2003, Xi'an Institute of Optics and Precision Mechanics had been ordered to prepare the satellite borne hyperspectral imager for "HJ-1-A Satellite" of our country. On September 6, the first star of A and B binary star of the environment and disaster monitoring and prediction minisatellite constellation has been successfully launched in Taiyuan Satellite Launch Center and smoothly entered into the preset orbit. One of the main loads prepared by Xi'an

Table 1.2 Main technical parameters of PHI

Working mode	Area array CCD detector push broom
Field of view	0.36 rad (21°)
Instantaneous field of view	1.0 mrad
Number of bands	224
Signal to noise ratio	300
Spectral resolution	<5 nm
Spectral range	400–850 nm
Number of pixels	367 pixel/line
Spectral sampling	1086 nm
Frame frequency	60 Fr/s
Data rate	7.2 Mb/s
Weight	9 kg

Table 1.3 Main technical parameters of OMIS

OMIS-I			OMIS-II		
Total number of bands		128	Total number of bands		128
Spectral range/μm	Spectral resolution/μm	Number of bands	Spectral range/μm	Spectral resolution/μm	Number of bands
0.46–1.1	10	64	0.4–1.1	10	64
1.06–1.70	40	16	1.55–1.75	200	1
2.0–2.5	15	32	2.08–2.35	270	1
3.0–5.0	250	8	3.0–5.0	2000	1
8.0–12.5	500	8	8.0–12.5	4500	1
Instantaneous field of view/mrad	3		1.5/3 optional		
Total field of view	>70°				
Scan rate/(Line/s)	5, 10, 15, 20 optional				
Number of pixels in a line	512		1024/512		
Data coding/bit	12				
Maximum data rate/Mbps	21.05				
Detector	Si, InGaAs, InSb, MCT linear array		Si linear array, InGaAs unit, InSb/MCT two-color		

Institute of Optics and Precision Mechanics for A star of the constellation, as the first hyperspectral imager prepared by the principle of spatial modulation interference spectrum imaging technique in our country, went into the space with the star. For more than a year, "HJ-1-A Satellite" hyperspectral imager has acquired the mass valid data products. As the important scientific achievement with the independent innovation, the successful preparation of "HJ-1-A Satellite" hyperspectral imager integrated the theoretical innovation, technical innovation, and integrated innovation. It not only created the first satellite borne spatial modulation interference hyperspectral imager with the business operation all over the world, but also firstly put forward the criteria of high throughput static interference spectrum in practice. It proposed and achieved the calibration method of the interference spectrum imager on the ground and satellite and the interference data compression and spectrum retrieval method. At the same time, through the preparation of the equipment, it further promoted the related technological development involving the optical imaging, signal processing, spectral data application, data compression, and CCD application research in our country. In March 2010, High-tech Research and Development Bureau of CAS hosted and convened "HJ-1-A Satellite Hyperspectral Imager" project appraisal meeting in Beijing, which has been assumed and prepared by Xi'an Institute of Optics and Precision Mechanics of CAS. The appraisal committee agreed that, the achievement had the theoretical breakthrough and technical invention, and also had the integrated innovation and successful application, which powerfully promoted the spectral imaging and related technological development of China, filled in the gap in the spaceflight hyperspectral remote sensing field of our country, and possessed the important significance.

1.1.4 Imaging Characteristics of HSI

During the spatial characteristic imaging on the target, the imaging spectrometer forms dozens of and even hundreds of narrow bands on each spatial pixel through the chromatic dispersion, for the purpose of the continuous spectrum coverage, and accordingly forms the hyperspectral image with the nanoscale spectral resolution ratio. While embodying the two-dimensional space vision information of the target image, the hyperspectral image can also reflect the spectral information of the one-dimensional representation pixel physical property of high resolution, i.e. unity of image and spectrum. Due to the enhancement of hyperspectral image resolution ratio, many insoluble problems under the multi-spectrum originally can be solved under the hyperspectrum at present. Through handling the spatial characteristics and spectral characteristics of the target image in the hyperspectral image, it can identify and distinguish the land object target by the higher credibility. It possesses the important application value in the aspects of the remote sensing image such as military reconnaissance, true/false target identification, and battlefield situation evaluation. Three outstanding features of the hyperspectral image can be summarized as follows. Firstly, the hyperspectral image possesses high spectral resolution, to enable it to solve many insoluble problems of multi-spectrum. Secondly, the adjacent bands had stronger relevance. This characteristic provides the possibility for the reduction of dimensionality (including the band selection and feature extraction etc.) and inter-spectrum compression. Thirdly, for the hyperspectral image, with the increase of dimensions, the hypercube volume focuses on the angle end, and the hyper-sphere and spheroid volume focuses on the shell. According to the characteristics of the hyperspectral image and the needs of the related technical processing, the information carried by the hyperspectral data generally adopts the following three space expression modes, i.e. image space, spectrum space and eigenspace (as shown in Fig. 1.1).

Different expressive modes emphasize different information, and suitable for different purposes and requirements. For the human visual system, the image space representation is the most natural and most intuitive expression mode. For each fixed wave length, equivalent to the photo of the ground scenery, two-dimensional image provides the geometrical relationship among the data samples. The image representation is very useful for the mutual positional relations among the overview land objects. For example, during the general training process, the selection of training samples should be made in the image space, therefore, the space geometric positional relations among the pixels are very important for some hyperspectral data processing. In the spectral space expression mode, the spectral response, as the

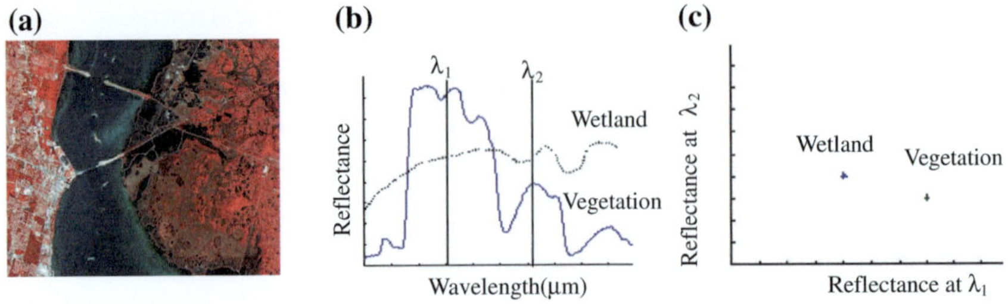

Fig. 1.1 Three expression methods of hyperspectral data. **a** Image space. **b** Spectrum space. **c** Eigenspace

function of the wave length, reflects the variation of the electromagnetic wave energy with the wave length. The spectral response curve includes the information as required by identification of land objects. The expression method is simple and effective, and provides the spectral information directly applied to unmixing. The eigenspace takes the samples in the spectrum space, and expresses the acquired data by the n dimensional vector, which is the other mode of the spectral response. The eigenspace expression mode is easy to understand from the concept, and is convenient from the perspective of mathematics. n dimensional vector contains all spectral information of the corresponding pixel. From the perspective of the information extraction, the eigenspace expression method among three expression methods is most suitable for the application of the mode recognition.

1.2 Classification Technique of HSI

Classification is an important way of acquiring the information, with the aim of making the classification of the land cover types or land cover topics of all pixels in the image automatically. The classification of hyperspectral remote sensing image can be divided into the supervised classifications and unsupervised classifications, parameter classifications and nonparameter classifications, crisp classifications and fuzzy classifications, and other classification methods (Liu 2005). For some classification methods, it may belong to one type as well as the other type.

1.2.1 Supervised Classifications and Unsupervised Classifications

The most common classification is to select the training samples depending on whether needing the priori knowledge of the known image to be classified, and the classification methods can be divided into supervised classifications and unsupervised classifications.

1. **Supervised classifications**

Before the supervised classifications, it is necessary to select the representative training sample region through the field measurement or the real land object map for each land object type, then count the average value, standard deviation, covariance matrix and other statistical characteristics of the training sample regions, and classify the unknown pixel into a certain category in accordance with the corresponding classification discrimination function with these statistical characteristics as the standard. At present, the representative supervised classifications are as follows.

Minimum distance classification: This is a relatively simple and practicable method. Firstly, it utilizes the training samples to calculate the average value and standard deviation of each type, and regards the average value as the central position of this type in the eigenspace, to calculate the distance between each pixel to the center of each type, and then classify each pixel into the type of the minimum distance. While possessing K sample pixels, the unknown pixels will be classified into the advantage type of K pixels. Or, in consideration of the weight, it will bestow on the weight depending on the distance between the training sample pixel and the unknown class (the weight is small if it is far, and the weight is large if it is near), and then the unknown pixels will be classified into the type of the largest weight. Minimum distance method generally adopts the distance discrimination functions, involving Mahalanobis distance, Euclidean distance and Taxi distance etc. This method supposes the equal covariance in all classes, so the classification speed is relatively fast, and the weakness is the unsatisfactory classification effect.

Fisher classification: This method deems the ratio between the inter-class average variance and the sum of intra-class variance as the maximum decision rule. It demands the maximum inter-class distance, i.e. the maximum inter-class average difference, while the inter-class discreteness is the minimum, i.e. the variance quadratic sum is the minimum. This method is consistent with the thought of the minimum distance classification. However, its decision criterion is the linear discrimination function, rather than the distance.

Maximum likelihood discrimination classification: The maximum likelihood method with the identification and classification for the remote sensing image according to Bayes code is the supervised classification method with the most extensive application at present, also known as Bayes classification method. This method is a nonlinear classification method, and makes the discrimination based on belonging to a certain kind of the principle of maximum or minimum probability misclassification. When the projects of various classes in any direction of the eigenspace can not be separated, adopting the processing method of the linear discrimination method does not work at all. Only after establishing the nonlinear classification boundary in the eigenspace, can it gain good effect. Therefore, the maximum likelihood method generally can gain preferable effect.

On basis of the above-mentioned basic supervised classification algorithm, Dundar et al. put forward the Bayes rules upon the kernel function. Landgrebe et al.

utilized the mixed model to express the class density, and the proposed classifier estimated the subclass and class statistics while selecting the best model. In 2002, Chang C et al. put forward Positive Boolean function (PBF) classification method based on Greedy modular eigenspace (GME), and made the simulation experiment. The experimental result indicated that, for the classification pretreatment, GME feature extraction method was applicable to nonlinear PBF multi-class classifier, and the final classification result was better than the traditional principal component analysis (PCA) method. Comprehensively, the current supervised method is mostly improved on the basis of the original method, so as to gain better classification effect.

2. **Unsupervised classifications**

For unsupervised classifications, without the priori knowledge, only according to the characteristics of the image data, it utilizes the similarity and difference of land object spectral characteristics represented by these data for the classification, thus, the unsupervised classification is also called as clustering. The representative unsupervised classifications are as follows.

K mean clustering method: The clustering code is to make the minimum quadratic sum of the distance between the multiple pixel point and the center of such class in each clustering. The basic thought is to move the centers of all classes successively through iteration, till gaining the best clustering result. Such algorithm result is impacted by the factors such as the quantity of the selected clustering center and the initial position, geometric property of the image distribution, and read-in sequence. Moreover, during the iterative process, there is not the measure of adjusting the class numbers. Therefore, it may make different initial classifications gain different results, which is the weakness of this method.

Iterative self-organizing data analysis techniques algorithm (ISODATA) is a dynamic clustering method. Firstly, it provides an initial clustering, and then adopts the iterative method to repeatedly modify and adjust clustering, so as to approach a correct clustering. It has two differences from K mean algorithm. (1) It calculates the mean value of various samples after adjusting all samples, rather than recalculating the mean value of various samples once while adjusting the class of a sample each time. (2) ISODATA algorithm not only can finish the clustering analysis of the sample through adjusting the class of the sample, but also can carry out the class "combination" and "splitting" automatically, and accordingly gain clustering result with the relatively reasonable number of classes.

The parallel pipeline method is based on the spectral characteristic curve of the land objects. Supposing the spectral curve similarity of the similar land objects as the discrimination standard, it sets up a similar threshold value. Thus, the similar land objects are manifested as "pipe" with the center of characteristic curve and the radius of the similar threshold value in the eigenspace. Such clustering method is substantially a heuristic method based on the nearest neighbor rule.

In addition to the above classical unsupervised classification algorithms, the researchers from various countries have also put forward the new algorithms in recent years. In 2004, Lee S et al. proposed the multi-stage grade clustering algorithm, which made the multi-stage grade clustering for the data after the band selection, involving two stages—local and global stages. In 2005, Jimenez et al. evolved ECHO supervised classification algorithm into unsupervised classification algorithm. The main feature of the algorithm was the reconstruction process of the simplified image spatial structure. Independent component analysis (ICA) was not applicable to the high data dimension of hyperspectral image, therefore, in 2004, Du Q et al. applied the image pre-processing into the hyperspectral image, to make ICA algorithm possess the common applicability. In the same year, Shah C et al. improved ICA single algorithm into ICA mixed model, and supposed the category distribution as non-Gaussian distribution. The classification effect of this algorithm on the land object coverage was better than K mean algorithm.

1.2.2 Parameter Classifications and Nonparameter Classifications

1. **Parameter classifications**

Parameter classifications suppose that the class possesses the specific probability distribution function and estimates the distribution parameter. The most typical parameter classification algorithm is the maximum likelihood classification. The nearest mean value method also belongs to the parameter classification, and may be regarded as the special case of the maximum likelihood classifier. The nearest mean value method supposes that the covariance and the priori probability of various training samples are equivalent, and the discrimination function only depends on the mean value of classes and the value of characteristic vector, approximating as linear function. It means that the decision boundary becomes the hyperplane, and becomes the straight line in two-dimensional plane, unlike the decision boundary in the maximum likelihood classifier being the curved surface or the curve in two-dimensional plane. Each pixel will be classified in accordance with the distance between the characteristic vector and the mean vector of various training samples. The precision of the nearest mean value method is not as high as that of the maximum likelihood classification.

The representative parameter classification algorithms proposed recently are as follows. In 2002, Jia X et al. proposed the simplified maximum likelihood classification algorithm, firstly make the principal component transformation for the related data, and then express the category by the clustering space. For the final category mark, the comprehensive decision decides the attribution of the category by the comprehensive decision. In 2004, Gomez-Chova L et al. put forward the comprehensive application of ISODATA, Learning Vector Quantization (VQ), maximum likelihood, neural network and other various methods.

2. **Nonparameter classifications**

The classification with some kind of probability distribution function and estimation of parameters without needing the assumption of the category is called as nonparameter classification. This kind of classifier mainly includes the following kinds.

Parallelepiped classification method: Also known as box classification method. The basic thought is to firstly find out the position and shape of each category in the eigenspace through the data in the training sample zone, and then regard the "box" containing this category as the discrimination function of this category. The discrimination rule is that, if the unknown vector $\boldsymbol{x}$ is fallen into this "box," $\boldsymbol{x}$ belongs to this category, otherwise, it is compared with other box again. This classification method has the misclassification phenomenon in the box overlapping region, and the misclassification probability is related to the comparison of the precedence order of the box.

Neural network classification: The neural network possesses the learning ability and should not make assumptions on the probability model, which is applicable to handling various problems of the spatial pattern recognition. Therefore, the neural network technique increasingly becomes the effective means of the hyperspectral remote sensing image classification processing. The greatest difference of the neural network and the traditional classification method lies in that the artificial neural network is not based on a certain assumed probability distribution, and it gains the network weight through learning the training samples, and forms the classifier. The merits of the neural network classification method is mainly embodied in: (1) it should not make the probability distribution assumption for the original category like the statistical pattern recognition, which is very effective for solving the land objects with the complicated categorical distribution and unclear background knowledge. (2) The discrimination function is nonlinear, and can form the complex nonlinear decision boundary in the eigenspace, and accordingly can solve the division of the nonlinear discriminant eigenspace. However, the relevant parameters of the network are many and need the constant adjustment, so at the learning stage, it should spend plenty of time in analyzing and calculating.

In 2005, Benediktsson J et al. applied the morphology into the hyperspectral image, and made the PCA on the image. These pre-processing algorithms made definite contributions to the final neural network classification. In 2003, Goel P et al. proposed the application of decision tree classification algorithm to distinguish different growth situations of the soybean field, and made the comparison of the final classification precision with the neural network algorithm. The merit of the decision tree method is the more simpleness and possessing the clearer classification rules. In 2004, Mercedes F et al. proposed applying the multilayer feedback network to 224 band hyperspectral image, and made comparison with other methods.

1.2.3 *Crisp Classifications and Fuzzy Classifications*

1.2.3.1 Crisp Classifications

1. **Crisp classifications**

The classification method of assigning some pixel of the image into a certain definite category is called as crisp classifications, also known as hard classifications. The traditional classification methods are crisp classifications.

2. **Fuzzy classifications**

The classification method that each pixel of the image can be assigned to two or more categories at the same time is called as fuzzy classifications. The thought of the fuzzy classifications is based on that the expression of the matters sometimes is not absolutely but has a uncertain fuzzy factor. The hyperspectral remote sensing image also has the fuzziness. Therefore, the fuzzy method can be used for the image clustering. For the fuzzy classifications, it is necessary to firstly select the initial classification center of each category, then carry out iteration by using the membership function, and repeatedly correct the category center and the degree of membership for each pixel vesting in each category. When the difference value of the previous and later iterations is smaller than some small number, stop the iteration, and complete the classification.

1.2.4 *Other Classification Methods*

Spectral angle mapper (SAM) is the spectral classification based on the physics. It refers to the measured standard spectrum in the laboratory or the average spectrum of the known point directly extracted from the image, determines the similarity between two spectrums, and valuates the generalized included angle for each pixel vector and reference spectrum vector in the image. The reference spectrum can be from the training area, and can also be from ASCII document of spectrum library. Each pixel is assigned to the class of the minimum included angle with the reference spectrum. SAM method lays stress on the shape features of the spectrum, and greatly reduces the feature information. In spite of this, for the mass data of hyperspectral image, the classification efficiency is not yet high.

The most typical one of the coding matching method is the spectrum binary coding technique. The algorithm carried out coding for the hyperspectral image through the presetting threshold value. The band brightness value of more than this threshold value is 1, or, it is 0. In this way, each pixel generates a binary coding curve, matches with the binary coding vector in the spectrum library, and calculates the matching coefficient, thus, it determines the land categories in the image. This coding technique can enhance the efficiency of the hyperspectral data analysis and processing, but sometimes it can not provide the reasonable spectrum division. Moreover, during the coding process, it will lose many details and spectral information.

Expert classifier is the comparatively new classification method. In view of one or more hypotheses, it sets up a system rule or decision tree, for the purpose of describing the numerical value or attribute of the variables. The hypothesis, rule, decision tree and variable constitute the knowledge base, and this working procedure is finished by the knowledge engineer. Then the knowledge classifier uses the knowledge base to finish and output the classification.

1.3 Endmember Extraction Technique of HSI

Before setting up the linear mixed model and conducting the spectral unmixing operation, it is very necessary to select the spectral endmember. The good or bad selection of the spectral endmember is the key to the mixed pixel spectral unmixing effect. The selection of the spectral endmember should be representative, and become the category component collection of most pixels in the image. During nearly 10 years, multiple automatic supervised hyperspectral image spectral endmember selection methods have been developed successively. The typical spectral endmember selection techniques include Pixel Purity Index (PPI), N-FINDR algorithm, iterative error analysis (IEA), optical real-time adaptive spectral identification system (ORASIS), and automated morphological endmember extraction (AMEE), etc. We will make the brief introductions to the above algorithms as follows.

The main features of PPI lie in its supervision. In N dimension data gained after MNF conversion, for each point in the image cube, generate L lines randomly, so as to calculate the pixel purity. All points in this data space are projected to these straight lines, and the points falling into the end point of each straight line are counted. After projecting to different random straight lines repeatedly, the pixel points with the count value exceeding the presetting numbers are affirmed as the pure pixels, i.e. spectral endmember. MEST (manual endmember selection tool) algorithm is a manual endmember selection tool proposed on basis of PPI thought. The first step is to determine the number of endmembers in the mixture by the PCA. If the PCA determines that the data can be expressed by N characteristic vectors, the number of endmembers is $N + 1$ regularly. Then the data are projected to N dimension space composed by the previous N characteristic vectors of the PCA. MEST provides a method of seeking the representative of the ground spectral model in $N + 1$ dimension space. They are included in the data set, and expanded to the geometry of the data set by the linear blending. Similarly, the supervision of this method is also a major weakness.

N-FINDR algorithm is aimed at seeking a group of pixels, to make the simplex composed by these pixels possess the maximum volume and accordingly make the pixels in the hyperspectral image fall into the simplex as much as possible. After the reduction of dimensionality of the original hyperspectral data, randomly select the specified number of the pixel points as the spectral endmember, and calculate the constitutive simplex volume by using these points. In order to gain the simplex with the larger volume, apply each pixel to successively replace the currently selected spectral endmember. If some replacement can gain the convex polyhedron with greater volume, such replacement will be retained as the valid replacement, or it will be eliminated as the invalid replacement. Repeat such basic process, till any replacement can not result in the volume increase of the convex polyhedron. PPI and N-FINDR algorithm will be introduced in details below.

IEA algorithm needs executing a series of constrained spectral unmixing, and selects the pixel set which can minimize the unmixing error as the spectral endmember each time. At the beginning of iteration, firstly initialize an error vector, and generally select the average spectrum of hyperspectral image data. Carry out the constrained linear spectral unmixing for the vector, and then gain the corresponding error image. Select three parameters, involving the desired number N of spectral endmembers, the number n of pixels, and angle value θ. n is the number of some pixels with larger error value from the error image. Find out the independent pixel with the maximum error, and obtain the corresponding spectral vector. Calculate the spectral vectorial angle θ between n pixels and the pixel with the maximum error, select the definite number of pixels under the given threshold value, and average their spectral vector as the new error vector. This process is always repeated till finding out the required number of the spectral endmembers.

ORASIS method has been researched by NRL for more than 5 years. This method also includes two main steps of pre-processing and sample selection. The later process is to eliminate the redundant spectrum through calculating the spectral angle distance (SAD) among the spectral vectors. Any vector will be removed from the data without separating under some threshold angle. The improved Gram-Schmidt method finds out a basic set with lower dimensionality from the original data. The sample spectrum is projected to these basic spaces, and finds out a simplex through the minimum distance conversion. Furthermore, some scholar puts forward an improved ORASIS algorithm, and reduces the data volume under the significance of the minimum loss of important information, which can be deemed as the single-pass learning VQ process. The sample is the expression of the entire data space. In other words, all spectrums of the original data at least have a sample within the scope of the threshold angle. However, the first discovered sample may not be the best spectral match in the entire set. Thus, if stopping the search after finding out a sample within the scope of the threshold angle for each spectrum in the scene, it will increase the unnecessary error. In the current ORASIS application, after finding out the first sample, the algorithm does not stop the search, but it strives to find a sample of the best match with the spectrum.

The input pf AMEE method is the entire original hyperspectral data cube without the reduction of dimensionality. This method is based on two parameters, such as the minimum value $S_{\min}$ and maximum value $S_{\max}$ of the space nuclear.

Firstly, make the minimum nuclear $K = S_{\min}$. Define a pixel space representing function $h(x,y)$ around each hyperspectral pixel. Upon utilizing the morphological operation, obtain the purest pixel and impure pixel of the spectrum in $h(x,y)$ neighborhood defined by K alternately. Through increasing the size of the nuclear gradually, repeat this operation on each pixel in the scene. The result is used for evaluating the component of each pixel in the space and spectrum. The iteration executes this algorithm, till $K = S_{\max}$. Define the morphology eccentricity index (MEI) by the morphology calculation formula. MEI value of each selected pixel is updated depending on the new value acquired from the next iteration, till generating the final MEI diagram. The selection process of the spectral endmember is full automatic.

From the above mentioned principle of the endmember selection algorithm, their existing problems can be seen. IEA is the iteration-based method. During this process, the pixel which can reduce the constrained spectral unmixing error is regarded as the spectral endmember. The weakness of this method is that the earlier the spectral endmember is selected, the poorer the foundation is. Moreover, once the pixel is selected, the spectral endmember can not be updated, and the inter-dependent relationship among the spectral endmembers can not be met to the maximum. ORASIS carried out the selection of the spectral endmember by learning and VQ, but this method is very sensitive to the threshold value parameters. While AMEE utilizes the morphology to select the spectral endmember, it also utilizes the space and spectral information, and the deficiency lies in the larger operand. PPI and N-FINDR are classic examples of the convex polyhedron search endmember selection in view of N dimension spectral space. N-FINDR is the full automatic method, and PPI is semi-automatic method. By contrast, N-FINDR is mostly used due to the merits such as full automation, parameter free, and better selection effect. However, this algorithm needs the pre-processing of the data dimension reduction, and contains a large number of volume calculations, which is the most time-consuming section. Also, the complexity of the volume calculation (i.e., mainly determinant calculation) will show the cubic increase with the increased number of the selected spectral endmembers, and accordingly result in the great reduction of the arithmetic speed of the algorithm. At present, some typical literatures have put forward the improvement program on N-FINDR algorithm. In 2002, Plaza A et al. introduced the concept of virtual dimensionality (VD) to determine the number of the spectral endmembers to be selected, which possessed some significance for the implementation of the algorithm, but it can not change the above two deficiencies of this algorithm. In 2008, Wu C et al. adopted the mode of the pixel pre-selection to reduce the complexity of the follow-up search, and also reduce the calculated amount of the algorithm from the side. In the feature selection, the progressive selection mode sometimes was used for replacing the joint selection mode. The progressive selection mode is far away from the essential features of N-FINDR algorithm. Once the pixel is selected, it can not be updated, so the interdependent relationship among the spectral endmembers can not be met to the maximum. In 2007, the method proposed by Tao X et al. could be directly conducted in the original data space free from the pre-processing of the dimension reduction, which broke through the traditional model of N-FINDR algorithm needing the pre-processing of the dimension reduction. This method also belonged to the progressive selection. In 2010, the method proposed by Chang C et al. could also be free from the pre-processing of the dimension reduction, but it was easy to have the ill-conditioned matrix equation solving problems and was not ideal. At the same time, implementing the random initialized endmember selection method over and over again also increased the calculated amount of the search process.

1.4 Spectral Unmixing Technique of HSI

Relative to the classification technique, the spectral unmixing, i.e. soft classification technique has a late start. Although the spectral resolution of the hyperspectral image is improved greatly, the spatial resolution of the corresponding land object target of the pixel is relatively low. For example, the spatial resolution of AVIRIS is 20 m × 20 m. Thus, a pixel may contain two or more land object targets, i.e. the pixel is mixed. When the interested target is not one pixel or several pixels, the object to be researched and analyzed will focus on the mixed pixel. If only one mixed pixel is assigned to a certain category, it will certainly bring about some classification error and result in the decline of the classification accuracy, and accordingly influence the follow-up application of the analysis result. Therefore, researching the spectral unmixing method of the hyperspectral image is the key technology to be solved in the high accuracy classification and recognition of the hyperspectral image.

During the period of multi-spectral remote sensing, the spectral unmixing problems of the mixed pixel are difficult to solve, because the spectral resolution of multi-spectral remote sensing is too low, and the acquired remote sensing radiation value is not enough to completely depict the spectral properties of the land objects. Moreover, during the period of hyperspectral remote sensing, due to the enhancement of the spectral resolution, the acquired pixel radiation value constitutes an approximately continuous spectral curve, on behalf of the spectral response characteristics of the land objects integrally. Thus, in the hyperspectral remote sensing, it completely may determine the land object composition category and proportion of the pixel by analyzing the composition of the spectral curve, and accordingly achieve the purpose of the spectral unmixing of the mixed pixel.

From the mixed mode, the pixel mixed model can be divided into two classes, linear mixed model and nonlinear mixed model. Moreover, for the unmixing mode of the linear spectral mixed model (LSMM), from the selected representative pixel of each class, i.e. the number of endmembers, it can also be divided into single-endmember unmixing mode and the multi-endmember unmixing mode. At present, people mostly research the LSMM—single-endmember unmixing mode.

1.4.1 Nonlinear Model

Relative the linear method, the nonlinear method study is less, but the typical models include Hapke model, Kubelk-Munk model, vegetation/soil spectral mixed model based on radiant flux density theory, and SAIL mdoel (Tong et al. 2006).

Hapke model is proposed on account of the planet surface. For the deficiencies, it is difficult to be applicable to the landmark with the vegetation cover, difficult to collect the data, and difficult to determine the scattering refraction coefficient, and phase function. The improved Hapke model can be applied to the spectral reflectivity research of the soil on earth. However, only when the reflection ratio of the single scattering for the asymmetry parameters of the soil particles is more than the specific threshold value, it will have better effect than the original Hapke model.

The application of Kubelk-Munk model is also confronted with many restrictions, such as needing the measurement of the hemisphere reflectivity, and the assumption of the isotropic reflection etc. This model has higher requirements on the surface reflectivity, resulting in the difficult applying to the remote sensing application of the earth surface. It can only convert the reflectivity into the quantity in proportion to the substance absorption coefficient within the limited range.

Vegetation/soil spectral mixed model based on radiant flux density theory can rapidly calculate the spectral radiation value of any level in the vegetation, and also can simulate very complex structure. However, the model still needs more detailed research in the theory and experiment, so as to gain the improved index and nonlinear decomposition method.

SAIL model is a mixed spectral model for calculating the vegetation leaf area index. This model makes the simple descriptions for the structure of crown, and makes rough estimate for the radiation conductivity equation. This model can not explain the isotropy, non Lambert reflector characteristic land object, and hot spot effect well. The improved SAIL model considers the hot spot effect meticulously, and considers the influence of the size and shadow of the blade while calculating the contribution of the secondary reflectivity on the single reflectivity.

In addition to the above typical nonlinear method, there are other types of nonlinear models, such as geometric model, mixed media model, mixed model, and computational simulation model etc.

The establishment and solution of the nonlinear mixed model are relatively difficult. Therefore, the spectral unmixing research based on the nonlinear mixed model is less. By contrast, the linear mixed model is easily established due to the explicit physical significance. The spectral unmixing based on the linear model is to suppose that the mixed pixel is the linear mixture by several classes of the pure land objects in accordance with the definite proportion. The spectrum of the mixed pixel is the linear combination of these land object spectrums. During the spectral unmixing, according to whether needing the priori knowledge relevant to the land object category, the spectral unmixing can be divided into the supervised method and unsupervised method. Besides, the spectral unmixing can be made under the constraint conditions, and also can be made without the constraint conditions. The spectral unmixing achieves the prediction and acquisition of the substance spectral information of less than a pixel, so it can be used for the target recognition processing of the known spectral characteristics.

1.4.2 Linear Model

The linear model is to suppose there is not the interaction between objects, and each photon can only "see" a substance and superpose the signal to the pixel spectrum. When multiple scattering occurs among the substances, it can be regarded as a process of the iteration product as well as a nonlinear process. The mixture of the objects and the spatial scale of the physical distribution decide the nonlinear degree.

In the early 1970s, the scholars in foreign countries began to research the land object mixture phenomena. In the research course of the spectral unmixing, the representative one was the thesis of Harsanyi J et al. from University of Maryland of America in 1994. After putting forward OSP method in 1994, Chang also developed and introduced a series of the mixed pixel researches based on this method in succession, and applied Kalman filter into the linear mixed model. This linear separation Kalman filter could detect the abrupt changes of various mixed component contents in the pixel, and could detect the useful target characteristics for the classification. In 1974, Adams J et al. firstly began to research utilizing the linear mixed model to analyze the geological mixture composed by the plant and rock, so as to determine the type and distribution of the minerals, also utilizing the shadow size in the image to link the Thematic Mapper response with the height and surface composition of the ground plant, and utilizing the shadow image to further explain the change of the forest structure. Since 1990s, the spectral unmixing research based on the LSMM has been developed gradually and extensively.

1. **Linear Spectral Mixture Analysis (LSMA)**

LSMA means the spectral unmixing method from the solution of LSMM by using the traditional optimization algorithm. LSMM becomes the most widely used model due to the relative simpleness and efficiency. Nevertheless, the traditional LSMM only uses an endmember to depict a class (known as single-endmember model). When intra-class spectral changes of this class are larger, this description is not accurate. On the other hand, LSMM is the linear model, and it is difficult to carry out the nonlinear promotion. The above situations result in the unsatisfactory unmixing effect. In the practical application, LSMA is generally attached with the non-negativity and normalization constraint (i.e., the valuated mixing ratio is non-negative and the total is 100 %, collectively referred to as full constraint) conditions under the least squares sense, while the traditional iteration solution process of the full constrained least squares LSMA (FCLS-LSMA) is very complex. In addition to the problem of the complexity, the above method fails to provide the reasonable thought in conformity with the least squares criterion for the full constraint adjustment of the mixing ratio result, and influences the unmixing accuracy.

On the other hand, some literatures proposed replacing the single-endmember mode by the multi-endmember mode, so as to overcome the intra-class spectral changes, which gained very good effect. However, these methods focused on the complex endmember selection from the supervised data, or possess the distinct pertinence, such as only applying to some specific territory and having relatively poor universality, while the unmixing process was still assisted by the traditional LSMA method. Bateson C et al. in 2000 and Roberts D et al. in 2010 proposed the typical multi-endmember spectral mixture analysis method. The superior point than the traditional LSMA method is to automatically expand each traditional fixed endmember to a reasonable range. However, such method selects less training samples, so the ability of considering the categorical distribution state is slightly poor. In 2000, Monte Carlo method used by Asner G et al. considered a good number of the training samples, but it only belonged to the statistical average of massive simple LSMA models, with bad effect. How to effectively enhance the solution speed and accuracy of the single-endmember FCLS-LSMA, consider the intra-class spectral changes and categorical distribution state better, and implement the multi-endmember unmixing mode more reasonably shall have the research prospect.

LSMA generally should be attached with the normalization and non-negativity constraint (collectively referred to as full constraint) conditions to meet the practical physical significance, and is to solve in accordance with the least squares sense. The normalization constraint condition means that the sum of mixing ratio of each pixel is 1. The non-negativity constraint condition means that the mixing ratio of each pixel is non-negative and less than 100 %. The traditional iterative optimization solution process of the FCLS-LSMA is very complex, and how to reduce the complexity is worthy of researching. In 2011, Heinz D et al. improved this, but the employed iteration solution method still had higher complexity, and the solution process was designed in accordance with the second best method.

In 2004, Xiurui Geng et al. set forth the geometrical significance of LSMM, and provided an intuitive geometric solution method based on the volume calculation, which conquered the problems of the excessive complexity to a certain extent. The volume calculation formula is the positive type (i.e., the value is positive). The volume calculation is mainly embodied in the determinant calculation, and the complexity is the cube of the endmember numbers and is still relatively high. On this basis, in 2008, Wenfei Luo et al. proposed a method of replacing the volume calculation with the distance calculation, with a certain effect. The complexity has been reduced than that of the volume calculation, but the total calculated amount has been still larger.

Another important aspect relevant to LSMM geometric solution is the full constraint handling problem of the unmixing component (geometric measurement).

When the mixed pixel falls inside the convex polyhedron formed by each endmember, the results from the above method meet the full constraint conditions. When the mixed pixel falls outside the convex polyhedron formed by each endmember, it can not completely meet full constraint conditions. It is easy to know that the positive and negative type measurement can only meet the normalization constraint, while the positive type measurement can only meet the non-negativity constraint. For the common full constraint adjusting way, in the case of the negative measurement (unmixing component), it will be set as 0. In the case of the measurement (unmixing component) of more than 1, it will be set as 1. Furthermore, divide each measurement (unmixing component) by the sum of all measurements (unmixing component). However, such adjusting way is merely pursuing the acquisition of the full constraint result, rather than conforming to the least squares criterion. How to establish low-complexity FCLS-LSMM geometric solution method completely in conformity with the least squares criterion becomes a problem with the practical significance.

2. **MVC-NMF (minimum volume constrained nonnegative matrix factorization) Algorithm**

MVC-NMF combined the least squares analysis and the convexity geometry by adding the volume restriction into NMF (Li et al. 2011). The proposed cost function includes two sections. One is to assess the observation data, endmember and abundance, and reestablish the approximate error among the data. The other is composed of the minimum volume restriction. These two sections are regarded as two forces. The external force (minimizing the approximate error) makes the estimated results move externally to the point cloud, and the internal force (minimizing the unit volume) makes the endmembers close up mutually as much as possible in the opposite direction. The specific process of the algorithm is:

1. Building the objective function

$$\begin{cases} \min f(\boldsymbol{A}, \boldsymbol{S}) = \frac{1}{2}||\boldsymbol{X} - \boldsymbol{A}\boldsymbol{S}||_F^2 + \lambda J(\boldsymbol{A}) \\ \boldsymbol{A}_{i,j} \geq 0, \boldsymbol{S}_{i,j} \geq 0, \boldsymbol{1}_p^T \boldsymbol{S} = \boldsymbol{1}_n^T \end{cases} \tag{1.1}$$

In the formula, $\boldsymbol{1}_p^T$ is p dimensional column vector with the element of 1. $\boldsymbol{1}_p^T$ is n dimensional column vector with the element of 1. $J(\boldsymbol{A})$ is the penalty term. Calculate the unit volume composed by the estimated endmember, $\lambda \in R$.

2. Initialization

Select p points from the point cloud data randomly, and constitute into the initial value of $\boldsymbol{A}$. $\boldsymbol{S}$ matrix can also be initialized randomly. In 2007, Miao L et al. initialized the matrix $\boldsymbol{S}$ into the zero matrix in the experiment.

3. Estimate the number of endmembers p by using the VD.
4. Stopping criterion: Provide the number of iterations and error threshold value.
5. According to the definite criterion, calculate the matrix $\boldsymbol{A}$ and $\boldsymbol{S}$ that can minimize the objective function. In the case of meeting the stopping criterion, the iteration stops. Or, update $\boldsymbol{A}$ and $\boldsymbol{S}$, and continue to seek the matrix of minimizing the objective function.

3. SPA (successive projection algorithm)

SPA algorithm is established on the basis of the convexity geometry and rectangular projection, and contains a restriction in the endmember candidate pixel space adjacency, so that it can reduce the influence of the outside point pixel and generate the practical endmember (Li et al. 2011). SPA algorithm does not need the data dimension reduction during the process of the endmember extraction form the hyperspectral data. It utilizes the spatial relationship between the spectral angle and the pixel in the image to restrict the selection of the candidate pixel on behalf of the endmember. Many objectives have the spatial continuity in the image, so the spatial restriction is beneficial to the endmember search. The additional achievements of SPA are the monomer volume ratio generated from the adjacent iteration during the process of the endmember extraction. It describes the influence of the new endmember on the data structure, and meanwhile provides the information for the convergence of the algorithm. It can provide the guidance for restricting the total number of the endmembers during the process of searching. The endmember set determined by the convexity geometry relevant to SPA algorithm is equipped with the following properties. ① The endmember stands for containing the pixel with the maximum volume. This property is applied to SPA, and decides whether the algorithm is convergent. ② The vector with the maximum Euclidean distance is located in some vertex of the monomer, which is the main step for SPA to determine the vertex pixel of the monomer. ③ For a given point of the monomer, the point with the maximum distance must be the vertex of the monomer. ④ The affine transformation of the monomer is still the monomer, and the endmember is still located in the vertex of the new monomer after transformation. It makes the orthogonal subspace projection as the core mechanism of SPA endmember extraction algorithm. SPA algorithm is based on two assumptions. ① The adjacent pixels of the space are more likely to have the similar spectral characteristics, on behalf of the same endmember. ② The probability of two adjacent pixels with the noise is very low. The specific steps are as follows. ① Set up the parameters: Set up three parameter values—the number of endmembers p to be sought, the spectral angle threshold value t_θ, and the space threshold value t_pixel. ② Extraction of the first endmember $\boldsymbol{e}_1$: Calculate the vector module of all image pixels, and the pixel with the maximum positioning module. According to the property 2, the pixel is located in a vertex of the monomer, and especially the pixel with the brightest image cube. ③ Extraction of the second endmember $\boldsymbol{e}_2$: Calculate the distance between all pixels and $\boldsymbol{e}$, and the pixel with the maximum positioning distance. According to the property 3, the pixel is located in the other end point of the monomer, and generally corresponds to the darkest object in the scene. ④ Extraction of rectangular projection and new endmember: The endmember matrix $\boldsymbol{E} = [\boldsymbol{e}_1, \boldsymbol{e}_2]$ has been built by the previously defined endmember. All pixels are projected to the subspace S_{proj}, and are orthogonal to the space expanded by $\boldsymbol{E}$. $\boldsymbol{P}_{(i,j)\_\text{proj}} = O\boldsymbol{P}_{(i,j)}$, $\boldsymbol{P}_{(i,j)\_\text{proj}}$ and $\boldsymbol{P}_{(i,j)}$ are respectively the projection vector and original vector in the position (i,j) of the image. O is the projection operator $O = \boldsymbol{I} - \boldsymbol{E}\boldsymbol{E}^{+}$. $\boldsymbol{I}$ is the unit matrix. $\boldsymbol{E}^{+}$ is the pseudo-inverse of $\boldsymbol{E}$,

$\boldsymbol{E}^{+} = (\boldsymbol{E}^{T}\boldsymbol{E})^{-1}\boldsymbol{E}^{T}$. In projecting the subspace S_{proj}, the effect of the endmember on the mixed pixel in $\boldsymbol{E}$ is eliminated. According to the property 4, the projection data of the new space still conform to the convex characteristics, i.e. the endmember is still in the vertex of the monomer. In projecting the subspace S_{proj}, the pixel with the maximum module corresponds to $\boldsymbol{e}_3$. The pixel is located in the farthest vertex of the monomer in the subspace expanded by the previously defined endmember $\boldsymbol{e}_1, \boldsymbol{e}_2$. ⑤ Finish the search of all end elements: At the moment, the matrix $\boldsymbol{E}$ has been updated to be $\boldsymbol{E} = [\boldsymbol{e}_1, \boldsymbol{e}_2, \boldsymbol{e}_3]$. Execute the step 4 repeatedly, and seek the new endmember, till meeting the previously determined number of endmembers p.

4. **Other methods**

The method that can be applied to the spectral unmixing also includes the multilayer perceptron, nearest neighbor classifier, and ICA, etc. The multilayer perceptron and nearest neighbor classifier method is the fuzzy classification method based on the supervision. This method is not applied widely due to the lower unmixing accuracy. ICA is the analysis method based on the signal high order statistical property, and has been widely concerned in the field of the signal processing. This method is to conduct a certain linear unmixing for the observed data, to make it become the component with the independent statistics. Here it needs using a hidden statistical variable model, indicating how the observed data are generated by the mixing of the independent component. If the model is reversible, the independent component can be obtained. The assumption of ICA algorithm is the statistic independence of various components and non-Gaussian distribution, supposing the unknown mixing matrix is the square matrix. Such assumption makes the practical application of ICA algorithm under restrictions. In recent years, some new methods of the spectral unmixing based on the nonnegative matrix factorization and grade Bayesian model occurred, and especially the former one became one of the hot spot unmixing techniques researched this year.

1.4.3 Multi-endmember Mode of Linear Model

In this model, one class is represented by one single spectral endmember. The spectral endmember can be acquired by various endmember extraction methods such as N-FINDR, PPI, IEA, and also can be acquired by the mean value of various classes of the supervised data. However, the large hyperspectral image space range results in the generally great intra-class spectral changes. Under this condition, one endmember spectrum is difficult to depict one class correctly. How to conquer the intra-class spectral changes becomes the important content of the spectral unmixing. In 1993, Roberts D et al. further carried out the spectral unmixing through analyzing the intra-class spectral changes of the residual spectrum. In 2000, Asner G et al. applied Monte Carlo to estimate the peak value and reliability section to conquer the spectral changes. In the method proposed by Bateson C et al. in 2000, multiple endmembers rather than one endmember was used for expressing one class, and the acquired maximum peak value and minimum peak value were used for estimating the expected result. At present, Roberts D et al. provides a method from the endmember selection, spectral unmixing to sub-pixel mapping through their website, and describes one class by multiple pure pixels, so as to reflect the intra-class spectral changes. The above typical multi-endmember spectral unmixing method plays a positive role in conquering the intra-class spectral changes, but it has the problems involving large calculated amount (a large number of unmixing averages), tedious endmember pre-selection, or different effects of various endmembers failing to be reflected reasonably.

In recent years, the feasibility, methodology and unique advantage of the support vector machine (SVM) applying to the spectral unmixing have gained the extending study. It is noted that the existing SVM model incorporated the hard classification error constraint conditions into the optimization function, while the general evaluation principle of the spectral unmixing was the unmixing error, i.e. soft classification error. Both had some difference. Chapter 4 of this book will focus on introducing the multi-endmember unmixing method based on SVM.

1.5 Sub-pixel Mapping Technique of HSI

In the late 1990s, some scholars gradually began the research of the sub-pixel mapping. In 1998, Foody G sharpened the low resolution image by using the high resolution image for two images with different spatial resolutions imaging simultaneously in the same place. In 1997, Gavin J et al. gained the high resolution mapping result by the Bayesian method and adding the priori knowledge of the true image into the random model. In 1998, the sub-pixel mapping method proposed by Gross H et al. also needed the image with higher spatial resolution as the additional information. Such additional information was often hard to acquire in the practical application. In 1997, Atkinson P came up with the earth surface cover category with the spatial correlation in the pixel and between the pixels, i.e. the natural land object distribution rules of the homogeneous aggregation and heterogeneous separation became the theoretical basis of the footstone in this field. In 2001, Atkinson P supposed that the spatial correlation in the land object spatial distribution could be determined by the distance scale, i.e. the similar pixels would have the similar values. Furthermore, in 2002, Verhoeye J et al. raised the spatial correlation (or spatial independence)-based sub-pixel mapping method, achieved by adopting the linear optimization technique, and converted the sub-pixel mapping thought into a linear optimization problem to maximize the spatial correlation, with a preferable effect. As a matter of fact, the spatial correlation theory became the core theory of the sub-pixel mapping since the twenty-first century. Some typical neural network-based mapping methods had been formed like this. In 2001, Tatem A et al. deemed Hopfield network model as the energy tool, and the spectral unmixing result as the constraint condition of the energy function. The network was finally converged into the minimum energy, and this state was the sub-pixel accuracy mapping result to be obtained finally, with a preferable effect. In 2008, Zhang J et al. put forward a BP neural network-based method, and it had good mapping effect. This method also needed the priori information to train the network. In 2006, Mertens K et al. utilized the space gravity model, to effectively realize the spatial correlation theory proposed by Atkinson. In the gravity model, the spatial correlation determined the sub-pixel spatial distribution through calculating the gravity between each sub-pixel and the corresponding neighborhood pixel in a pixel. This model directly considered the correlation between each sub-pixel and the original low resolution pixel in the neighborhood, and could gain the sub-pixel spatial distribution result without the iteration, which was a real-time yet effective sub-pixel mapping method. However, it considered the correlation among all sub-pixels less and failed to fully apply the spatial correlation theory, which became the major defect of such method. Thus, the spatial correlation-based space gravity model occupied the dominant position in the sub-pixel mapping without dispute, and meanwhile this model had the structure defects and promotion space. The effective means of "sublation" became the key to break through the bottleneck of the sub-pixel mapping technique.

The above are sub-pixel mapping methods suitable for the soft classification result diagram. Generally the complexity of the original hyperspectral image was considerably higher than the soft classification result diagram, so the above methods were difficultly applied to the mapping treatment of the original image effectively. In 1993, Schneider W et al. came up with a sub-pixel mapping method of the original image. However, this method could only be applied to the image with the straight line margin feature under some spatial resolution, and the employed model could not be realized when the image contained more than two land object classes. In 2001, the method of Aplin P et al. could also be applied to the original image. But the harsh condition of needing the hyperspectral image with high resolution made this method have few practical values. Till now, there is not yet the sub-pixel mapping method that can be applied to the original image effectively. To sum up, the existing sub-pixel mapping methods need the unavailable additional information, or fail to fully implement the spatial correlation theory, or fail to apply to the complex original image limited to the mapping treatment of the soft classification result diagram (unmixing component diagram).

In 2009, Atkinson P divided the sub-pixel mapping methods into two categories. First, make the maximization treatment for the spatial correlation of the sub-pixels and obtain the distribution situation of the sub-pixel scale inside the mixed pixel. This method divides each sub-pixel into the appointed land object class, so as to gain the spatial distribution map. This method is applicable to the example of H-type resolution, i.e. the situation of the land object type greater than the pixel size. Second, match the correlation of the sub-pixels by some priori model. These priori models include the space covariance or variation function model. This method is applicable to L type situation, i.e. the area of the land object class is smaller than the region size covered by the mixed pixel. In the past, many people researched in these aspects, and meanwhile many methods were proposed for the application in the sub-pixel classification.

In recent years, the methods used by the domestic and overseas researchers have been summed up as four main approaches, involving utilizing the increasing of the spatial correlation auxiliary information, utilizing spatial geostatistics method to provide the auxiliary model, utilizing the neural network to make the sub-pixel distribution simulation, and gaining the best match by the exchange among sub-pixels (Ren and Ge 2011).

1.5.1 Spatial Correlation-Based Sub-pixel Mapping

The sub-pixel mapping by using the property of the spatial correlation should play the role of the spatial correlation to the maximum. In each land object class, the inherent scale of the spatial variability is equal to or more than the pixel space sampling scale, supposing that the position of the land object class in the pixel will depend on the position of the land object class in the neighborhood pixel to some extent. That is to say, on the precondition that the spatial variable scale is larger than the remote sensing image pixel scale, the spatial distribution of different intra-pixel and inter-pixel land objects has the correlation. The spatial dependence, generally referred to as spatial correlation or spatial autocorrelation, means that the observed value with the similar space possesses more similar trend than the observed value with the longer space distance for a given attribute. In other words, in the mixed pixel and among different pixels of the remote sensing image, the sub-pixel with the shorter distance is more likely to belong to the same type compared with the sub-pixel with the relatively long distance. On the basis of increasing the spatial autocorrelation auxiliary information, many domestic and overseas researchers made the research in the aspect of the sub-pixel mapping. Aplin P utilized the land line digital vector data of the Ordnance Survey, and adopted territory-based Perfield method rather than the traditional pixel-based land cover classification method to make the territory boundary mapping in the sub-pixel scale. Also, it utilized the vector data to make the pixel segmentation, and the segmented pixel fragments are arranged as per the area. These areas are marked as some land object class through the typicality of the land object class. Nevertheless, in most cases in reality, it is very difficult to acquire the available accurate vector data set. In 1993, Schneider W put forward the knowledge-based analysis technique, depending on the knowledge related to the straight line boundary characteristics in Landsat T scene, so as to achieve the automatic positioning of the territory boundary with the sub-pixel accuracy. In 1998, Foody G adopted the method based on the simple regression, utilized the sharpening image with higher spatial resolution, and sharpened the soft classification output result of the image with coarser resolution, so as to refine the sub-pixel land object type map. In 2002, Jan V et al. brought the soft information of the pixel in the mixed model in the manner similar to the spectral mixed model, realized the hard classification in the sub-pixel scale by the means of the least squares estimation, and acquired the sub-pixel distribution diagram. This linear solution enables each sub-pixel class to predict through the neighborhood information in the surrounding pixel level. However, the classification result had the obvious linear effect. On account of the assumption that the adjacent pixels are more likely to belong to the same land object class than other pixels, in 2005, Kasetkasema T et al. proposed a super resolution drawing model based on Markov random field (MRF). It eliminated a large number of misclassification pixels manifested as the discrete pixels, to gain the high resolution classification image. But this method would result in the loss of the distributed information for the small interested objectives. Domestically, in 2005, Feng Ling came up with a new cellular automata model for simulation of the spatial sub-pixel mapping. Upon the corresponding adjustment of the original cellular automata, the cellular automata of solving the sub-pixel mapping problems can be obtained. According to the current state of the cellular and the neighbor status, determine the dynamic function of the cellular state at the next moment, build the evolution rules, and finally get the sub-pixel distribution diagram. The experiment result indicated that the result of utilizing the cellular automata model for the sub-pixel mapping has been obviously superior to the result of utilizing MLC method for the classification in the visual sense. In 2007, Chang Yi et al. put forward a neural network-based cellular automata model under the multi-scale. Transplant the cellular automata theory into the evolution of different spatial scales, establish the neural network-based multi-scale cellular automata model, and extract the super resolution information of the urban land in Haidian District, Beijing by this model. The result showed that, this method can effectively express the spatial autocorrelation among the image pixels. In 2009, Ge Y et al. raised an algorithm. The sub-pixel spatial distribution simulation algorithm was mainly used for solving the spatial distribution problem of the basic component in the mixed pixel with the faceted distribution mode in accordance with the spatial autocorrelation characteristics of the remote sensing image. After the accuracy evaluation on the experiment result, the acquired sub-pixel distribution image had been greatly improved in the classification accuracy and vision than MLC classification result. In 2009, Ke Wu et al. proposed a brand new remote sensing image sub-pixel mapping method based on Agent evolutionary theory, provided the evolutionary mechanism and design flow of Agent, and utilized two different sets of experiments to make the comparative analysis on the performance of this method. The experiment result showed it could conduct the mapping on the sub-pixel in the mixed pixel rapidly and effectively. This method was built on the basis of the spatial correlation, combined the assumption relations of the spatial neighborhood, utilized two behavior patterns of duplication and diffusion of Agent, made the random dynamic selection, and provided an accurate position for each sub-pixel. It was suitable for the local area with the connected consistency area, and Agent points in different areas could be handled at the same time. Moreover, this method was easily described and realized. But how to keep Agent point to finish the given task in the fastest way and determine the reproduction and diffusion direction of Agent point still needed the further research. In a word, the spatial correlation-based algorithm is the main direction for the sub-pixel spatial mapping drawing now, so combining different simulation algorithms can gain good experiment results. Although the above methods seem like diversity, they have some common characteristics. Such type of algorithm mainly utilizes the soft information of the neighborhood pixel scale to predict the distribution situation inside the mixed pixel. The algorithm is generally fast, simple and efficient, and also can provide the experiment results under different magnification factors on request, and provide the higher resolution boundary information relevant to the land object class. The weakness is to only utilize the information of the surrounding neighborhood pixel, but neglect the structural features of some land object distributions. It can make the relatively accurate prediction for the distribution situation of the faceted land objects on the sub-pixel scale, but the treatment for the linear land objects and some dotted land objects (L type land objects) of less than the mixed pixel is not satisfactory. Besides, the assumption of the maximum spatial correlation of the pixel is applicable in most cases, but this assumption may have some error in some cases due to the complexity of the practical situation. Therefore, how to gain better expression way for the pixel spatial distribution rules is crucial for enhancing the accuracy of the sub-pixel mapping, and needs the deeper research.

1.5.2 Spatial Geostatistics-Based Sub-pixel Mapping

The previous geostatistical method for handling the mixed pixel problems was the semi-variable function based on the traditional two-point geostatistics, but the semi-variable function could only seize half of the information of two-point histogram gray level co-occurrence matrix, with scarce handling capacity for the complex spatial structure, such as connectivity etc. In 2006, Boucher A et al. proposed utilizing the geostatistics-based method such as indication Kruger and indication random simulation as a priori structural model from the perspective of the probability theory inverse problems. They firstly utilized land object cover situation in the high resolution image to get the indication variation function model, then deemed this model as a priori probability, made the downscaling positioning for the low resolution image, and accordingly got the land object distribution situation in the higher resolution. In 1993, Guardiano F et al. proposed the multipoint geostatistics conquering the defects of the traditional two-point geostatistics, which produced the new hop for solving this problem. In the multipoint geostatistics, it replaced the variation function by the training image to express the structural information and spatial autocorrelation of the land objects, so it could conquer the deficiency of the traditional geostatistics of failing to reproduce the geometrical morphology of the complex land objects. At the same time, this method still regarded the pixel as the simulation unit and adopted the sequential non-iterative algorithm. It included a search tree structure for storing the frequency of encountering the patterning in the training image. It could be loyal to the hard data, and the speed was superior to the objective-based random simulation algorithm. In 2008, Atkinson P came up with the algorithm of the sub-pixel mapping by the downscaling collaborative Kruger. The collaborative Kruger considered the autocorrelation and cross-correlation of the image, and could handle in allusion to the pixels of various sizes. Also, it could consider the point spread function of the sensor with the prediction coherence, and meanwhile support the combination with other auxiliary data for handling. However, the biggest problem was that the auto-covariance and cross covariance as required by the super resolution Kruger could not be gained by the experience. In 2006, Boucher A put forward a more complex priori model in the form of the training image based on the previous work, and carried out the sub-pixel mapping in combination with the sequential non-iterative algorithm. Kruger-based geostatistics utilized the indication variation function to simulate the correlation between two points in various categorical distribution situations of the space. The training image is a grid image for expressing the land object distribution mode. Generally speaking, the training image can be used for expressing the distribution mode that the variation function can not express, such as the curved land object class. Compared with the spatial autocorrelation, the spatial geostatistics-based method can be used for the modeling of the complex geometrical morphology land objects well and also has faster speed. But this algorithm has the disadvantages, such as excessively considering the simulation by the structural information of the training image, neglecting the consideration in the aspect of the spatial autocorrelation information, different results realized each time, and low degree of accuracy, etc.

1.5.3 Neural Network-Based Sub-pixel Mapping

The neural network has also been one of the mainstream techniques for the domestic and overseas researchers to carry out the sub-pixel mapping in recent years. Hopfield Neural Network (HNN) is generated from the mutual combination of massive simple neural processing units, with the constraints involving symmetry, indirect self feedback, and asynchronous motion. HNN is often used as an optimization tool, for minimizing the energy function. HNN strategy is close to the method of constantly repeating the iteration of the previous sub-pixel classification situation, and especially is applicable to the situation of H type mixed pixel that the land object target class is greater than the pixel size. Initially, HNN is designed for solving the two-value situation (i.e., only two land object types, objective land objects and backgrounds, one or the other), but this method is also applicable to the situation of multiple land object classes. From 2001 to 2003, Tatem A et al. proposed the sub-pixel classification by HNN technique. It utilized the output result of soft classification to constrain the HNN. The spatial clustering function encoding enters into the neural network. It carries out mapping on the land types inside each pixel by utilizing the information included in the surrounding pixel. Later, it expands this method to the sub-pixel scale mapping with various land types through adding the new constraint in the energy function. Furthermore, it adds the priori information relevant to the special land object class typical spatial arrangement into the energy function as the semi-variation constraint, to solve the problems of the target spatial pattern of less than the ground resolution of the sensor. In the follow-up research, in 2003, Tatem A et al. applied HNN to match the spatial correlation between the sub-pixel and some specific priori model (such as semi-variance diagram). It utilizes the true Landsat TM crops image to test the proposed method. This method is more applicable to the situation of L type mixed pixel that the land object target is less than or equal to the pixel size, but the algorithm is relatively time-consuming. Compared with the previous method, HNN method can generate more accurate spatial distribution situation of the land object target in the sub-pixel scale. Moreover, the clustering target and the matching target can be made at the same time in allusion to different land object classes in the same image. In addition to HNN, the other neural network model for the sub-pixel mapping frequently is the BP neural network. The BP network is often used for building the training model to describe the relationship between the mixed pixel and the neighborhood pixel as well as the intra-pixel spatial distribution, and further gets the distribution of the mixed pixel in the sub-pixel scale by the super resolution model. The experiment result proves that, the algorithm can gain the fairly good drawing result, and the algorithm complexity is low, which can handle the original hyperspectral image.

1.5.4 Pixel-Swapping Strategy-Based Sub-pixel Mapping

Compared with the above-mentioned neural network method, the pixel-swapping-based strategy is simple and fast. Therefore, two pixel-swapping-based methods are also developed. First, it is designed for the H-type mixed pixel situation. The algorithm was originally proposed in allusion to the situation of two land objects (two-value). It allows the sub-pixel classes in the same pixel to be swapped. Thus, it can ensure that the proportion value of the initial soft information inside the mixed pixel is not changed, while making the classes in the sub-pixel scale distribute in accordance with the most correct position distribution situation. So the so-called correct trend makes the spatial correlation in the sub-pixel scale reach the optimum. Upon the repetitive execution of the algorithm in allusion to different classes for several times, it can make this method expand, and also is applicable to the situation of multiple classes. But this method is only applicable to H-type mixed pixel. For L-type mixed pixel, in 2008, Atkinson P proposed a new two-point histogram-based method for realization. The two-point histogram is the set of all swapping probabilities between the selected mixed pixel and the neighborhood pixel in the vector distance formulated apart. In remote-sensing science, it is similar to the gray level co-occurrence matrix. This algorithm replaces the variation diagram (and covariance function) by the two-point histogram, because the variation diagram only expresses half of the available information (i.e., distinction, without the direction). Therefore, this pixel-swapping-based method depends on the two-point histogram from the training image. In this way, this method needs a small region of high resolution image (image with good classification) as the training image, and then applies it to the large region in the interested low resolution remote-sensing image, so the distribution situation in the sub-pixel scale can be gained. The remote-sensing image sub-pixel mapping is a new field in the quantitative remote-sensing research. As the follow-up analysis method of the mixed pixel decomposition, this technique can effectively eliminate the spatial uncertainty of the mixed pixel decomposition result, which not only possesses very important theoretical research value, but also possesses very powerful practical significance in the related field.

From the development history of the sub-pixel mapping research, it can be seen that, from the proposal of different model solution algorithms at the early stage of research, to different description methods of spatial correlation and the related research of model improvement, and the analysis on the uncertainty and error of the model at present, the related theoretical research has been developed quickly. However, as a new field, many unknown problems need to be explored deeply. In the practical application, the sub-pixel mapping technique is also applied to the ground control-point selection, landscape index calculation, change detection, and other related fields, in addition to the successful application in the aspects of the land-cover drawing, the lake boundary extraction, and the coastline extraction. But relatively speaking, the practical application is fewer. How to make the deeper and more effective application in the related field and accordingly discover and solve the new problems also needs further research.

In general, the sub-pixel mapping has many problems to be solved from the theoretical practical application view. We think the following aspects will be the main problems that the sub-pixel mapping research is confronted with now and the further research direction in the future.

1. Multiscale comprehensive model: In allusion to two relationships between the land object size and pixel resolution, now it is being handled by using different sub-pixel mapping models. However, in the practical work, the same research area generally will have these two situations at the same time, and cannot be simply summed up as a certain scale relation, while needing to adopt different research methods on account of different land object classes or different spatial positions. At present, there is no theoretical models that can consider these two situations at the same time, so how to make the comprehensive analysis on the models of two scales needs further research.
2. Description of spatial correlation: The key to the sub-pixel mapping is the spatial distribution characteristic of land objects. Although many methods are used for describing this characteristic from the initial spatial correlation maximum to the subsequent multipoint statistics, they have some deficiencies. How to describe the complex land object space correlation more effectively is a problem needing continuous research for the sub-pixel mapping. In addition, on account of some special problems, it discovers and proposes some special description methods of the spatial correlation. For example, the rivers and roads

are mostly the linear land objects, and the shadow distribution is mostly relevant to the terrain. Accordingly, a priori information of the land object distribution is embedded in the sub-pixel mapping algorithm. How to apply more remote sensing and auxiliary data into the sub-pixel mapping is a matter of further research direction in the future.

3. Model selection and comparison: At present, many sub-pixel mapping models and algorithms are proposed, so how to select the algorithm in the practical application and what advantages and disadvantages various algorithms have will be a problem to be confronted. Atkinson P put forward the suggestion of the sub-pixel mapping algorithm comparison. In the case of the algorithm comparison, many basic problems of the sub-pixel mapping should be discussed, involving the test data set selection, result accuracy evaluation parameters, and uncertainty analysis, etc.
4. Fusion with other two problems of the mixed pixel: As previously mentioned, the "mixed pixel" of remote-sensing image includes three problems, such as endmember selection, pixel decomposition, and sub-pixel mapping. Different endmember selection results will gain different mixed pixel decomposition results, while different mixed pixel decomposition results will gain different sub-pixel mapping results. Hence, these three problems are correlative. In the practical application, generally it abides by the sequence of selecting the endmember, then decomposing the mixed pixel, and conducting the sub-pixel mapping. So it should not independently consider the sub-pixel mapping problems away from the other two problems, but it should build a comprehensive analysis model of the "mixed pixel".

1.6 Super Resolution Technique of HSI

The image resolution is simply a measurement of the imaging system on the image detail resolution capability, as well as an index of the objective subtle degree in the image. It indicates the level of the scenery information details. In the remote-sensing technique with rapid development today, there are higher and higher requirements on the resolution of the remote-sensing image, but the existing imaging device is far from meeting the requirements of all aspects. Consequently, enhancing the resolution of the imaging system effectively by other means, or gaining the high-resolution remote-sensing image have the practical value and important significance. The resolution enhancement method based on the hardware should be restricted by the technological level and other factors. Thus, enhancing the resolution by the image processing technique becomes a very active task in the remote-sensing field currently.

The traditional interpolation methods such as the nearest neighbor interpolation, bilinear interpolation, and cubic spline interpolation can increase the number of pixels of the output image. However, this method does not increase the information quantity of the original image, and also does not really enhance the resolution of the image in the strict sense, but it is fuzzy to the margin of the image.

In 1960s, the super resolution was proposed by Harris and Goodman in the concept and method of the single image restoration. Soon afterwards, many people researched and successively came up with various restoration methods, such as linear extrapolation method, superposition sine template method, and prolate ellipsoidal function method etc. The above methods achieved preferable simulation results, but they did not gain the ideal effect in the practical application. In the early 1980s, Huang T et al. firstly put forward the super resolution reconstruction problems based on the sequence image, and provided the reconstruction method based on the frequency domain approximation. In the late 1980s, many valuable methods were proposed and developed, such as projections onto convex sets method, continuous reduction of energy method, and Bayes analysis method etc. In the meantime, the super resolution image reconstruction by the sequence image became a hot spot of research by the people. It made the most of the similar yet different information among the sequence images with the mutual displacement, so it possessed relatively good super resolution restorability. However, in most cases, it was difficult to get the remote-sensing image of the time sequence, so its application was restricted greatly. Such a method could be collectively referred to as super resolution restoration method of sequence image.

Since 1970s, as an effective resolution enhancement method, the multi-information fusion has been widely researched. The multi information fusion means the process of handling the automatic detection, interconnection, correlation, estimation, and combination of the data and information from multiple sources, as well as the process of the information enrichment. In 1997, Wald L et al. enhanced the spatial resolution of the satellite remote-sensing image by the high-frequency component calculation in the multivariate data and in the mode of fusion, and gained preferable effect. In 1996, Yocky D et al. introduced the wavelet transformation and pyramid algorithm into the super resolution algorithm, which became the model of the multi-resolution analysis. In 1999, Aiazzi B et al. applied the generalized Laplacian Pyramid algorithm to conduct the effective research on the multi-information fusion with the non-integer resolution. Such a method could be collectively referred to as the resolution enhancement method based on the multi-information fusion.

With the constant in-depth research on the mixed pixel interpretation technique, in the 1990s, some scholars began the research of the mixed pixel resolution enhancement method gradually. Due to the prevailing mixed pixel in the remote-sensing image, it seriously impacted the spatial resolution capability of the image. The mixed pixel problem not only was the important barrier of the remote-sensing technique towards the quantification development, but also seriously influenced the application of the remote-sensing technique in various fields. The resolution enhancement method of the mixed pixel is to split each pixel into several sub-pixels, and then determine the gray level (or other characteristic values) of each sub-pixel, so it can enhance the image resolution to the sub-pixel level. The resolution enhancement methods of the mixed pixel include two links of mixed pixel interpretation and super resolution processing. Creating Repetitive Information (CRI) is the simplest sub-pixel mapping method, i.e., making the super resolution for the low resolution pixel into several similar sub-pixels under the state of keeping the constant numerical value. Obviously, this method could not better reflect the marginal information of the image. In 1998, Foody G et al. sharpened the low resolution image by using the high resolution image for two images with different spatial resolutions imaging simultaneously in the same place. In 1998, the sub-pixel mapping method proposed by Gross H et al. also needed the image with higher spatial resolution as the additional information. Such additional information was often hard to acquire in the practical application. In 1997, Atkinson P supposed that the land object spatial distribution could be determined by the distance scale, i.e., the similar pixels would have the similar values. According to this assumption, Verhoeye J et al. in 2002, and Jan V et al. in 2000 raised the spatial dependence (SD)-based sub-pixel mapping method, and achieved by adopting the linear optimization technique, with a preferable effect and fast operating rate. In 2004, Jiao Li researched the super resolution processing of the neural network-based interpretation component diagram, and obtained a preferable effect, but the calculated amount was larger. The super resolution technique of the mixed pixel becomes a very active issue in the remote-sensing field at present. The development status shows the following three outstanding features. Firstly, many methods need the unavailable additional information. Secondly, most methods have larger calculated amount. Thirdly, most methods are only limited to applying the super resolution processing of the relatively simple spectral interpretation component diagram. Such a method could be collectively referred to as the sub-pixel mapping method.

On the premise of the limited original image information, that increases the number of pixels of the output image by the interpolation method. At present, there are many classic interpolation methods, and the most common ones are various transformations of the nearest neighbor interpolation, bilinear interpolation, and the cubic spline interpolation. In 1996, the thesis of Schultz and Stevenson made a brief summary for various complex interpolation methods currently, involving the improved algorithm of the cubic spline interpolation, rule-based method, edge preserving method, and Bayes method. In the edge preserving-based interpolation method, it is worth mentioning of the spatial self-adaptive interpolation by the local structure, the iterative method based on the projections onto the convex sets, and edge direction-based interpolation, etc. Jensen K et al. applied the interpolation method of the second-order statistical model in the stationary random process, gained a preferable effect on the edge preserving, and reduced the calculated amount in the mode of mixing the linear interpolation method while realizing. In 2002, Leizza R et al. put forward the local area self-adaptive nonlinear interpolation method. For a point to be interpolated, through calculating the local standard deviation, compare result with the presetting threshold value, and accordingly determine the way of finishing the interpolation calculation of this point. Such a method could be collectively referred to as interpolation-based spatial resolution enhancement method.

The research on the resolution enhancement method in our country started late, but more and more research institutes and researchers paid close attention to this research. Peking University, Harbin Institute of Technology, Beijing Institute of Technology, Beijing Normal University, Wuhan University, Harbin Engineering University, Institute of Remote Sensing Applications CAS, and other units have done much work. In 2002, Junping Zhang utilized the auxiliary high-resolution optical image and enhanced the spatial resolution of the hyperspectral image by the means of fusion. In 2005, Jinzong Li et al. made the super resolution method research of the sequence image. In 2006, Liu Y et al. researched the mixing super resolution method based on the projections onto convex sets and the maximum posterior probability, and gained some effect. Wuhan University has done much work in the super resolution drawing method based on the spectral component image and has continued all to work till date.

In the above four typical methods, most resolution enhancement methods should be completed by the unavailable auxiliary information. For example, the fusion-based method generally needs the high-resolution optical image, while the super resolution restoration method of the sequence image needs utilizing the image with the spatial motion compensation relation. Various spectral sections of the hyperspectral image do not belong to the unrelated independent image, and also differ from the sequence image with the spatial motion compensation, but they are the image group with the inter-spectrum complementary information. Starting from this fact, building the specific super resolution model and researching the hyper spectral image resolution enhancement methods independent of the auxiliary information which makes the most of the inter-spectrum complementary information and meets the special application demands are highly anticipated. In 2005, Akgun T et al. made the groundbreaking research on such resolution enhancement methods, and gained the definite achievements. However, due to the changes of the continuity conditions of the spectrum in the transformation domain, the establishment of the imaging model in this method is short of the reasonability. In addition, the problems of the algorithm on the special application demands and the application of MAP algorithm here have not been researched.

1.7 Anomaly Detection Technique of HSI

Anomaly detection mainly depends on calculating the statistical changes of the local area to detect the anomaly target. The classic anomaly detection algorithm is derived from the RX algorithm of the hyperspectral image. The RX algorithm had been firstly proposed by Reed I et al. in 1990, and the RX algorithm was a constant false alarm rate detection method based on the generalized likelihood ratio test, for detecting the target object with the known spatial pattern and unknown spectral characteristic under the conditions of the Gaussian background statistical property and spatial whitening. The detector in the progressive sense simplified the original algorithm into the Mahalanobis distance from the sample to the ensemble average vector. The RX algorithm was the likelihood ratio detection operator constructed under some simplified assumed conditions. Handling the hyperspectral image by using RX algorithm directly will generate higher false alarm probability, which is mainly caused by two reasons. Firstly, the local statistical model adopted by the RX algorithm assumes that the data are the space independent or the spatial whitening, and the data are requested to obey the local normal distribution. Such assumption can not fully describe the situation of the real scenes. Secondly, when the RX algorithm is directly used for the hyperspectral image processing, it is necessary to calculate the covariance matrix of the sample, while the dimensionality of the covariance matrix will rapidly increase with the increase of the band numbers, which will result in the enormous calculated amount. Thereafter, Chang C et al. researched on the basis of the RX algorithm, put forward a series of the improved detection operator, and effectively enhanced the performance of the hyperspectral anomaly detection. In 1998, Ashton E et al. combined the LSMM, and used the RX detection anomaly upon the background suppression through the endmember extraction. In 2004, Riley R et al. constructed the weighted Euclidean distance by using the noise covariance matrix, combined it with the RX detector in the Lagrange multiplier method, and constructed the new detector. The detector effectively suppressed the noise disturbance while maintaining the sensitivity of the RX detector for the anomaly.

In 1998, Schweizer S et al. used the three-dimensional Gaussian MRF for modeling of the data, which could directly acquire the estimation of the inverse matrix of the band covariance matrix and avoid the difficulty during the inversion process. Also, constructed the generalized likelihood ratio detector for detection on the basis of the spatial correlation and inter-spectrum correlation. In 2004, Yver R et al. came up with an interested target anomaly detection algorithm based on the maximum posterior probability. It eliminated the false target by the Markov rule, but the energy function in the algorithm was only converged to the local optimal solution. Moreover, the parameter estimation of the multiple MRF process was comparatively complex. In 2003, Plaza A et al. expanded the traditional morphological filtering method to the high-dimensional spectral space, and achieved the joint target detection of the spatial dimension and spectral dimension. In 2007, Broadwater J et al. united multiple detectors for the detection of the sub-pixel target, considered the physical information and statistical information of the image while constructing the linear mixed model, with a preferable effect. In 2003, Kwon H et al. put forward the sub-space isolation self-adaptive anomaly detection algorithm based on the double-window data projection. The inner window data stand for the target, and the outer window data stand for the background. The mean values of the inner window and outer window data are projected toward the specific direction, and the degree of separation of the projected inner window and outer window data are deemed as the criterion of anomaly. Caefe et al. of the Air Force Research Laboratory firstly utilized the histogram of the principal component for the segmentation of the image, and then utilized the local statistics of the pixel neighborhood and segmentation area to construct the local anomaly measurement, so as to detect the point anomaly target.

In recent years, with the rapid development and progressive perfection of the statistical learning theory, it is possible to apply the kernel machine learning algorithm to handle the hyperspectral image. The hyperspectral data are mapped to the high-dimensional eigenspace for handling, to fully excavate the implicit non-linear information and further improve the separation performance between the hyperspectral small target and background. In 2007, the foreign scholars Kwon H et al. improved the traditional hyperspectral small target detection algorithms (such as RX, ASD, and CEM), mapped the original spatial spectral signal to the high-dimensional eigenspace by the kernel function, and accordingly realized the anomaly detection of the small target. By comparing with the detection performance of the traditional hyperspectral small target detection algorithms, it concluded that the kernel algorithm has been significantly improved in the aspects of enhancing the target detection probability and reducing the false alarm. In 2007, Goldberg H et al. achieved the anomaly detection of the small target in the eigenvalue decomposition method in the high-dimensional eigenspace, and meanwhile gained the detection result superior to the traditional method, but it was greatly impacted by the background interference. In 2007, Banerjee A raised an anomaly target detection algorithm based on Support Vector Data Description (SVDD), to avoid the inverse operation of the covariance matrix during the detection process and effectively enhanced the operating rate of the algorithm. However, these algorithms used the single form of the kernel function, and had a relatively poor adaptation. The parameters were determined through the constant attempts in the experiment, and are relatively tedious. The anomaly detection algorithm after the band fusion and dimensionality reduction by the kernel method was another train of thought, such as the band fusion method based on the support vector machine and the band fusion algorithm of selecting the kernel PCA. But the complex of the calculation and the last choice of the characteristics suitable for the anomaly detection still needed further research.

The domestic research on the hyperspectral image anomaly small target detection started relatively late. Institute of Remote Sensing Applications CAS, National University of Defense Technology, Harbin Institute of Technology, Northwestern Polytechnical University, The PLA Information Engineering University, and other units made greater progress in the research of the hyperspectral image target detection. In 2005, Xiurui Geng made in-depth research in the aspects of hyperspectral image feature extraction, unsupervised classifications, endmember extraction, and anomaly detection, and put forward the spectral rearrangement-based feature extraction method and the weighted sample covariance matrix-based target detection algorithm. In 2004, Zhiyong Li et al. researched the basic theory of the hyperspectral anomaly small target detection, analyzed the detection performance of different models, and automatically finished the hyperspectral image small target detection without the a priori knowledge. In 2004, Wei Lu et al. applied the combination of the projection pursuit (PP) and genetic algorithm to the hyperspectral image anomaly small target detection, and effectively projected the hidden target structure information of the high-dimensional data into the low-dimensional eigenspace centrally, but the genetic algorithm optimizing process was easily caught in the local optimum. In 2008, He L et al. gained a series of the research achievements in the multi-spectral/hyperspectral target detection, and effectively restrained the noise and enhanced the detection performance through making full use of the spatial information and scale information of the hyperspectral image. In 2002, the algorithm proposed by Yanfeng Gu et al. utilized the kernel principal component transformation for the nonlinear feature extraction, and carried out the small target detection after reducing the hyperspectral image data dimension. In 2007, Lina Xun et al. came up with an endmember extraction-based anomaly small target detection algorithm. In 2008, Derong Chen improved SVDD anomaly detection algorithm by the sample segmentation optimization selection, enhanced the operation efficiency of the algorithm on the basis of maintaining the detection performance, and also researched the low probability detection based on the compression and vertex component analysis with good effects. Qingbo Li et al. started from the spectral dimension, and utilized the Mahalanobis distance of the spectral angle matching for the anomaly detection, with higher operating rate. Yaxin Ji et al. from Harbin Engineering University put forward the anomaly target detection algorithm based on the secondary curvelet transformation and pulse coupling neural network fusion, and effectively achieved the hyperspectral data dimensionality reduction and feature extraction in allusion to the anomaly detection.

1.8 Dimensionality Reduction and Compression Technique for HSI

1.8.1 Dimensionality Reduction: Band Selection and Feature Extraction

For the band selection on the hyperspectral data, it is necessary to firstly conform to the criteria of the follow-up processing and determination of the band selection. The existing criteria mainly include the following classes.

1. From the perspective of information theory, the information quantity of the selected band or band combination should be kept to the maximum.
2. From the perspective of mathematical statistics, the selected bands should have weak correlation, to keep the independence and effectiveness of various bands.
3. From the perspective of the spectroscopy, the spectral characteristic difference of the land objects to be recognized in the research area should be the maximum.
4. From the perspective of the classification, the land object classes to be discriminated should have the strongest class discrimination in the selected band combination.

Those bands with large information quantity, small correlation, big land object spectral differences, and good discrimination will be the best bands to be selected.

1. **Best band selection based on information quantity**

Best band selection based on information quantity mainly considers the factors involving the inter-band combination entropy, covariance matrix, optimum index factor (OIF), self-adaptive band selection, and other methods. Such a method mainly involves the entropy and combination entropy, covariance matrix eigenvalues method, OIF, and self-adaptive band selection, etc.

2. **Best band selection based on inter-class discrimination**

In a general way, the inter-class object discrimination can calculate in allusion to the single band as well as the multi-band combination image. Accordingly, the processing result of the inter-mean standard distance model indicates that the processing result of the class object in the discrimination of each band, dispersion, Bhattachryya distance and Jeffries-Matusita distance, mixed distance, spectral angle mapping method, spectral correlation coefficient, and other models can reflect the class discrimination of the class object in the multi-band combination image. Such a method mainly involves the inter-mean standard distance method, dispersion method, B distance method, inter-class mean discrimination method, and the fractal method, etc.

3. **Fractal-based band selection**

The traditional band selection method is generally made on the basis of some basic statistics, such as variance, maximum, minimum, variation coefficient, etc. The maximum, minimum and standard deviation of the spectral band are simple and applicable, and can provide some information for understanding the feature of each band, but they do not reflect the spatial change information. The variable coefficient can not measure the spatial change component. The spatial structure information is the key to the remote sensing application. The structure determines the function. The spatial structure of the band determines the available information from it. Therefore, selecting the band from the spatial configuration should be another importance of the hyperspectral remote-sensing band selection method. Fractal is a new tool for exploring the spatial structure and space complexity, while the fractal dimension is the quantification expression. The fractal dimension as an index of the hyperspectral remote-sensing image band selection can make up the defect that the tradition-based statistics can not reflect the spatial information of the image. The differences of the fractal dimension reflect the undetectable spatial structure information differences in the visual method and the traditional method. When two bands are difficult to select, the size of the fractal dimension can be regarded as the scale for the selection. In addition, the fractal dimension differences of the vegetation and sand also reflect the spatial changes of the spectrum of different land object classes, which lays a new foundation for selecting the band in accordance with the discrimination among different land objects. Therefore, the combination of the traditional method and the fractal dimension will provide the new theoretical and technical supports for the hyperspectral remote-sensing band selection.

In addition to the above typical band selection methods, there are some methods worthy of reference for the readers: (I) the index method of the band selection, (II) the band selection based on genetic algorithm, (III) the band selection based on support vector machine, etc.

The feature extraction is the transformation-based dimensionality reduction. Compared with the band selection method, this method generally can carry more energies of the original signal, so it has relatively extensive application.

4. **PCA-based feature extraction**

PCA is to try to recombine the original numerous indexes with some correlation into a group of new irrelevant aggregative indexes instead of the original indexes. In order to make these new indexes reflect the information of the original indexes as much as possible, the projection variance of the original data point in the index is required to be large, if possible. Moreover, the information included in the previous principal component need not appear in the later principal component. PCA transformation is the linear transformation of meeting the following two conditions. First, remove the correlation between the transformation coefficients, i.e., the covariance matrix of the transformation coefficient is the diagonal matrix. Second, make the variance of the transformation coefficient centralize highly, i.e., the transformed energy mainly focuses on the previous M item, to guarantee that the mean square error upon removing the later several items will be the minimum during the feature extraction. Thus, the evaluation of the transformation matrix is the key. In terms of mathematics, this process is equal to the transformation of coordinates. The sequence of the principal component selection is determined by the order of the eigenvalues. Furthermore, these principal components constitute the transformation matrix Φ of the required solution. The specific implementation steps of PCA can be made in accordance with the following processes. (1) Firstly, calculate the covariance matrix $\boldsymbol{\Sigma}_X$ of the given image. (2) Then, evaluate the eigenvalue of $\boldsymbol{\Sigma}_X$ and the corresponding characteristic vector. (3) Determine the transformation matrix Φ from the characteristic vector. (4) Make the orthogonal transformation on the image by Φ.

Under different conditions or requirements, the partitioning PCA or noise regulation PCA (NAPCA) can replace the original PCA for the feature extraction. The other common transformation similar to PCA is MNF (see also Sect. 3.1.1 of this book), which can also be used for the feature extraction, with the details omitted.

5. **Projection pursuit (PP)-based feature extraction**

In 1999, Jimenez L et al. put forward the hyperspectral data feature extraction algorithm based on the projection method. This method utilized the projection matrix, and divided the adjacent bands into groups, and then project n_i feature bands in each group into one feature. It effectively reduced the data dimension through the projection, to provide the convenience for the subsequent analysis and processing. Later, they also developed the projection method of the supervision and parameters, and calculated the low-dimensional subspace by a priori knowledge.

The advantage of this method is to consider the true land object information, but the process of gaining the projection matrix is relatively complex.

6. **Best-bases (BB)-based feature extraction**

In 2001, Shailesh K et al. proposed a series of the best-bases feature extraction algorithms. This method is simple, fast, and very effective for the hyperspectral image feature extraction. It intelligently turned the subset composed by the adjacent bands into the subset with less features through the feature extraction. It included two algorithms of top-down and bottom-up. Top-down algorithm regressively divided the band into two (not always the same size) subsets, and replaced the finally obtained subset by the mean value. Bottom-up algorithm constructed the adjacent high-correlation bands into an aggregated tree, and extracted more effective features. These two algorithms converted the original *N* classes of problems into a class 2 problem of $\binom{C}{2}$.

7. **Discriminant Analysis Feature Extraction (DAFE)**

DAFE is often used for the hyperspectral image dimensionality reduction. Because this method uses each class of the mean vector and covariance matrix, it is also known as the parameter feature extraction algorithm. The purpose is to find out a transformation matrix $\boldsymbol{A}$, to make the class discrimination of the transformation data $\boldsymbol{Y}$ is the maximum. For the advantage of DAFE, it does not necessarily know the distribution form. Its weakness lies in the effectiveness only in the normal class distribution. When the class shows the multi-model mixed distribution, the effect of DAFE is unsatisfactory.

8. **Decision Boundary Feature Extraction (DBFE)**

Lee and Landgrebe came up with the feature extraction by the decision boundary, and applied it to the neural network and statistical classifier, respectively. They proved that the sequence of Decision Boundary Feature Matrix (DBFM) was capable of acquiring the demanded minimum dimension of the same classification accuracy as the original eigenspace. In other words, for some classifier, the minimum feature as required by the classification can be gained by DBFM. However, in this method, the estimation of the class statistical parameters as required by the classification should be made in the entire data space. For the high-dimensional data, it needs a large number of training samples, and the calculation time is longer.

9. **Nonparameter weighting-based feature extraction**

Due to the weakness of DAFE, in 2002, Kuo B et al. put forward the nonparameter weighting-based feature extraction. This method is based on the full rank scatter matrix. Compared with the parameter DAFE method, this method appoints the feature number of the ideal quantity, and reduces the influence of the singular problems. Assign the large weight to the sample in the decision boundary, and increase the classification accuracy greatly.

Besides, there have been the discriminant criterion-based feature extraction method and the nonlinear criterion-based feature extraction method in the works of Tong et al. (2006).

1.8.2 Compression: Lossy Compression and Lossless Compression

Two types of the remote-sensing image compression mode are the lossless compression (distortionless compression) mode, and the lossy (limited distortion compression) mode (Wan et al. 2010).

With the continuous development of the imaging spectrum technique, the cognitive ability of the people on the remote-sensing image is deepened constantly. Meanwhile, the new problems are generated, and resulted in very great difficulties for the hyperspectral data application, i.e., the mass data storage and transmission problems. How to conduct the effective hyperspectral image data compression and diminish the signal bandwidth is the urgent problem in the imaging spectrum technique, and also becomes one of the most challenging issues in communication technology.

From the perspective of information theory, all compression techniques achieved the purpose of the compression by eliminating the redundancy. The lossless compression can recover the original image from the compressed data inerrably. However, the compression ratio of the lossless compression is limited, and it may exceed the upper limit of the channel capacity after compression. At this time, it is necessary to abandon some data, and adopt the mode of the limited distortion lossy compression to gain the larger compression ratio. For the lossy compression and lossless compression modes, the main implementation methods can be divided into three classes, involving prediction-based method, transformation-based method, and the VQ-based method.

The prediction-based method mainly utilizes the correlation among the pixels in the image, i.e., utilizing the spatial adjacent pixels of the current pixel to joint predict, then subtracting the predicted value from the current pixel value to get the predicted residuals, and adopting the corresponding entropy coding method to compress the residual data. The most basic linear prediction method is the differential pulse code modulation (DPCM), which minimizes the difference value between the output signal predicted value of the predictor, and the current actual value through the selection of the predicted coefficient.

The transformation-based method is also one of the methods in the field of the hyperspectral image lossless compression. In the image spatial domain, the redundant information is distributed in the larger scale of the spatial pixel set, and is difficult to make the decorrelation directly. Moreover, the image is mapped to the transformation domain from the spatial domain, and the energy in the image is centralized. A few transformation coefficients with larger amplitude stand for the most energy in the image, while the vast majority of the transformation coefficients stand for some unimportant detail components of the image. Describe the principal energy component represented by the coefficient with larger amplitude through the fewer code words, while quantizing the detail components represented by the transformation coefficients with the smaller amplitude, to enable and obtain a higher compression ratio. At present, the common transformation methods mainly include PCA also known as the Karhunen-Loeve transform (KLT), discrete cosine transform (DCT), and discrete wavelet transform (DWT). While making the lossless compression in the transformation-based method, it is necessary to use the integer transformation form, i.e. the data before and after transformation are the integer forms, so as to achieve the purpose of the lossless compression.

The principle of VQ is to quantize the data block directly, without the decorrelation pre-processing. VQ regards the high-order entropy of the information as the lower limit, and obtained the optimal compromise between the high-compression ratio and average minimum distortion. The corresponding spectrum of any pixel in the hyperspectral image can be expressed by a vector, and each element in the vector corresponds to a specific wave length. Because the similar earth's surface possesses the similar spectral curve, the VQ is the ideal method of the hyperspectral image compression.

In the above three compression methods, the VQ-based compression method can obtain a higher compression effect, but the overlarge calculated amount restricts the application in practice. The prediction-based method has a long history and the compression effect is the best, so it arouses the wide concerns. The transformation-based method has the common compression performance and the algorithm complexity is not high, but it is mostly applied to the lossy compression of the hyperspectral image.

Although the lossless compression of the hyperspectral image can achieve the complete reconstruction of the data, but numerous pre-processing means adopted in the lossless compression increase the calculated amount of the algorithm to a large extent, and it is difficult to finish the real-time coding and decoding of the hyperspectral data. On the other hand, the compression ratio of the lossless compression of the hyperspectral data is generally lower. In the case of the existing communication bandwidth, it does not yet meet the demand of the real-time transmission of the hyperspectral data. In some occasions with higher requirements on the data timeliness, such as the battlefield situation, the lossless compression method is seemingly unable to do as well as one would wish. For the hyperspectral data acquired from the satellite borne imaging spectrometer, restricted by the storage capacity of the satellite, and the satellite link transmission bandwidth, in the case of realizing the real-time transmission of the hyperspectral data, it is necessary to adopt the efficient lossy compression method to reduce the data volume.

The lossy compression of the hyperspectral image can be divided into the above three methods.

In addition to two compression methods of the lossy compression and lossless compression, the classification/clustering method is also applied to the compression. The classification/clustering method is the land object type recognition classification in the hyperspectral image, and only records the class mark of each pixel in the image while storing. Such a method can gain the highest compression ratio, but the information loss is great and the application range is very limited. In order to enhance the compression ratio of the hyperspectral image substantially in the case of keeping the useful information as much as possible, people applied the region of interest (ROI) compression thought into the hyperspectral image compression, and made the lossless compression or high fidelity compression for ROI, and the compression with high compression ratio for non-ROI, to achieve the combination of the lossless compression and lossy compression of the image. In some cases, using one method alone is difficult to take effect, and the compression train of thought combining multiple methods becomes the trend of the compression technological development. In the practical application, it is necessary to select different compression modes as per the specific application demands.

References

Li ES, Zhang BM, Song LH, Yu WJ, Tang DJ (2011) A review on spectral unmixing algorithms based on linear mixing model. Sci Surv Mapp 36(5):42–44

Liu CH (2005) Research on dimensional reduction and classification of hyperspectral remote sensing image. Harbin Engineering University, Nangang

Ren W, Ge Y (2011) Progress on sub-pixel mapping methods for remotely sensed images. Remote Sens Technol Appl 26(1):33–44

Sun JB (2003) Remote sensing principle and application. Wuhan University Press, Wuhan, pp 210–211

Tong QX, Zhang B, Zheng LF (2006) Hyperspectral remote sensing: principle, technology and application. Higher Education Press, Beijing

Wan JW, Nian YJ, Su LH, Xin Q (2010) Research progress on hyperspectral imagery compression technique. Sig Process 26(9):1397–1407

Chapter 2
Classification Technique for HSI

Classification is one of the most basic and most important research contents of the hyperspectral data processing (Richards and Jia 2006). Classification is an analytical technique of describing the land object target or class, with the main task of a process of giving a class mark to each pixel point of the data volume to generate the thematic map. It is one of the important ways for people to extract the useful information from the remote-sensing image. The thematic map upon classification can clearly reflect the spatial distribution of the land objects, so that people can know and discover the rules and the hyperspectral remote-sensing image possesses the real use value and is effectively put into the practical application. After introducing several typical classification methods and evaluation criteria, this chapter focuses on the burgeoning SVM (Vapnik 2000) based classification method.

2.1 Typical Classification Methods

1. **Spectral angle match**

Spectral Angle Match (SAM) (Sohn and Rebello 2002) is an angle-based hyperspectral image classification method. It automatically compares the image spectrum with various spectrums or spectral libraries. According to the physical basis of remote sensing, the reflection spectrum of the land objects can determine the land object types to a great extent, and accordingly lead out SAM classification. Thereby finishing the transformation from the measurement space to the eigenspace through mapping the measurement spectral vector into a series of angle values on behalf of the similarity of this vector and the reference spectral vector. Calculating the spectral angle between the two spectrums can determine the level of their similarity. The dimension of the spectral vector is the number of bands. The similarity α between the unknown spectrum $\boldsymbol{t}$ and reference spectrum $\boldsymbol{r}$ shall be determined by the following formula.

$$\alpha = \cos^{-1}\frac{\langle \boldsymbol{t}, \boldsymbol{r} \rangle}{\|\boldsymbol{t}\| \cdot \|\boldsymbol{r}\|} \tag{2.1}$$

L. Wang and C. Zhao, *Hyperspectral Image Processing*,
DOI 10.1007/978-3-662-47456-3_2

With the standard spectrum measured in the laboratory or the average spectrum of the known point directly extracted from the image for reference, take the generalized included angle α for each pixel vector and reference the spectral vector in the image. The smaller α is, the greater the level of their similarity is. In the general application, usually select the known type of the area from the image, classify with the sample center of the average spectrum, take the included angle between each pixel of the image and the center of each class, and then put the pixel into the class of the corresponding minimum included angle.

2. **Maximal likelihood classification**

Maximal likelihood (ML), also known as Bayes code (Chen and Tu 1996; Jia and Richards 1994), discrimination function is the parametric method of the statistical pattern recognition. This method should employ various a priori probabilities $p(\omega_i)$ and conditional probability density functions $p(\boldsymbol{X}/\omega_i)$. The priori probabilities $p(\omega_i)$ are generally given in accordance with all kinds of a priori knowledge (practical situation of the specific problems, and information accumulated in history etc.) or are supposed to be equal. Moreover, $p(\boldsymbol{X}/\omega_i)$ firstly determines the distribution form, and then estimates the parameters used in this form by the training field. The estimation of the distribution form has multiple methods such as the maximum entropy method and the polynominal method. In the remote-sensing problem, the assumption of the normal distribution is reasonable, i.e., some non-normal problems can be converted into the normal problems for handling in the mathematical method.

Given $p(\boldsymbol{X}/\omega_i)$ is the $i(i = 1, 2, \ldots, N)$ class of the probability density function in d dimension characteristic data space, and $p(\omega_i)$ is the i class of the occurrence probability in the data set, the decision $\boldsymbol{X}$ is ω_i class, rather than ω_j class, equivalent to

$$p(\boldsymbol{X}/\omega_i)p(\omega_i) \geq p(\boldsymbol{X}/\omega_j)p(\omega_j) \tag{2.2}$$

In the practical application, the probability density function is often assumed to be the normal or Gaussian distribution, and then the class probability density function is expressed as

$$p(\boldsymbol{X}/\omega_i) = \frac{1}{(2\pi)^{N/2}|\boldsymbol{\Sigma}_i|^{1/2}} \exp\left[-\frac{1}{2}(\boldsymbol{X} - \boldsymbol{\mu}_i)^{\mathrm{T}} \boldsymbol{\Sigma}_i^{-1} (\boldsymbol{X} - \boldsymbol{\mu}_i)\right] \tag{2.3}$$

where, $\boldsymbol{\mu}_i$ is the class mean value vector, and $\boldsymbol{\Sigma}_i$ is the covariance matrix. In this case, the class mean value vector and the class covariance matrix can be estimated only if selecting the appropriate samples.

If the assumption of Gaussian distribution is correct, the decision function can be simplified further. Then, for all $j = 1, 2, \ldots, N$, if the formula (2.2) is correct, then

$$\ln[p(\boldsymbol{X}/\omega_i)p(\omega_i)] \geq \ln\left[p(\boldsymbol{X}/\omega_j)p(\omega_j)\right] \tag{2.4}$$

It is also correct, so the decision function can be expressed as

$$g(\boldsymbol{X}) = \ln[p(\omega_i)] - \frac{1}{2}\ln|\boldsymbol{\Sigma}_i| - \frac{1}{2}(\boldsymbol{X} - \boldsymbol{\mu}_i)^{\mathrm{T}}\boldsymbol{\Sigma}_i^{-1}(\boldsymbol{X} - \boldsymbol{\mu}_i) \tag{2.5}$$

We make the classification and recognition mainly in accordance with the decision criterion in the above formula.

3. **Fisher discriminant analysis**

Fisher discriminant analysis is a supervised classification method, and the main idea is to conduct the linear combination for the multivariate observed value to set up the new discriminant amount, and maximize the specific value of the inter-class variance and intra-class variance of the new discriminant amount.

Given unmixing the Nc classes, and each class has Ntr_i training samples and the number of bands for each training sample are ND, so the training sample of each class constitute a matrix of ND $\times$ Ntr_i. Meanwhile, given the training samples are expressed by $x_1, x_2, x_3, \ldots, x_{\mathrm{Ntr}_i}$.

Each sample mean $\overline{\boldsymbol{m}_i}$ shows the mean value of each sample.

$$\overline{\boldsymbol{m}_i} = \frac{\sum_{p=1}^{\mathrm{Ntr}_i} \boldsymbol{x}_p}{\mathrm{Ntr}_i} \quad i = 1, 2, 3, \ldots, \mathrm{Nc} \tag{2.6}$$

The inter-class sample mean $\bar{\boldsymbol{m}}$ shows the total mean value of all samples.

$$\bar{\boldsymbol{m}} = \frac{\sum_{q=1}^{\mathrm{Nc}} \sum_{p=1}^{\mathrm{Ntr}_i} \boldsymbol{y}_{qp}}{\sum \mathrm{Ntr}_i} \quad i = 1, 2, 3, \ldots, \mathrm{Nc} \tag{2.7}$$

In formula (2.7), $\boldsymbol{y}_{qp}$ shows the p training sample in the q class.

For the sample intra-class dispersion matrix $\boldsymbol{S}_i$ and total intra-class dispersion matrix $\boldsymbol{S}_w$, $\boldsymbol{S}_i$ stands for the internal differences of the i class training, while $\boldsymbol{S}_w$ stands for the total internal differences of all training samples.

$$\boldsymbol{S}_i = \sum_{p=1}^{\mathrm{Ntr}_i} (\boldsymbol{y}_{ip} - \overline{\boldsymbol{m}_i})(\boldsymbol{y}_{ip} - \overline{\boldsymbol{m}_i})^{\mathrm{T}} \quad i = 1, 2, 3, \ldots, \mathrm{Nc} \tag{2.8}$$

$$\boldsymbol{S}_w = \sum \boldsymbol{S}_i \quad i = 1, 2, 3, \ldots, \mathrm{Nc} \tag{2.9}$$

The inter-class dispersion matrix $\boldsymbol{S}_b$ stands for the total inter-class dispersion, which can stand for the differences among all classes.

$$\boldsymbol{S}_b = \sum_{\mathrm{i}=1}^{\mathrm{Nc}} (\overline{\boldsymbol{m}_i} - \bar{\boldsymbol{m}})(\overline{\boldsymbol{m}_i} - \bar{\boldsymbol{m}})^{\mathrm{T}} \quad (2.10)$$

Considering the linear combination

$$y = \boldsymbol{Ux} \quad (2.11)$$

where, $\boldsymbol{U}$ is the matrix of $1 \times \mathrm{ND}$, indicating some linear combination operation on the original spectrum. Then the degree of separation upon the matrix $\boldsymbol{U}$ transformation is:

$$J = \frac{\boldsymbol{US}_b\boldsymbol{U}^{\mathrm{T}}}{\boldsymbol{US}_w\boldsymbol{U}^{\mathrm{T}}} \quad (2.12)$$

In the formula (2.12), J is the dispersion. When $\boldsymbol{U}$ maximizes J, $\boldsymbol{U}$ will minimize the intra-class distance of the sample and maximize the inter-class distance. Thus, such $\boldsymbol{U}$ is the linear transformation which we seek. Such $\boldsymbol{U}$ can be obtained by calculating the characteristic vector of $\boldsymbol{S}_w^{-1}\boldsymbol{S}_b$. The characteristic vector enabling J to gain the maximum vector is called the first discriminant vector, and the characteristic vector enabling J to gain the second is called the second characteristic vector. Similarly, we can gain multiple discriminant vectors, and distinguish Nc classes to find the Nc − 1 discriminant vector.

2.2 Typical Assessment Criterions

The pixel level-based accuracy assessment of the hyperspectral image classification results is obtained on the basis of the classification confusion matrix. The form of the classification confusion matrix is as follows:

$$\boldsymbol{M} = \begin{bmatrix} m_{11} & m_{12} & \cdots & m_{1\mathrm{Nc}} \\ m_{21} & m_{22} & \cdots & m_{2\mathrm{Nc}} \\ & \cdots & \cdots & \\ m_{\mathrm{Nc}1} & m_{\mathrm{Nc}2} & \cdots & m_{\mathrm{NcNc}} \end{bmatrix} \quad (2.13)$$

where, m_{ij} shows the number of pixels that should belong to the i class sample in the experimental area and are assigned to the j class, and Nc is the number of classification classes. In the confusion matrix, the larger the element numerical value of the diagonal is, the higher the reliability of the classification results is. On the contrary, it shows the more serious phenomena of the mis-classification.

According to the classification confusion matrix, it can calculate the overall accuracy OA, user's accuracy CA_i_user, and producer's accuracy CA_i_producer:

$$\begin{aligned} \mathrm{OA} &= \frac{1}{\mathrm{Nte}} \sum_{i=1}^{\mathrm{Nc}} m_{ii}, \\ \mathrm{CA}_i\_\mathrm{user} &= \frac{m_{ii}}{\mathrm{Nte}_i}, \\ \mathrm{CA}_i\_\mathrm{producer} &= \frac{m_{ii}}{\mathrm{Nte}_i^*}, \quad i = 1, 2, \ldots, \mathrm{Nc} \end{aligned} \tag{2.14}$$

In the formula, Nte is the total test samples, Nte_i is the total test samples of the i class, and Nte_i^* is the total pixels assigned to the i class. Thus, m_{ii} is the number of samples of the i class correct classification. When the number of various samples is equal, the overall accuracy is equivalent to the average user's accuracy.

The other accuracy analytical method is to make the quantification assessment for the overall effective performance of the classifier on the basis of the classification confusion matrix, and the most common one is Kappa coefficient.

$$\mathrm{Kappa} = \frac{\mathrm{Nte} \sum_{i=1}^{\mathrm{Nc}} m_{ii} - \sum_{i=1}^{\mathrm{Nc}} m_{i+} m_{+i}}{\mathrm{Nte}^2 - \sum_{i=1}^{\mathrm{Nc}} m_{i+} m_{+i}} \tag{2.15}$$

where, + shows the summation of the line or row, and Nte is the total test samples. Thus, this calculation uses each element in the classification matrix. The significance of Kappa coefficient can be interpreted as, if Kappa value of the classification result is 0.8, the classification method should be superior to 80 % of the method which randomly gives each point to a class. In the practical application, it often adopts the following form:

$$\mathrm{Kappa} = \frac{\theta_1 - \theta_2}{1 - \theta_2} \tag{2.16}$$

where,

$$\theta_1 = \frac{\sum_{i=1}^{\mathrm{Nc}} m_{ii}}{\mathrm{Nte}}, \theta_2 = \frac{\sum_{i=1}^{\mathrm{Nc}} m_{i+} m_{+i}}{\mathrm{Nte}^2}.$$

This book mainly adopts the overall accuracy to express the classification results, and if necessary, makes assessment by the user's accuracy of each class (referred to as classification accuracy of each class) and the mean value (referred to as the mean classification accuracy).

2.3 SVM-Based Classification Method

SVM is the new generation machine learning theory developed on the basis of the statistical learning theory, and seeks the best compromise between the model complexity and learning ability in accordance with the limited sample information, in the hope of gaining the best generalization ability.

2.3.1 *Theory Foundation*

SVM is the epitome of several standard techniques in the field of the machine learning, which integrates multiple techniques such as maximum margin hyperplane, Mercer kernel, convex quadratic programming, sparse solution, and slack variable. Here, we will introduce the important theoretical basis of SVM—VC dimension theory, and the footstone of SVM algorithm—structural risk minimization principle.

1. **VC dimension** (Karpinski and Werther 1989)

The statistical learning theory defines a series of the performance indexes relevant to the function set learning, and the most important one is the VC dimension. VC dimension of an indication function set $Q(z, \boldsymbol{\alpha}), \boldsymbol{\alpha} \in \Lambda$ means the maximum number h that can be divided into two classes of the vectors $z_1, z_2, \ldots, z_h$ in all possible 2^h modes by the centralized function of the function, i.e., the maximum number of the vector that can be scattered by the function set. If the set of an n vector for any natural number n can be scattered by the function set $Q(z, \boldsymbol{\alpha}), \boldsymbol{\alpha} \in \Lambda$, VC dimension of the function set is infinity. Illustrate VC dimension below.

VC dimension of the linear indication function set $Q(z, \boldsymbol{\alpha}) = \sum_{i=1}^{d} \alpha_i z_i + \alpha_0, \alpha_0, \ldots, \alpha_d \in (-\infty, \infty)$ in d dimension coordinate space $Z = \{z_1, z_2, \ldots, z_d\}$ is $d + 1$, because the function in this set can scatter $d + 1$ vectors to the maximum. VC dimension reflects the learning ability of the function set. The larger the VC dimension, the more complex the learning machine is. As shown in Fig. 2.1, the linear function set of two-dimensional space can make eight possible second-class classifications for three data points, and VC dimension is 3.

2. **Structural risk minimization principle**

The statistical learning theory systematically researches the relationship between the empirical risk and actual risk for various kinds of the function sets, i.e., the promotional boundary. In regard to two classification problems, for all functions in the indication function set, the empirical risk $R_{\mathrm{emp}}(\alpha)$ and actual risk $R(\alpha)$ meet the following relations by at least $1 - \eta$ probability.

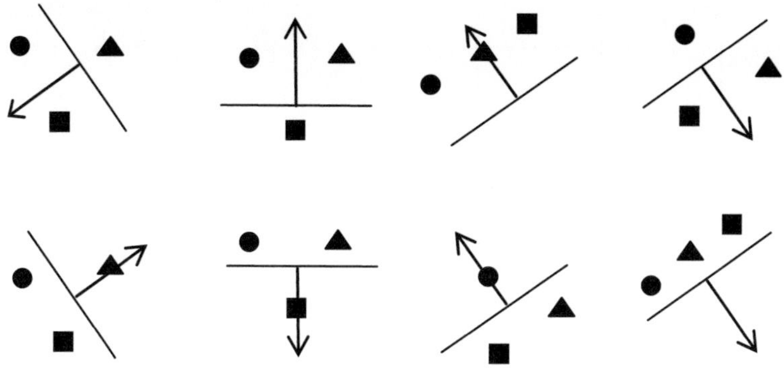

Fig. 2.1 2 VC dimension of dimensional space linear function set

$$R(\alpha) \leq R_{emp}(\alpha) + \sqrt{\frac{h[\ln(2n/h) + 1 - \ln(\eta/4)]}{n}} \tag{2.17}$$

where, h is the VC dimension of the function set, and n is the number of samples.

This conclusion in theory indicates that the actual risk of the learning machine is composed of two sections, involving the empirical risk (training error) and confidence risk (VC confidence), and reflects the generalization ability of the learning machine gained in accordance with the empirical risk minimization principle, so-called as the promotional boundary. It shows that, under the limited training sample, the higher VC dimension of the learning machine is, the greater the confidence risk is. Accordingly, it results in the larger differences between the actual risk and the empirical risk. It is the reason that it has the overfitting phenomenon.

In the traditional method, the process of selecting the learning model and algorithm is the process of adjusting the confident range. If the model is relatively applicable to the existing training sample, it can gain the preferable effect. However, because of lack of theoretical guidance, such selection can only depend on the priori knowledge and experience, and result in the overdependence of the neural network and other methods on the user's skills. When the training sample is applicable to the existing model, the expected risk is close to the value of the empirical risk. In this case, the smaller empirical risk can guarantee the smaller expected risk. For the sample with the number of n, if the specific value n/h is smaller (generally subject to 20 times), we deem that the number of the samples is small, i.e., we deem that the sample set is the small sample. If the training sample is the small sample, a small empirical risk value cannot guarantee the small expected risk value. In this case, to minimize the actual risk value, the learning ability (VC dimension) of the function set must become a controllable variable. The statistical learning theory puts forward a new strategy, i.e., constructing the function set into a function subset sequence, making each subset arrange in accordance with the size of VC dimension, seeking the minimum empirical risk, in each subset, compromising the consideration of the

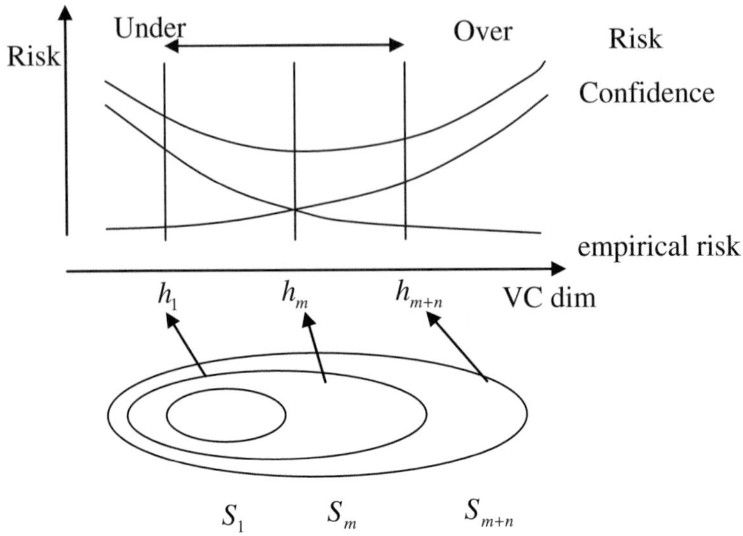

Fig. 2.2 Schematic diagram of structural risk minimization

empirical risk and the confidence risk among the subsets, and obtaining the minimum of the actual risk. This idea is called the structural risk minimization (SRM), as shown in Fig. 2.2.

2.3.2 *Classification Principle*

The original SVM theory is used for handling two types of classification problems. The classification principle can be summarized as seeking a classification hyperplane, enabling two types of sample points in the training sample to be separated, and having the distance from the plane as far as possible. While for the problems of the linear inseparability, we map the data of the low-dimensional input space into the high-dimensional space through the kernel function, and accordingly transform the linear inseparability problems of the original low-dimensional space into the linear separability problem of the high-dimensional space. Before the specific introduction, we firstly provide the basic definition and theorem in several optimization theories.

Define the original problem in the domain $\Omega \subseteq R^n$:

$$\begin{aligned} \min \quad & f(\boldsymbol{w})\boldsymbol{w} \in \Omega \\ \text{s.t.} \quad & g_i(\boldsymbol{w}) \leq 0 \quad i = 1, \ldots, k_1 \\ & h_i(\boldsymbol{w}) = 0 \quad i = 1, \ldots, k_2 \end{aligned} \tag{2.18}$$

Then the generalized Lagrange form of the original problem (2.18):

$$\begin{aligned} L(\boldsymbol{w},\boldsymbol{\alpha},\boldsymbol{\beta}) &= f(\boldsymbol{w}) + \sum_{i=1}^{k_1} \alpha_i g_i(\boldsymbol{w}) + \sum_{i=1}^{k_2} \beta_i h_i(\boldsymbol{w}) \\ &= f(\boldsymbol{w}) + \langle \boldsymbol{\alpha}, g(\boldsymbol{w})\rangle + \langle \boldsymbol{\beta}, h(\boldsymbol{w})\rangle \end{aligned} \tag{2.19}$$

Furthermore, the Lagrange dual problem of the original problem (2.18) can be expressed as:

$$\begin{aligned} &\max \quad \theta(\boldsymbol{\alpha},\boldsymbol{\beta}) = \inf_{\boldsymbol{w}\in\Omega} L(\boldsymbol{w},\boldsymbol{\alpha},\boldsymbol{\beta}) \\ &\text{s.t.} \quad \boldsymbol{\alpha} \geq 0 \end{aligned} \tag{2.20}$$

Kuhn–Tucker theorem (Cristianini and Shawe-Taylor 2004): Given an optimization problem (2.18) defining in the convex domain $\Omega \subseteq R^n$, where f is the continuous convex function, and g_i and h_i are affine functions. Generally, the necessary and sufficient condition that a point $\boldsymbol{w}^*$ is the optimal point will have $\boldsymbol{\alpha}^*$ and $\boldsymbol{\beta}^*$, meeting:

$$\begin{aligned} \frac{\partial L(\boldsymbol{w}^*,\boldsymbol{\alpha}^*,\boldsymbol{\beta}^*)}{\partial \boldsymbol{\alpha}} &= 0 \\ \frac{\partial L(\boldsymbol{w}^*,\boldsymbol{\alpha}^*,\boldsymbol{\beta}^*)}{\partial \boldsymbol{\beta}} &= 0 \end{aligned} \tag{2.21}$$

$$\begin{aligned} &\alpha_i^* g_i(\boldsymbol{w}^*) = 0 \quad i = 1,\ldots,k_1 \\ &g_i(\boldsymbol{w}^*) \leq 0 \quad i = 1,\ldots,k_1 \\ &\alpha_i^* \geq 0 \quad i = 1,\ldots,k_1 \end{aligned} \tag{2.22}$$

where, the relational expression $g_i(\boldsymbol{w}^*) \leq 0 \quad i = 1,\ldots,k_1$ is called the KKT complementary condition.

1. **Optimal classification hyperplane**

For two types of classification problems of the linear separability, one of the key techniques is to seek the optimal classification hyperplane, i.e., determining the optimal linear discriminant function. Given $\boldsymbol{x}_i \in R^d$ is the sample data, and $y_i \in \{+1, -1\}$ is the corresponding class mark, $i = 1,\ldots,$ Ntr. The general form of the linear discriminant function is $g(\boldsymbol{x}) = \langle \boldsymbol{w}, \boldsymbol{x}\rangle + b$, and the corresponding classification plane is $\langle \boldsymbol{w}, \boldsymbol{x}\rangle + b = 0$. In the formula, $\boldsymbol{x}$ is the d dimension characteristic vector, and $\boldsymbol{w}$ is called the weight vector. It can be expressed as $\boldsymbol{w} = [w_1, w_2,\ldots,w_d]^{\mathrm{T}}$. b is the constant, called as threshold weight. For the linear classifier with two types of problems, it can adopt the following decision rules:

$$\begin{cases} g(\boldsymbol{x}) > 0 \Rightarrow \boldsymbol{x} \in \omega_1 \\ g(\boldsymbol{x}) < 0 \Rightarrow \boldsymbol{x} \in \omega_2 \\ g(\boldsymbol{x}) = 0 \Rightarrow \boldsymbol{x} \in \omega_1 \text{ or } \omega_2 \end{cases} \tag{2.23}$$

The equation $g(\boldsymbol{x}) = 0$ defines a decision surface, and divides the points of belonging to different classes. The decision hyperplane is recorded as H.

$g(\boldsymbol{x})$ can be regarded as an algebra measurement of the distance from some point $\boldsymbol{x}$ to the hyperplane H in the eigenspace. If $\boldsymbol{x}$ is expressed as

$$\boldsymbol{x} = \boldsymbol{x}_p + r\frac{\boldsymbol{w}}{\|\boldsymbol{w}\|} \tag{2.24}$$

In the formula, $\boldsymbol{x}_p$ is the projection vector of $\boldsymbol{x}$ in H, r is the vertical distance from $\boldsymbol{x}$ to H, $\frac{\boldsymbol{w}}{\|\boldsymbol{w}\|}$ is the unit vector in the $\boldsymbol{w}$ direction. In combination with the two formulas of (2.23) and (2.24), it can obtain

$$g(\boldsymbol{x}) = \left\langle \boldsymbol{w}, \left(\boldsymbol{x}_p + r\frac{\boldsymbol{w}}{\|\boldsymbol{w}\|}\right)\right\rangle + b = \langle \boldsymbol{w}, \boldsymbol{x}_p\rangle + b + r\frac{\langle \boldsymbol{w}, \boldsymbol{w}\rangle}{\|\boldsymbol{w}\|} = r\|\boldsymbol{w}\| \tag{2.25}$$

or indicating it as

$$r = \frac{g(\boldsymbol{x})}{\|\boldsymbol{w}\|} \tag{2.26}$$

If $\boldsymbol{x}$ is the original point, then

$$g(\boldsymbol{x}) = b \tag{2.27}$$

In combination with the two formulas of (2.26) and (2.27), the distance from the original point to the hyperplane is obtained:

$$r_0 = \frac{b}{\|\boldsymbol{w}\|} \tag{2.28}$$

Thus, in order to separate the samples to be separated as well as possible, we demand the maximum geometric margin (i.e., the projection of two classes of the minimum distance segment in the direction perpendicular to the classification hyperplane), equivalent to minimizing $\|\boldsymbol{w}\|$. Figure 2.3 provides the graphical representation of the SVM maximization margin attribute in two cases.

Maximizing the geometric margin is equivalent to minimizing $\|\boldsymbol{w}\|$. Therefore, seeking the optimal classification plane is transformed into the following optimal problem:

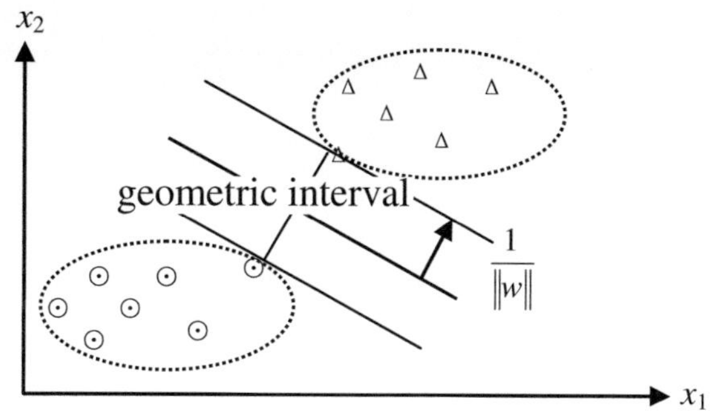

Fig. 2.3 Classification hyperplane of maximization geometric margin

$$\begin{aligned} &\min \quad \frac{1}{2}\|\boldsymbol{w}\|^2 \\ &\text{s.t.} \quad y_i(\langle \boldsymbol{w}, \boldsymbol{x}_i \rangle + b) - 1 \geq 0, \quad i = 1, 2, \ldots, \text{Ntr} \end{aligned} \tag{2.29}$$

Construct Lagrange function by the expression (2.29):

$$L(\boldsymbol{w}, b, \boldsymbol{\alpha}) = \frac{1}{2}\|\boldsymbol{w}\|^2 - \sum_{i=1}^{\text{Ntr}} \alpha_i [y_i(\langle \boldsymbol{w}, \boldsymbol{x}_i \rangle + b) - 1] \tag{2.30}$$

Here, Lagrange multiplier (support value) is $\alpha_i \geq 0$. By taking the derivative for the corresponding $\boldsymbol{w}$ and b, the following relational expression can be obtained.

$$\begin{aligned} \frac{\partial L(\boldsymbol{w}, b, \boldsymbol{\alpha})}{\partial \boldsymbol{w}} &= \boldsymbol{w} - \sum_{i=1}^{\text{Ntr}} \alpha_i y_i \boldsymbol{x}_i = 0 \\ \frac{\partial L(\boldsymbol{w}, b, \boldsymbol{\alpha})}{\partial b} &= \sum_{i=1}^{\text{Ntr}} \alpha_i y_i = 0 \end{aligned} \tag{2.31}$$

i.e.,

$$\begin{aligned} \boldsymbol{w} &= \sum_{i=1}^{\text{Ntr}} \alpha_i y_i \boldsymbol{x}_i \\ \sum_{i=1}^{\text{Ntr}} \alpha_i y_i &= 0 \end{aligned} \tag{2.32}$$

Substitute this formula into the Lagrange function (2.30), and obtain the dual problem of the original problem, i.e., maximizing the following objective functions:

$$\begin{aligned} & L(\boldsymbol{w}, b, \boldsymbol{\alpha}) = \sum_{i=1}^{\mathrm{Ntr}} \alpha_i - \frac{1}{2}\sum_{i,j=1}^{\mathrm{Ntr}} \alpha_i\alpha_j y_i y_j \langle \boldsymbol{x}_i, \boldsymbol{x}_j\rangle \\ \text{s.t.} \quad & \sum_{i=1}^{\mathrm{Ntr}} \alpha_i y_i = 0, \quad \alpha_i \geq 0, \quad i = 1, 2, \ldots, \mathrm{Ntr} \end{aligned} \tag{2.33}$$

This dual problem is generally easier to handle than the original problem. According to Kuhn–Tucher theorem, the optimal solution meets:

$$\alpha_i[y_i(\langle \boldsymbol{w}, \boldsymbol{x}_i\rangle + b) - 1] = 0, \quad i = 1, 2, \ldots, \mathrm{Ntr} \tag{2.34}$$

Given $(\boldsymbol{\alpha}^*, b^*)$ is the optimal solution of the maximization (2.33), the corresponding discrimination functional expression is:

$$f(x) = \mathrm{sgn}\{\langle \boldsymbol{w}^*, \boldsymbol{x}_i\rangle + b^*\} = \mathrm{sgn}\left\{\sum_{i=1}^{\mathrm{Ntr}} \alpha_i^* y_i K(\boldsymbol{x}_i, \boldsymbol{x}) + b^*\right\} \tag{2.35}$$

where, the vector is $\boldsymbol{w}^* = \sum_{i=1}^{\mathrm{Ntr}} \alpha_i^* y_i \boldsymbol{x}_i$. It is noted that the value of b does not occur in the dual problem, and the optimal value b^* can be inferred (the form is not sole) as by Kuhn–Tucher theorem:

$$b^* = -\frac{\max_{y_i=-1}(\langle \boldsymbol{w}^*, \boldsymbol{x}_i\rangle) + \min_{y_i=+1}(\langle \boldsymbol{w}^*, \boldsymbol{x}_i\rangle)}{2} \tag{2.36}$$

2. **Generalized optimal classification hyperplane**

When we are confronted with an inversion problem, i.e., needing inferring the unknown reason from the known result, it is necessary to consider the theoretical exposition of the ill-posed problem. The ill-posed problem is not only the mathematical phenomenon, but also exists extensively in the practical problems. The regularization theory is just proposed in allusion to this problem. The important content is that the minimization functional cannot obtain a very good solution in solving the problem of the operator equation of defining the ill-posed problem. On the contrary, we should adopt the unobvious solution, i.e., minimizing "deteriorated" (regularization) functional, to solve. Constructing the SVM with the generalized optimal classification hyperplane is just the embodiment of this idea.

While handling the linear inseparability problem, introduce the slack variable e_i, $i = 1, 2, \ldots, \mathrm{Ntr}$, the constraint condition in formula (2.29) becomes:

$$y_i[\langle \boldsymbol{w}, \boldsymbol{x}_i\rangle + b] \geq 1 - e_i, \quad i = 1, 2, \ldots, \mathrm{Ntr} \tag{2.37}$$

Meanwhile, introduce the penalty factor γ to make the condition control for the misclassification sample. The corresponding objective function becomes:

$$J(\boldsymbol{w}, \boldsymbol{e}) = \frac{1}{2}\|\boldsymbol{w}\|^2 + \frac{\gamma}{2}\sum_{i=1}^{\mathrm{Ntr}} e_i \tag{2.38}$$

Moreover, the constraint condition corresponding to $\alpha_i \geq 0, \quad i = 1, 2, \ldots, \mathrm{Ntr}$ becomes $\gamma \geq \alpha_i \geq 0, \quad i = 1, 2, \ldots, \mathrm{Ntr}$. When the class division has the error, the corresponding slack variable is more than 0. Therefore, the sum of the slack variable is the upper bound of the classification error in the training set.

3. **Nonlinear problems**

While handling the nonlinear problems, SVM maps the data point without the linear separation by the original space into the linear separable point in the transformation space through introducing the nonlinear mapping ϕ, as shown in Fig. 2.4. Under this condition, $\boldsymbol{x}_i$ in the optimal expression should be correspondingly replaced with $\phi(\boldsymbol{x}_i)$, while the inner product $\langle \boldsymbol{x}_i, \boldsymbol{x}_j \rangle$ is replaced with

$$\boldsymbol{K}(i,j) = K(\boldsymbol{x}_i, \boldsymbol{x}_j) = \langle \phi(\boldsymbol{x}_i), \phi(\boldsymbol{x}_j) \rangle \tag{2.39}$$

Here, K is called as the kernel function operator, which is an inner product algorithm in the transformation space. $\boldsymbol{K}$ is the kernel function matrix of the sample set under the effect of the kernel function operator. In case that it is not confused, both in this book are often referred to as kernel function.

The nonlinear mapping ϕ is generally difficult to construct, and usually the dimension of the corresponding transformation space is very high and even infinity, resulting in great difficulties for the analysis. It is noted that above-discussed optimal and generalized linear classification functions, and the final classification discrimination function only includes the inner product of the support vector in the samples to be classified and the training sample, i.e., kernel function operation. Meanwhile, the solution process also involves the kernel function operation among the training samples. Thus, if solving the optimal linear classification problems in an eigenspace, it is necessary to only know the inner product operation in this space. If the inner product in the transformation space can be directly calculated by

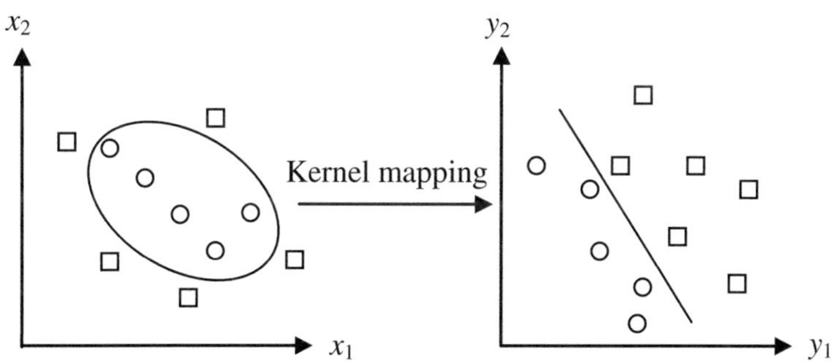

Fig. 2.4 Kernel mapping transforming nonlinear problem into linear problem

the variable in the original space through the kernel function, the calculation complexity of solving the problem of the optimal classification plane will not be increased much even if the dimension of the transformation space is increased much. In this way, the quotation of the kernel function skillfully solves the problem of constructing and disposing the nonlinear mapping.

The statistical learning theory indicates that, according to Hilbert–Schmidt principle, only if an operation meets the Mercer conditions, it can be used as the kernel function here. Mercer conditions state that, for any symmetric function $K(\boldsymbol{x},\boldsymbol{x}')$, the necessary and sufficient condition as the inner product operation in the eigenspace is, for any $\phi(\boldsymbol{x}) \neq 0$ and $\int \phi^2(\boldsymbol{x})\mathrm{d}\boldsymbol{x} < \infty$, then

$$\iint K(\boldsymbol{x},\boldsymbol{x}')\phi(\boldsymbol{x})\phi(\boldsymbol{x}')\mathrm{d}\boldsymbol{x}\mathrm{d}\boldsymbol{x}' > 0 \tag{2.40}$$

Such a symmetric function $K(\boldsymbol{x},\boldsymbol{x}')$ can be regarded as the kernel function.

The kernel function is closely related to the performance of SVM. How to construct the kernel function relevant to the practical problem has always been the main content of SVM research. The selection of the kernel function does not have some theoretical guidance, and the parameter selection still adopts the empirical mode. The several common kernel functions now are shown in Table 2.1. From a large number of experiment results, the classification result of the Gaussian radial basis kernel function is relatively good.

Below we replace the dot product in the optimal classification plane by the kernel function (inner product) $K(\boldsymbol{x},\boldsymbol{x}')$, equivalent to transforming the original eigenspace into a certain new eigenspace. Then, seek the optimization problem as shown in (2.41) transformed from the optimal classification plane:

$$\begin{aligned} \min \quad & \frac{1}{2}\|\boldsymbol{w}\|^2 \\ \text{s.t.} \quad & y_i[\langle \boldsymbol{w}, \phi(\boldsymbol{x}_i)\rangle + b] - 1 \geq 0 \\ & i = 1, 2, \ldots, \mathrm{Ntr} \end{aligned} \tag{2.41}$$

Then, Lagrange function is as follows:

Table 2.1 Several common Kernel functions

Kernel function name	Kernel function expression
Linear kernel function	$K(\boldsymbol{x},\boldsymbol{y}) = \langle \boldsymbol{x},\boldsymbol{y}\rangle$
Polynominal kernel function	$K(\boldsymbol{x},\boldsymbol{y}) = [\langle \boldsymbol{x},\boldsymbol{y}\rangle + 1]^d$
Gaussian radial basis kernel function	$K(\boldsymbol{x},\boldsymbol{y}) = \exp[-\|\boldsymbol{x}-\boldsymbol{y}\|^2/2\sigma^2]$
Index radial basis kernel function	$K(\boldsymbol{x},\boldsymbol{y}) = \exp[-\|\boldsymbol{x}-\boldsymbol{y}\|/2\sigma^2]$
Sigmoid kernel function	$K(\boldsymbol{x},\boldsymbol{y}) = \tanh(k\langle \boldsymbol{x},\boldsymbol{y}\rangle - \delta)$

$$L(\boldsymbol{w}, b, \boldsymbol{\alpha}) = \frac{1}{2}\|\boldsymbol{w}\|^2 - \sum_{i=1}^{\mathrm{Ntr}} \alpha_i[\langle \boldsymbol{w}, \phi(\boldsymbol{x}_i)\rangle + b - y_i] \tag{2.42}$$

(2.41) can be transformed into the objective function under the maximization in the same mode as the linear problem:

$$L(\boldsymbol{w}, b, \boldsymbol{\alpha}) = \sum_{i=1}^{\mathrm{Ntr}} \alpha_i - \frac{1}{2}\sum_{i,j=1}^{\mathrm{Ntr}} \alpha_i \alpha_j y_i y_j K(\boldsymbol{x}_i, \boldsymbol{x}_j) \tag{2.43}$$

Then, the corresponding discrimination function is:

$$f(x) = \mathrm{sgn}\{\langle \boldsymbol{w}^*, \phi(\boldsymbol{x}_i)\rangle + b^*\} = \mathrm{sgn}\left\{\sum_{i=1}^{\mathrm{Ntr}} \alpha_i^* y_i K(\boldsymbol{x}_i, \boldsymbol{x}) + b^*\right\} \tag{2.44}$$

After introducing the concept of the high-dimensional space inner product (i.e., kernel function), the basic idea of SVM can be simply summarized as firstly transforming the input space into a high-dimensional space by the nonlinear transformation, and then gaining the optimal linear classification plane in this new space, while the nonlinear transformation is realized by defining the proper inner product function.

4. **Expression form of basic theory**

Maximizing the classification interval is actually controlling the generalization ability, which is one of the core ideas of SVM. The statistical learning theory indicates that, in d dimension space, given the sample is distributed in the hyper-sphere range with the radius of R, VC dimension of the indication function set $f(\boldsymbol{x}, \boldsymbol{w}, b)$ composed by the regular hyperplane in conformity with the condition of $\|\boldsymbol{w}\| \leq k$ will meet the following boundary:

$$h \leq \min(R^2k^2, d) + 1 \tag{2.45}$$

Therefore, minimizing $\|\boldsymbol{w}\|^2$ is the minimum classification, i.e., minimizing the upper bound of VC dimension, and accordingly realizing the selection of the dual function complexity in SRM code. The optimal classification plane and generalized optimal classification plane are actually dividing the classification function set $S = \{(\langle \boldsymbol{w}, \boldsymbol{x}\rangle + b)\}$ into some normalized subsets in accordance with the module of the weight (classification interval in the case of the linear separability). Each subset is as follows:

$$S = \left\{(\langle \boldsymbol{w}, \boldsymbol{x}\rangle + b) : \|\boldsymbol{w}\|^2 \leq c_k\right\} \tag{2.46}$$

For the linear separability, the optimal classification plane is to seek the normalized subset with the minimum bound of the expected risk on the premise of the fixed empirical risk of 0. Moreover, in the case of the linear inseparability, the generalized optimal classification plane is to seek the minimum bound of the expected risk under the condition of controlling the misclassification sample. Therefore, they are optimal in the sense of the bound of the expected risk, and are the specific embodiment of the structural risk minimization principle. For the linear function in d dimension space, VC dimension is $d+1$. However, from the above discussion, under the constraint condition of $\|w\| \leq k$, VC dimension may be reduced greatly. The smaller VC dimension function set can also be gained even in very high-dimensional space, to guarantee the relatively good promotion. At the same time, we can see that, the complexity of the calculation depends on the number of samples, especially the support vector number in the sample, rather than the spatial dimension, through transforming the original problem to the dual problem. These characteristics make it possible for SVM to effectively handle the high- dimensional problems.

2.3.3 *Construction of Multi-class Classifier with the Simplest Structure*

At present, there are two most common multi-class classifiers, respectively, 1-a-r (1-against-rest) multi-class classifier and 1-a-1 (1-against-1) multi-class classifier. Figure 2.5 provides the structure of these two traditional multi-class classifiers. These two classifiers have the relatively complex structural design and very large calculated amount. Take N class problem as an example. The algorithm 1-a-r is to construct N two-class target subclassifiers. The k subclassifier regards the training sample in the k class as the positive training sample, and the others are the negative training samples. For some input sample, the classification result is the subclassifier output value as the maximum corresponding class. The algorithm 1-a-1 is proposed by Knerr, which is to construct each two classes in N class into a subclassifier, and

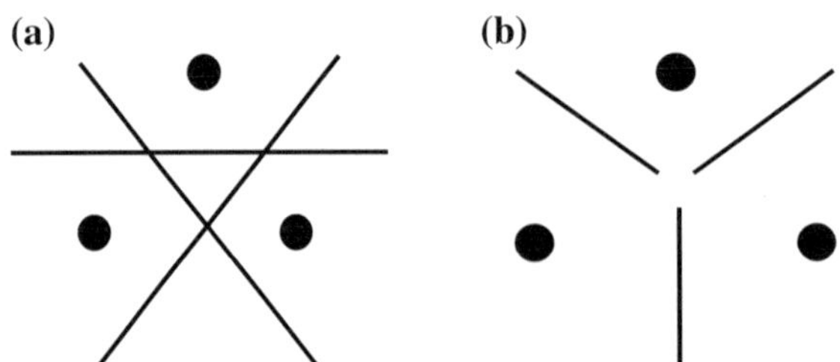

Fig. 2.5 Structure of two classic multi-class classifier. **a** 1-a-r structure. **b** 1-a-1 structure

accordingly needs constructing $N(N - 1)/2$ subclassifiers, combining these sub-classifiers, and determining the classification result by the voting method.

The common weakness of these two methods is unbounded promotion error, and a large number of classifiers, resulting in the low speed of decision. On account of the problems of the complex classifier structure due to the excessive subclassifiers, this section puts forward a method of simplifying the structure of the multi-class target classifier.

1. **Construction of multi-class classifier with the simplest structure**

For the convenience of the description, it can firstly describe the construction of the classifier structure with eight classes of problems as an example. Given all sample sets as P, the construction process of the classifier is as follows:

1. Firstly, P is equally divided into two sample sets by category, respectively, noted as P_1 and P_2. Also it is noted as $A_1 = P_1$ and $B_1 = P_2$, and the corresponding class marks of A_1 and B_1 are reset as +1 and −1. Then train A_1 and B_1 as two classes of targets, and gain the first two-class target sub-classifier C_1.
2. Firstly, P_1 is equally divided into two sample sets by category, respectively, noted as P_{11} and P_{21}. Then P_2 is equally divided into two sample sets by category, respectively, noted as P_{12} and P_{22}. Also it is noted as $A_2 = P_{11} \cup P_{12}$ ($\cup$ is the set union operator), $B_2 = P_{21} \cup P_{22}$, and the corresponding class marks of A_2 and B_2 are reset as +1 and −1. Finally, train A_2 and B_2 as two classes of targets, and gain the second two-class target sub-classifier C_2.
3. Firstly, P_{11} is equally divided into two sample sets by category, respectively, noted as P_{111} and P_{211}. Then P_{21} is equally divided into two sample sets by category, respectively noted as P_{121} and P_{221}. P_{12} is equally divided into two sample sets by category, respectively noted as P_{112} and P_{212}. P_{22} is equally divided into two sample sets by category, respectively noted as P_{122} and P_{222}. Also it is noted as $A_3 = P_{111} \cup P_{121} \cup P_{112} \cup P_{122}$, $B_3 = P_{211} \cup P_{221} \cup P_{212} \cup P_{222}$, and the corresponding class marks of A_3 and B_3 are reset as +1 and −1. Finally, train A_2 and B_2 as two classes of targets, and gain the third two-class target sub-classifier C_3.
4. Three subclassifiers of C_1, C_2 and C_3 are combined into a multi-class target classifier C.

Thus, the sample to be decided can be decided as the only one class through the decision intersection of three subclassifiers. Figure 2.6 provides the construction

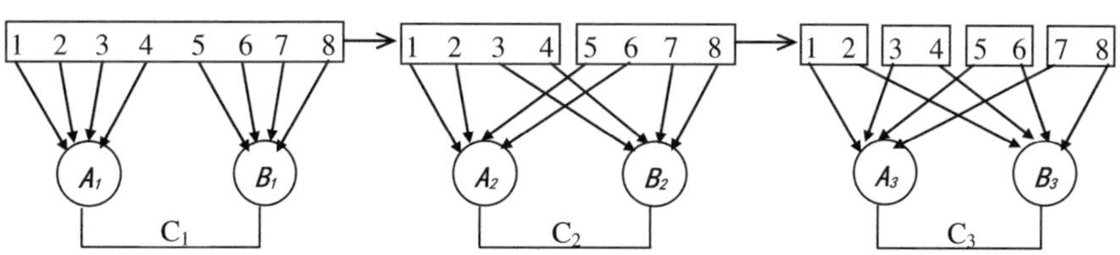

Fig. 2.6 Construction of three subclassifiers in eight classes

schematic diagram of three subclassifiers in this example. The rectangle box shows the original class set and the division situation. The circular box shows the original class mark set assigned into a class in some step. It is noted that the class dividing mode of each step is arbitrary in the above construction process of the classifier.

Generally, for 2^N type of problem, it can be finished by the following descriptive process.

1. 2^{N-1} class sample in the original sample is combined into a class sample set, and the remaining is the other class sample set. Thus, train and gain the first two-class target sub-classifier. The original sample is divided into two sets by category.
2. Take 2^{N-2} (totaling to 2^{N-1}) class samples respectively in the two divided sets from the above step, to combine into a class sample set, and the remaining is the other class sample set. Thus, train and gain the second two-class target sub-classifier. The original sample is divided into four sets by category.
3. In the k step, take 2^{N-k} (totaling to 2^{N-1}) class samples respectively in the 2^{k-1} divided class set from the $k-1$ step, to combine into a class sample set, and the remaining is the other class sample set. Thus, train and gain the k two-class target sub-classifier. The original sample is divided into 2^k sets by category.
4. Continue, gain N different subclassifiers, and combine them into a multi-class target classifier. Each sample can be decided as the only one class through the output value of N subclassifiers.

If the number of classes is between 2^{N-1} and 2^N, the number of classes can be transformed into "2^N" in the method of adding the virtual classes, and then the virtual class sample is removed from N subclassifiers constructed finally. In fact, the added virtual class is the formal participation without the practical participation. For example, when the number of classes is 6, it can add the 7th and 8th classes. So, we will gain the same construction form of the classifier as Fig. 2.6. The difference is that the design result removes the 7th and 8th classes of samples in the final A_1, B_1, A_2, B_2, A_3 and B_3.

It is easy to analyze that the number of subclassifiers from the new method is far less than two typical methods of 1-a-1 and 1-a-r. Table 2.2 is the comparison of the required number of subclassifiers by three methods under different classes (including $K = 2^N$).

If failing to consider the complexity of the design, the time for classification can be used for measuring the complexity of the classifier. For training, the complexity

Table 2.2 Comparison of required number of subclassifiers by structure of three classifiers

Number of subclassifiers		Structure of classifier		
		1-a-1	1-a-r	New method
Number of classes	4	6	4	2
	16	120	16	4
	K	$K(K-1)/2$	K	N

Table 2.3 Relative frequency of Kernel function calculation by three classifier structures in test

Frequency of Kernel function calculation		Structure of classifier		
		1-a-1	1-a-r	New method
Number of classes	4	4	3	2
	16	16	15	4
	K	K	K–1	N

of the subclassifiers under different structures may be different, and various subclassifiers in the same structure have the specific relations, so the number of the subclassifiers cannot only be used for measuring the complexity of the classifier. By contrast, the test process of the classification is not restricted by the above factors, and the time consuming is mainly used for the kernel function operation. Thus, the frequency of the kernel function operation as required for processing can be used for the measurement index of the complexity. Under the same conditions, for the same $K = 2^N$ class of the training sample, it can calculate the relative value for the frequency of the kernel function operation as required for processing by the classifier with different structures in the test (see Table 2.3).

For the essence, relative to the two traditional methods, the new method abandons the massive redundant information among various subclassifiers, and gains the simplification of the classifier structure, and the greater improvement of the classification speed.

2.3.4 *Least Squares SVM and Its SMO Optimization Algorithm*

In recent years, there have been many developed and transformed SVM types. During these development types, the least squares SVM has been widely applied due to the efficient classification and regression functions. What is more, the mathematical model of the least squares SVM is only an optimization problem of the error cost function sum of squares with the equality constraint, and the solution can be made in the linear system. This book focuses on adopting such type of SVM.

1. **Least Squares SVM** (Suykens et al. 2002)

The optimization problem expression of the least squares SVM is as follows:

$$\begin{aligned} &\min_{\boldsymbol{w},b,\boldsymbol{e}} \quad J(\boldsymbol{w},\boldsymbol{e}) = \frac{1}{2}\|\boldsymbol{w}\|^2 + \frac{\gamma}{2}\sum_{i=1}^{\mathrm{Ntr}} e_i^2 \\ &\text{s.t.} \quad y_i = \langle \boldsymbol{w}, \phi(\boldsymbol{x}_i)\rangle + b + e_i, \\ &\qquad i = 1, 2, \ldots, \mathrm{Ntr}, \ \gamma > 0. \end{aligned} \tag{2.47}$$

where, $\boldsymbol{x}_i \in R^d$ is the sample data, and $y_i \in \{+1, \ -1\}$ and e_i are, respectively, the class mark and discrimination error. $i = 1, \ldots, \mathrm{Ntr}$. The corresponding dual problem is:

$$\begin{aligned} \min_{w,b,e,\alpha} \quad & L(\boldsymbol{w}, b, \boldsymbol{e}, \boldsymbol{\alpha}) \\ & = J(\boldsymbol{w}, \boldsymbol{e}) - \sum_{i=1}^{\mathrm{Ntr}} \alpha_i \{\langle \boldsymbol{w}, \phi(\boldsymbol{x}_i)\rangle + b + e_i - y_i\} \end{aligned} \tag{2.48}$$

The optimal KKT condition is as follows:

$$\begin{cases} \dfrac{\partial L}{\partial \boldsymbol{w}} = 0 \rightarrow \boldsymbol{w} = \sum_{i=1}^{\mathrm{Ntr}} \alpha_i \phi(\boldsymbol{x}_i) \\ \dfrac{\partial L}{\partial b} = 0 \rightarrow \sum_{i=1}^{\mathrm{Ntr}} \alpha_i = 0 \\ \dfrac{\partial L}{\partial e_i} = 0 \rightarrow \alpha_i = \gamma e_i, \quad i = 1, 2, \ldots, \mathrm{Ntr} \\ \dfrac{\partial L}{\partial \alpha_i} = 0 \rightarrow \langle \boldsymbol{w}, \phi(\boldsymbol{x}_i)\rangle + b + e_k - y_i = 0, \quad i = 1, 2, \ldots, \mathrm{Ntr} \end{cases} \tag{2.49}$$

Upon eliminating $\boldsymbol{w}$ and e by the elimination method, the above formula can be further expressed as:

$$\begin{bmatrix} 0 & 1_v^T \\ 1_v & \boldsymbol{K} + \boldsymbol{I}/\gamma \end{bmatrix} \begin{bmatrix} b \\ \boldsymbol{\alpha} \end{bmatrix} = \begin{bmatrix} 0 \\ \boldsymbol{y} \end{bmatrix} \tag{2.50}$$

where, $\boldsymbol{y} = [y_1, y_2, \ldots, y_{\mathrm{Ntr}}]^{\mathrm{T}}$, $1_v = [1, 1, \ldots, 1]^{\mathrm{T}}$, $\boldsymbol{\alpha} = [\alpha_1, \alpha_2, \ldots, \alpha_{\mathrm{Ntr}}]^{\mathrm{T}}$.

2. **SMO optimization algorithm of least squares SVM**

We know that the least squares SVM can be directly and conveniently solved into the linear system. However, when the number of the training samples is too large, the direct solution becomes very difficult. Therefore, it is necessary to promote the efficient SMO algorithm (Shevade et al. 2000) to a such type of SVM solution, as the effective replacement of the linear solution.

The dual form of the formula (2.48) is as follows:

$$L = \frac{1}{2}\|\boldsymbol{w}\|^2 + \frac{\gamma}{2}\sum_{i=1}^{\mathrm{Ntr}} e_i^2 + \sum_{i=1}^{\mathrm{Ntr}} \alpha_i [y_i - \langle \boldsymbol{w}, \phi(\boldsymbol{x}_i)\rangle - b - e_i] \tag{2.51}$$

Upon applying the Wolfe duality theory, we can obtain the following form of the optimization problem:

$$\begin{aligned} \max \quad & f(\boldsymbol{\alpha}) = -\frac{1}{2}\sum_{i=1}^{\mathrm{Ntr}}\sum_{j=1}^{\mathrm{Ntr}} \alpha_i\alpha_j\tilde{K}(\boldsymbol{x}_i,\boldsymbol{x}_j) + \sum_{i=1}^{\mathrm{Ntr}} \alpha_i y_i \\ s.t. \quad & \sum_{i=1}^{\mathrm{Ntr}} \alpha_i = 0 \end{aligned} \tag{2.52}$$

where,

$$\tilde{K}(\boldsymbol{x}_i,\boldsymbol{x}_j) = K(\boldsymbol{x}_i,\boldsymbol{x}_j) + \frac{1}{\gamma}\delta_{i,j}, \quad \delta_{i,j} = \begin{cases} 1, & i = j \\ 0, & i \neq j \end{cases} \tag{2.53}$$

The Lagrange form of the formula (2.52) is as follows:

$$\bar{L} = -\frac{1}{2}\sum_{i=1}^{\mathrm{Ntr}}\sum_{j=1}^{\mathrm{Ntr}} \alpha_i\alpha_j\tilde{K}(\boldsymbol{x}_i,\boldsymbol{x}_j) + \sum_{i=1}^{\mathrm{Ntr}} \alpha_i y_i + \beta\sum_{i=1}^{\mathrm{Ntr}} \alpha_i \tag{2.54}$$

Defining

$$F_i = -\frac{\partial f}{\partial \alpha_i} = \sum_{i=1}^{\mathrm{Ntr}} \alpha_i\tilde{K}(\boldsymbol{x}_i,\boldsymbol{x}_j) - y_i, \quad i = 1,2,\ldots,\mathrm{Ntr}. \tag{2.55}$$

From KKT conditions of the formula (2.54), we can get

$$\frac{\partial \bar{L}}{\partial \alpha_i} = \beta - F_i = 0 \Rightarrow F_i = \beta, \quad i = 1,2,\ldots,\mathrm{Ntr}. \tag{2.56}$$

This formula explains that the necessary and sufficient condition of the support value vector $\boldsymbol{\alpha} = [\alpha_1,\alpha_2,\ldots,\alpha_{\mathrm{Ntr}}]^{\mathrm{T}}$ as the optimal solution is:

$$\max_i \{F_i\} = \min_i\{F_i\} \tag{2.57}$$

Thus, we can provide the iterative method of the solution optimal $\boldsymbol{\alpha} = [\alpha_1,\alpha_2,\ldots,\alpha_{\mathrm{Ntr}}]^{\mathrm{T}}$. Note

$$\begin{aligned} i_{\max} &= \arg\max_i\{F_i\} \\ i_{\min} &= \arg\min_i\{F_i\} \end{aligned} \tag{2.58}$$

$$\begin{aligned} \tilde{\boldsymbol{\alpha}} &= [\tilde{\alpha}_1,\tilde{\alpha}_2,\ldots,\tilde{\alpha}_{\mathrm{Ntr}}]^{\mathrm{T}} \\ \tilde{\alpha}_i &= \begin{cases} \alpha_i - t, & i = i_{\max} \\ \alpha_i + t, & i = i_{\max} \\ \alpha_i, & \text{other } i \end{cases} \end{aligned} \tag{2.59}$$

For the given $\boldsymbol{\alpha} = [\alpha_1, \alpha_2, \ldots, \alpha_{\mathrm{Ntr}}]^{\mathrm{T}}$, if it fails to meet the optimum condition (2.57), $\alpha_{i_{\max}}$ and $\alpha_{i_{\min}}$ are, respectively, replaced by $\alpha_{i_{\max}} - t$ and $\alpha_{i_{\min}} + t$, i.e., replacing $\boldsymbol{\alpha}$ by $\tilde{\boldsymbol{\alpha}}$. The selection of the parameter t needs maximizing $f(\tilde{\boldsymbol{\alpha}})$, and the optimal value is provided by the following formula:

$$\begin{aligned} &\frac{\partial f(\tilde{\boldsymbol{\alpha}}(t))}{\partial t} = 0 \Rightarrow t = t^* = (F_{i_{\min}} - F_{i_{\max}})/\eta \\ &\eta = \{2 * \tilde{K}(\boldsymbol{x}_{i_{\max}}, \boldsymbol{x}_{i_{\min}}) - \tilde{K}(\boldsymbol{x}_{i_{\max}}, \boldsymbol{x}_{i_{\max}}) - \tilde{K}(\boldsymbol{x}_{i_{\min}}, \boldsymbol{x}_{i_{\min}})\}. \end{aligned} \tag{2.60}$$

After getting $\tilde{\boldsymbol{\alpha}}$, the new iterative process starts from here. Thus, the least squares support vector machine theory is promoted completely.

2.3.5 *Triply Weighted Classification Method*

SVM shows good performance in the hyperspectral image classification, but how to further improve the classification performance is still a researchable content. During the process of the hyperspectral image classification, the generalization performance of SVM is sensitive to the outliers point and noise interference pixel (collectively referred to as anomaly pixel) during the training process, while they inevitably exist extensively in the hyperspectral data, and influence the correctness of the model. The modeling method of SVM excessively depends on the training samples, and is very sensitive to the existence of the anomaly pixel. Generally, the introduction of a few anomaly pixels may fully destroy the generalization performance of the model.

Suykens et al. (2002) came up with LSSVM weighted method, to make the pixel and outliers point seriously suffering from the noise interference in the hyperspectral image control effectively, and accordingly gain better Robust feature and generalization ability. Such a weighted method includes the complete preliminary training. Moreover, the calculated amount required by the training is generally larger, especially when the training samples are more, this method will be time consuming, Due to this reason, the method is not popularized effectively.

The existing hyperspectral image classification weighted method is generally implemented on account of the training samples, while little literature considers the following two situations. Firstly, the influence of different characteristics (or band and spectral section) of the hyperspectral image on the class separability is different, i.e., their effect on the classification is different. Thus, they should not be treated equally in the classifier design. Secondly, in the practical application, the remote-sensing data classes are numerous, while the significance of different classes on the hyperspectral data analysis is often different, or the researchers have different degree of interests on them. Thus, it is necessary to consider in the classifier design. For this purpose, this section introduces a triply weighted method in the LSSVM theory-based hyperspectral image classification problem, so as to further enhance the classification analysis effect.

1. **Pixel weighting in the hyperspectral image classification**

The optimization problem expression of LSSVM is shown in (2.47). In order to make the samples with different anomaly degrees embodying in the classification model, their corresponding classification error should be distributed with different weights in the cost function, i.e., gaining the weighting training model of LSSVM. Given e_i corresponds to the weight v_i, this formula becomes:

$$\begin{aligned} \min_{w,b,e} \quad & J(w,e) = \frac{1}{2}\|w\|^2 + \frac{\gamma}{2}\sum_{i=1}^{\mathrm{Ntr}} (v_i e_i)^2 \\ \text{s.t.} \quad & y_i = \langle w, \phi(x_i) \rangle + b + e_i \\ & i = 1, 2, \ldots, \mathrm{Ntr}, \quad \gamma > 0. \end{aligned} \tag{2.61}$$

Thus, how to reasonably determine the weight v_i becomes the key problem in the sample weighting. Due to longer relative distance between the anomaly sample and the corresponding class center in the training sample, the anomaly degree can be measured by the distance scale (Song et al. 2002). In this way, the smaller weight can be distributed for the sample with the larger anomaly degree to weaken the adverse effects. On the other hand, because of the differences of the intra-class spectrum, the pure sample cannot concentrate on the corresponding class center, but has a relatively small deviation. In view of this, while calculating the distance, we can subtract a correction constant from the distance obtained previously. For this purpose, we can firstly determine the class center as the center of the circle, including the minimum radius of the specified proportional sample point in the class. Further this radius is given as the above correction constant.

Given the class center corresponding to the sample $\boldsymbol{x}_i$ is $\boldsymbol{x}_0$, while the circle with $\boldsymbol{x}_0$ as the center and the radius of r includes the minimum circle of the specified proportional sample in the class. The uncorrected distance from the sample $\boldsymbol{x}_i$ to $\boldsymbol{x}_0$ is expressed by $\hat{D}(\boldsymbol{x}_i, \boldsymbol{x}_0)$, and then the calculation formula of $\hat{D}(\boldsymbol{x}_i, \boldsymbol{x}_0)$ is as follows:

$$\begin{aligned} \hat{D}(\boldsymbol{x}_i, \boldsymbol{x}_0) &= \|\phi(\boldsymbol{x}_i) - \phi(\boldsymbol{x}_0)\| \\ &= \left(K(\boldsymbol{x}_i, \boldsymbol{x}_i) + K(\boldsymbol{x}_0, \boldsymbol{x}_0) - 2K(\boldsymbol{x}_i, \boldsymbol{x}_0)\right)^{1/2} \end{aligned} \tag{2.62}$$

Accordingly, the correction distance $D(\boldsymbol{x}_i, \boldsymbol{x}_0)$ from $\boldsymbol{x}_i$ to the class center $\boldsymbol{x}_0$ can be stipulated as:

$$D(\boldsymbol{x}_i, \boldsymbol{x}_0) = \hat{D}(\boldsymbol{x}_i, \boldsymbol{x}_0) - r, \quad i = 1, 2, \ldots, \mathrm{Ntr}. \tag{2.63}$$

Noting

$$\begin{aligned} D_{\max} &= \max_i(D(\boldsymbol{x}_i, \boldsymbol{x}_0)) \\ D_{\min} &= \min_i(D(\boldsymbol{x}_i, \boldsymbol{x}_0)) \end{aligned} \tag{2.64}$$

And $\mathrm{Nor}D(\boldsymbol{x}_i, \boldsymbol{x}_{y_i})$ is used for expressing the normalization form of $D(\boldsymbol{x}_i, \boldsymbol{x}_{y_i})$, i.e.,

$$\mathrm{Nor}D(\boldsymbol{x}_i, \boldsymbol{x}_{y_i}) = D(\boldsymbol{x}_i, \boldsymbol{x}_{y_i})/D_{\max}, \quad i = 1, 2, \ldots, \mathrm{Ntr}. \tag{2.65}$$

The weight factor can be obtained by the following formula:

$$v_i = 1 - \mathrm{Nor}D(\boldsymbol{x}_i, \boldsymbol{x}_{y_i})^2 + (D_{\min}/D_{\max})^2 \quad i = 1, 2, \ldots, \mathrm{Ntr}. \tag{2.66}$$

It is easy to verify $0 < v_i \leq 1$. The original error term $\{e_i\}_{i=1}^{\mathrm{Ntr}}$ is replaced by the weighted form $\{v_i e_i\}_{i=1}^{\mathrm{Ntr}}$, and the sample weighting type LSSVM as shown in the formula (2.61) can be obtained.

2. **Feature weighting in the hyperspectral image classification**

The key to the feature weighting is to find out an appropriate weighting matrix. This matrix can enhance the effective feature, and weaken the feature with the poorer class separability. Fisher linear discriminant analysis is a widely used classification technique, and has been extensively applied in the pattern recognition. The inverse matrix of the intra-class divergence matrix can reflect different contributions of different features for the classification effect well (Ji et al. 2004), and the effect has been verified in the spectral separation (Chang and Ji 2006). Thus, it can be applied to the feature weighting of the hyperspectral image classification, and the specific methods are as follows.

Given Ntr training sample vectors are used for the classification, $\boldsymbol{\mu}_j$ is the mean value of the j class of the sample ($j = 1, 2, \ldots, \mathrm{Ntr}$), i.e.,

$$\boldsymbol{\mu}_j = \frac{1}{n_j} \sum_{x_i \in C_j} \boldsymbol{x}_i \tag{2.67}$$

C_j and n_j, respectively, stand for the j class of the sample set and the number of samples, and accordingly the intra-class divergence matrix $\boldsymbol{S}_W$ can be defined as follows:

$$\boldsymbol{S}_w = \sum_{j=1}^{\mathrm{Nc}} \boldsymbol{S}_j \tag{2.68}$$

Here

$$\boldsymbol{S}_j = \sum_{x \in C_j} (\boldsymbol{x} - \boldsymbol{\mu}_j)(\boldsymbol{x} - \boldsymbol{\mu}_j)^{\mathrm{T}} \tag{2.69}$$

$\boldsymbol{S}_w$ is the real symmetric matrix, consequently, the orthogonal matrix $\boldsymbol{U}$ transforms the opposite angle into the matrix $\boldsymbol{B}$:

$$\boldsymbol{U}^{\mathrm{T}} \boldsymbol{S}_w \boldsymbol{U} = \boldsymbol{B} \tag{2.70}$$

Further, it can be inferred:

$$\boldsymbol{S}_w{}^{-1} = (\boldsymbol{U}\boldsymbol{B}\boldsymbol{U}^{\mathrm{T}})^{-1} = (\boldsymbol{U}\boldsymbol{B}^{-1/2})(\boldsymbol{U}\boldsymbol{B}^{-1/2})^{\mathrm{T}} \tag{2.71}$$

Noting $\boldsymbol{G} = (\boldsymbol{U}\boldsymbol{B}^{-1/2})^{\mathrm{T}}$, then $\boldsymbol{G}$ can be used for the feature weighting matrix in the classification problem.

3. **Class weighting in the hyperspectral image classification**

The matrix equation of LSSVM is rewritten as follows:

$$\begin{bmatrix} 0 & 1_v^T \\ 1_v & \boldsymbol{K} + \boldsymbol{I}/\gamma \end{bmatrix} \begin{bmatrix} b \\ \boldsymbol{\alpha} \end{bmatrix} = \begin{bmatrix} 0 \\ \boldsymbol{y} \end{bmatrix} \tag{2.72}$$

$\boldsymbol{I}$ is the unit matrix of Ntr $\times$ Ntr. When $\boldsymbol{I}$ is the unit matrix, it shows that the training process equally considers each training sample. According to the thought of Suykens et al. (2002), if different classes of the training samples are treated differently in the optimization model, it will directly reflect as the different diagonal element assignments of $\boldsymbol{I}$ in the corresponding linear Eq. (2.72). In other words, the diagonal element assignment of $\boldsymbol{I}$ can embody the emphasis on each training sample. The larger some weight of $\boldsymbol{I}$ is, the less indifference the training process has on the corresponding sample, and vice versa. The class weighting means the diagonal element value through resetting the corresponding position of some class sample in $\boldsymbol{I}$, rather than the original equivalent setting, for changing the emphasis on each class, accordingly protecting the class of interest, and restraining the non-essential class. Thus, reducing the weight corresponding to the training sample in the class of interest properly and increasing the weight corresponding to the training sample in the class of non-interest properly can effectively enhance the classification accuracy of the class of interest.

The above three weighting methods can be used independently, and also can be used in combination by any check mode. Figure 2.7 is the relationship diagram for mapping the class center distance of some real hyperspectral data sample into the weighting value. Figure 2.8 provides the schematic diagram for the operation interface of the check weighting classification of the hyperspectral image.

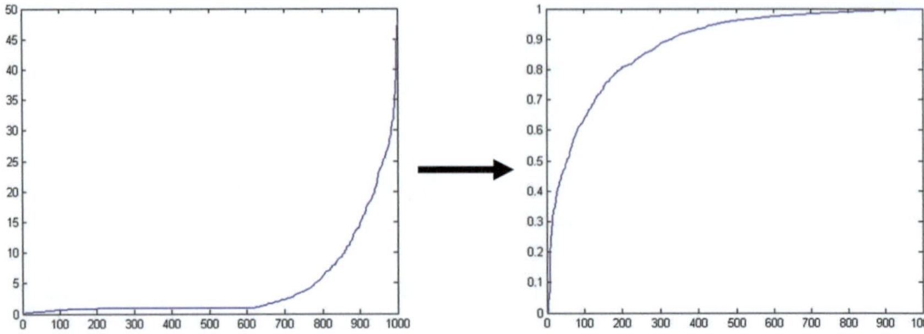

Fig. 2.7 Mapping from uncorrected distance to weight (X-coordinate sample is reciprocal correspondence relation)

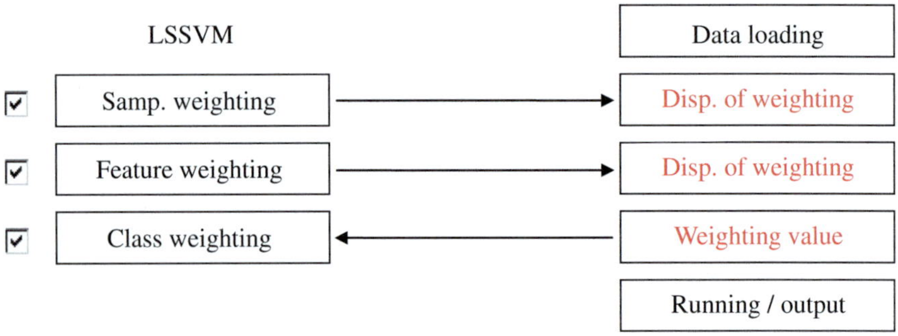

Fig. 2.8 Operation interface of least squares SVM check weighting

2.4 Performance Assessment for SVM-Based Classification

One of the hyperspectral remote-sensing images mainly used in this book was from one section of the Indian agriculture and forestry hyperspectral remote-sensing experimental area in the northwest of Indiana of America shot in June 1992. After removing some bands with larger noise effect, select 200 bands as the research object from the original 220 bands. The images are supervised, and the land object classes represented by the supervision class mark 0–17 are successively Background, Alfalfa, Corn-notill, Corn-min, Corn, Grass/ Pasture, Grass/ Trees, Grass/ pasture-mowed, Hay-windrowed, Oats, Soybeans-notill, Soybeans-min, Soybean-clean, Wheat, Woods, Bldg-Grass-, ree-Drives, and Stone-steel towers. Figure 2.9 provides the false color composite image by the bands of 50, 27 and 17. The number of various land object pixels and the image data characteristics are, respectively, shown as Fig. 2.10 and Table 2.4.

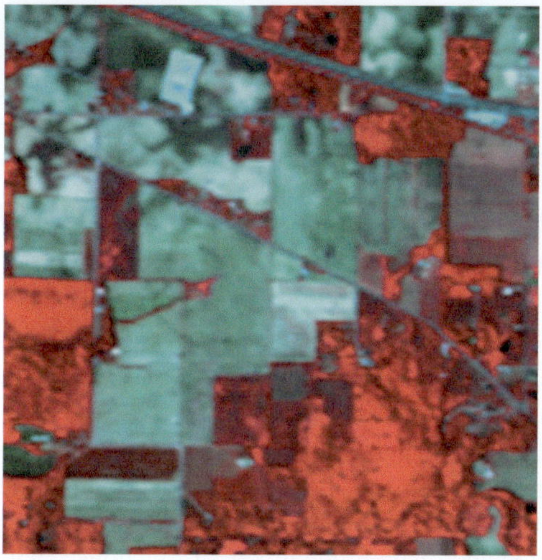

Fig. 2.9 False color composite image by Bands of 50, 27 and 17 as RGB channel

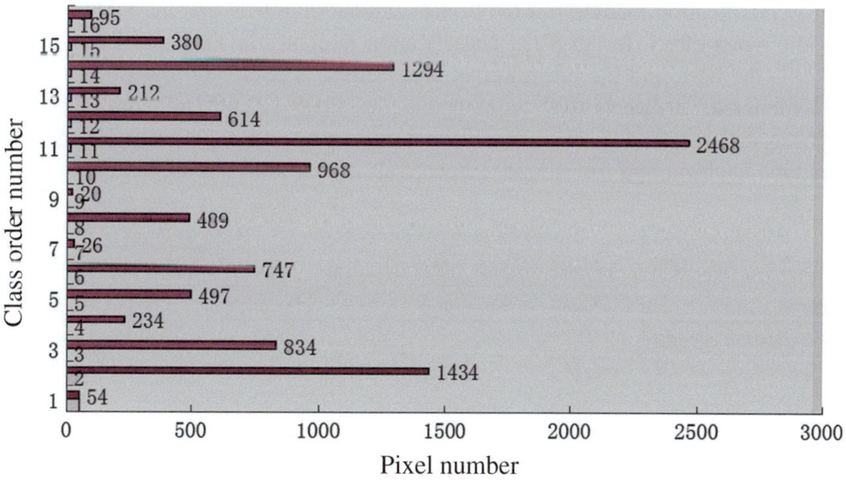

Fig. 2.10 Quantity statistics of pixels included in each class of image

Table 2.4 Data characteristics of experimental image

Number of bands	220 units
Wave length range	400–2500 nm
Spectral resolution	≈10 nm
Spatial resolution	20 × 20 m
Image size	144 × 144
Flight height	20 km (NASA ER-2 airplane)

This chapter focuses on comparing the SVM classification method and two common methods i.e., spectral angle matching method and maximum likelihood method by this image.

2.4.1 Performance Assessment for Original SVM-Based Classification

The experiment will make the detailed comparison on the classification performance of various methods by transforming the training numbers and sample dimension. In Experiment 1, the selected number of classes is six classes, involving corn, soybean, grass, forest land, hay, and wheat. The total number of the training samples is 1031, and the number of the inspection samples is 5144. SAM classification method and ML classification method are used for comparing with the SVM classification method with different kernel functions. Table 2.5 and Fig. 2.11 provide the result of SVM (Gaussian kernel)-based classification method and the thematic mapping-based image representation in details. The experimental result comparison of various methods is shown in Table 2.6. The result shows that the SVM classification method has the best classification effect, and the SAM method has the worst effect. In the SVM classification method, the Gaussian kernel SVM effect is the best, and the linear kernel SVM effect is relatively low.

Experiment 2 selects 400 training samples (100 samples of corn, meadow, soybean and forest land, respectively) and 320 test samples (80 samples of corn, meadow, soybean and forest land, respectively) from four land object classes, and

Table 2.5 Classification accuracy of each class (%)

Classes	Corn	Soybean	Grass	Forest land	Hay	Wheat
Classification accuracy	97.7	99.0	96.2	99.54	99.59	99.06

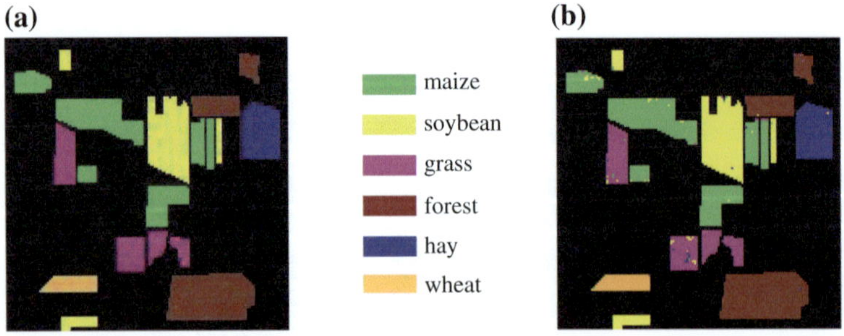

Fig. 2.11 Real land object chart and classification chart. **a** Real class. **b** SVM (Gauss)-based classification result

Table 2.6 Comparison of classification accuracy

Classification accuracy (%)		Classification method				
		SAM	ML	Linear kernel SVM	Polynomial kernel SVM	Gaussian kernel SVM
Number of samples	1031, 5144	79.35	95.63	96.64	97.65	98.46
	400, 320	71.88	82.81	88.13	93.44	96.56
	50, 320	71.25	–	82.19	83.44	85.31

then reduce the dimension to 50 in the wavelet fusion method. The experiment gains the similar result to Experiment 1 (see Table 2.6). In Experiment 3, the training sample in Experiment 2 is reduced to 50. In this case, the classification effect of SVM method still keeps the best, while ML method cannot be implemented due to less training samples.

The experiment shows that the classification accuracy gained from the maximum likelihood classification method is generally higher than SAM method, but the number of the training samples cannot be insufficient (theoretically, it should be more than the spectral dimension, while actually it is requested to be more). The classification accuracy gained from the SVM-based classification method is generally higher than the maximum likelihood classification method. In SVM, the efficiency of Gaussian kernel function is generally the maximum, while the efficiency of the linear kernel function is relatively poor. The experiment result shows clearly the excellent performance of SVM.

2.4.2 *Performance Assessment for Multi-class Classifier with the Simplest Structure*

It still adopts the agriculture and forestry remote-sensing area in Indiana of America. Four types of the land objects in the real chart are selected for the classification experiment. There are 400 pairs of training samples, and 320 pairs of test samples. The experiment adopts the least squares support vector machine of Gaussian kernel function and the efficient SMO algorithm. The iterative operation does not store the kernel function. At the same time, it adopts the 1-a-r method and 1-a-1 method for future reference. Table 2.7 provides the training time and test time used under the same iteration termination standard and different methods, with the time unit of second. The classification accuracy gained from the proposed method is 93.75 %, and the classification accuracy of two reference methods is, respectively,

Table 2.7 Comparison of training time and test time under structure of three classifiers

	1-a-r	1-a-1	Proposed method
Training time	81.7500	59.7970	37.5780
Test time	14.4702	11.21870	7.4288

94.69 and 94.37 %. The experiment result shows that the time as required during the training and testing by the algorithm constructed on the basis of the classifier structure proposed in this book is less than that of the two traditional methods, while the classification accuracy is reduced for less than 1 %. The experiment result fully verifies the previous theoretical analysis.

2.4.3 *Performance Assessment for Triply Weighted Classification*

The first group of the experimental samples is combined by the land object data of three classes of 3, 8, and 11 (the number of pixels is successively 834, 489, and 2468) in the agriculture and forestry remote-sensing image in Indiana. Extract the spectral characteristics of some pixels as the training sample, and the entire class data as the test sample. Adopt the unweighted, sample weighting, feature weighting, class weighting mode, and triply weighting mode successively for the effect test. The classification result is shown successively in Fig. 2.12a–e. In the experiment,

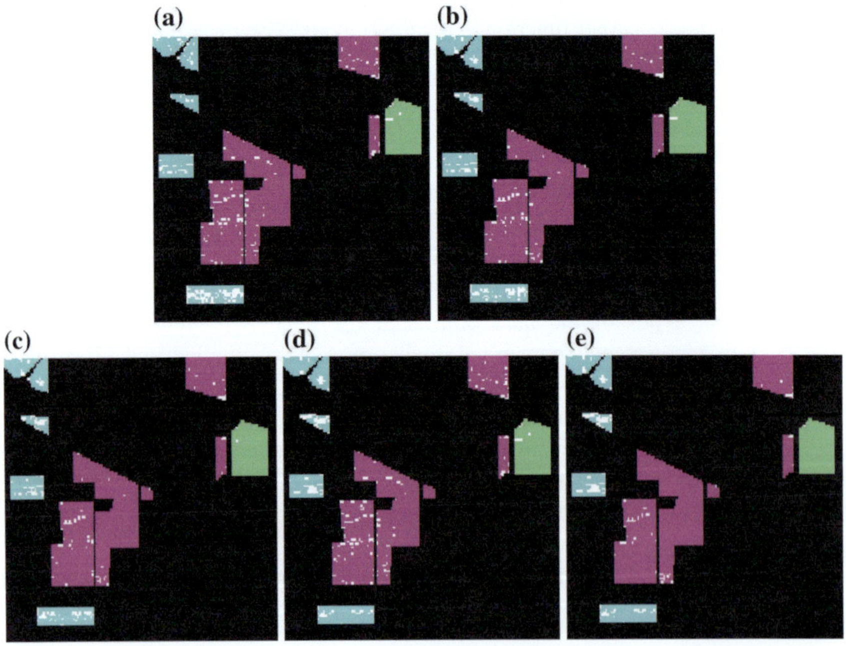

Fig. 2.12 Classification result chart under different weighting conditions in the first group of classification experiment. **a** Unweighted classification result chart. **b** Classification result chart of sample weighting. **c** Classification result chart of feature weighting. **d** Classification result chart of class weighting. **e** Classification result chart of triply weighting

SVM adopts the Gaussian kernel function, and the training samples are taken from the front 100 pixels of each class. In the class weighting experiment, the weights of three classes are successively set as 1, 5, and 10, i.e., focusing on considering the classification effect of Class 3. In the classification result, the above three classes are successively marked by different colors. The pixel of the classification error in the image is displayed by the white dot. The experiment result indicates that, using the sample weighting and feature weighting methods can enhance the entire classification accuracy to varying degrees, while the class weighting method can improve the analysis effect of the class corresponding to the relatively small weight (meanwhile reduce the analysis effect of the class corresponding to the relatively large weight). Although the triply weighting classification analysis effect of Class 3 is not as good as the effect of independently applying the class weighting, using three weighting methods at the same time can reach better analysis effect in general.

The second group of the experiment selects the land objects in three classes of 2, 6, and 10, and the number of pixels is successively 1434, 747, and 968. The experimental mode is ditto, and the classification result is shown in Fig. 2.13. The objective evaluation indexes of the above two groups of the experiments are respectively shown in Tables 2.8 and 2.9. The soft classification error means the absolute error mean statistics between the SVM decision result without the

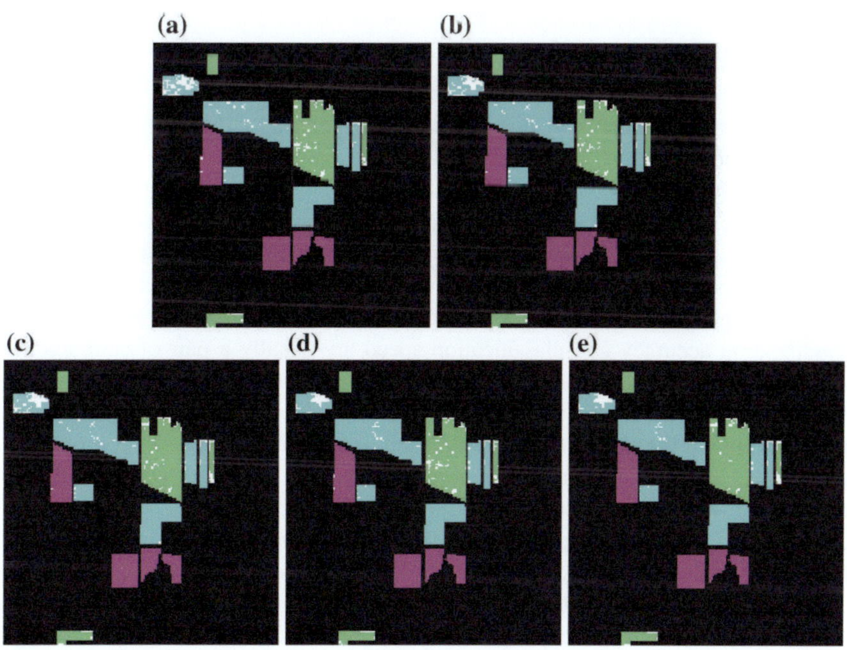

Fig. 2.13 Classification result chart under different weighting conditions in the second group of classification experiment. **a** Unweighted classification result chart. **b** Classification result chart of sample weighting. **c** Classification result chart of feature weighting. **d** Classification result chart of class weighting. **e** Classification result chart of triply weighting

Table 2.8 Number of misclassification pixels in the first group of classification experiment

Land object class	Unweighted	Sample weighting	Feature weighting	Class weighting	Triply weighting
Class 3	165/0.2248	145/0.1992	138/0.1914	130/0.1880	133/0.1897
Class 8	7/0.2248	6/0.2248	2/0.2248	7/0.2248	1/0.2248
Class 11	136/0.2248	87/0.2248	70/0.2248	147/0.2248	54/0.2248

Table 2.9 Number of misclassification pixels in the second group of classification experiment

Land object class	Unweighted	Sample weighting	Feature weighting	Class weighting	Triply weighting
Class 2	114/0.125	107/0.125	104/0.125	83/0.125	82/0.125
Class 6	2/0.125	1/0.125	1/0.125	3/0.125	0/0.125
Class 10	89/0.125	82/0.125	79/0.125	90/0.125	82/0.125

two-value quantization and the supervised classification result. This mode is more accurate than the traditional hard classification accuracy statistics. The experimental result further affirms that the weighting method has the effect.

2.5 Chapter Conclusions

One of the main contents in this chapter is to put forward a classifier with the simplest structure, which can greatly reduce the complexity of the classifier. There are many advantages, such as reducing the training time, reducing the test time, lowering the complexity of programming, and reducing the number of the sub-classifiers, to make the independent parameter adjustment in each decision function possible. It is noted that the advantage from the proposed method is at the cost of sacrificing the smaller classification accuracy. In the target classification problem, classification accuracy and classification speed are usually a pair of contradictory indexes. During solving the practical problem, which classifier is adopted should be determined as per the requirements of the user. For the problem with higher requirements on the classification speed such as the real-time application of SVM, the method proposed in this book is very effective. In the case of comprehensively considering the classification accuracy and classification speed, the traditional classifier and the classifier proposed in this book can be combined into the mixed classifier to coordinate the demand contradiction between the two.

The other main content in this chapter is the proposed multiple weighting classification method, to map the nonlinear distance into the corresponding weight to finish the sample weighting in accordance with the relationship between the sample anomaly degree and the distance of the sample deviating from the class center. According to the weighting characteristics of the intra-class divergence matrix on the linear spectral separation problem, it is promoted to LSSVM classification problem to finish the feature weighting. According to the special meaning of the unit matrix diagonal element in the system of LSSVM linear equation, it is set as different numerical values of reflecting the importance of the class, to finish the class weighting. In three weighting methods, the sample weighting is the way specially implemented on the training sample, and the feature weighting is the operation on all data, while the class weighting is the resetting of the matrix diagonal element during the training process. Three weighting methods can be used independently, and also can be used in combination by any check mode. In the practical application, it can be selected as per the specific demand.

In addition, for the mass data problem, the dominant solution method of LSSVM is perplexed by the aspects of the data storage. SMO optimization solution algorithm proposed in this chapter, as the substitute way, is conducive to solving this problem.

References

Chang C-I, Ji B (2006) Weighted abundance constrained linear spectral mixture analysis. IEEE Trans Geosci Remote Sens 44:378–388
Chen C-H, Tu T-M (1996) Computation reduction of the maximum likelihood classifier using the winograd identity. Pattern Recognit 29(7):1213–1220
Cristianini N, Shawe-Taylor J (2004) Support vector machine introduction. In: Guozheng L, Meng W, Huajun Z (eds). Electronic Industry Press, Beijing
Emami H Introducing correctness coefficient as an accuracy measure for sub pixel classification results. http://www.ncc.org.ir/articles/poster83/H.Emami.pdf
Ji B, Chang Chein-I, Jensen JO, Jensen JL (2004) Unsupervised constrained linear Fisher's discriminant analysis for hyperspectral image classification. In: 49th annual meeting, SPIE international symposium optical science and technology, vol 49. Imaging spectrometry IX (AM105), Denver, CO, pp 2–4
Jia XP, Richards JA (1994) Efficient maximum likelihood classification for imaging spectrometer data sets. IEEE Trans Geosci Remote Sens 32(2):274–281
Karpinski M, Werther T (1989) VC dimension and uniform learnability of sparse polynomials and rational functions. SIAM J Computing. Preprint 8537-CS, Bonn University
Richards JA, Jia XP (2006) Remote sensing digital image analysis, 3rd edn. Springer, Berlin
Shevade SK, Keerthi SS, Bhattacharyya C et al (2000) Improvements to the SMO algorithm for SVM regression. IEEE Trans on Neural Networks 11(5):1188–1193
Sohn Y, Rebello NS (2002) Supervised and unsupervised spectral angle classifiers. Photogram Eng Remote Sens 68(12):1271–1280
Song Q, Hu WJ, Xie WF (2002) Robust support vector machine with bullet hole image classification. IEEE Trans Systems Man and Cybern Part C 32:440–448
Suykens JAK, Brabanter JD, Lukas L, Vandewalle J (2002) Weighted least squares support vector machines: robustness and sparse approximation. Neurocomputing 48(1–4):85–105
Vapnik VN (2000) The nature of statistical learning theory. In: Zhang X (ed). Tsinghua University press, Beijing

Chapter 3
Endmember Extraction Technique of HSI

Before establishing the linear mixed model and conducting the spectral unmixing operation, it is very necessary to extract the spectral endmember, which acquires the essential priori information for the spectral unmixing. The spectral endmember extraction should be representative, and become the class component set of most pixels in the image. In the past ten years, multiple automatic and supervised hyperspectral image endmember extraction methods have been developed one after another. By contrast, the N-FINDR algorithm based on the convex polyhedron geometrical characteristics has been mostly used due to the advantages of the full automation, parameter free, and preferable extraction effect. This chapter will expand focusing on the promotion method of this algorithm.

3.1 Endmember Extraction Method: N-FINDR

Before the implementation of N-FINDR spectral endmember extraction algorithm, it is generally necessary to firstly make the dimension reduction pre-processing for the spectral data. For this purpose, it firstly introduces two common dimension reduction transformations, and then introduces the PPI endmember extraction algorithm for the pre-processing of constructing the fast N-FINDR algorithm below.

3.1.1 Introduction of Related Theory

1. **MNF transformation**

Maximum noise fraction (MNF) transformation was proposed by Green et al. (1988). The purpose of the transformation is to sequence the transformation image in accordance with the image quality.

The hyperspectral data with ND bands are arranged band by band to form the two-dimensional matrix $\boldsymbol{Z}_i(x)$, $i = 1, 2, \ldots, \text{ND}$. Given $\boldsymbol{Z}(x) = \boldsymbol{S}(x) + \boldsymbol{N}(x)$,

L. Wang and C. Zhao, *Hyperspectral Image Processing*,
DOI 10.1007/978-3-662-47456-3_3

including $\mathbf{Z}^{\mathrm{T}}(x) = \{\mathbf{Z}_1(x), \ldots, \mathbf{Z}_{\mathrm{ND}}(x)\}, \boldsymbol{S}(x)$ and $\boldsymbol{N}(x)$ are the irrelevant signal and noise components of $\mathbf{Z}(x)$, then their covariance matrixes possess the following relations:

$$\mathrm{Cov}\{\mathbf{Z}(x)\} = \boldsymbol{\Sigma} = \boldsymbol{\Sigma}_S + \boldsymbol{\Sigma}_N \tag{3.1}$$

The noise component of the i band is defined as the specific value of the noise variance and population variance of the band:

$$\mathrm{Var}\{\boldsymbol{N}_i(x)\}/\mathrm{Var}\{\mathbf{Z}(x)\} \tag{3.2}$$

MNF transformation selects the linear transformation $\boldsymbol{a}_i$:

$$\boldsymbol{Y}_i(x) = \boldsymbol{a}_i^{\mathrm{T}}\mathbf{Z}(x), \quad i = 1, \ldots, \mathrm{ND} \tag{3.3}$$

While meeting the conditions of being orthogonal to $\boldsymbol{Y}_k(x)$, $k = 1, 2, \ldots, i$, maximize the noise component of $\boldsymbol{Y}_i(x)$. From the principal component analysis principle, $\boldsymbol{a}_i$ is the characteristic vector of $\boldsymbol{\Sigma}_{\mathrm{ND}}\boldsymbol{\Sigma}^{-1}$ starting from the left side. $\mu_i(\mu_1 \geq \mu_2 \geq \cdots \geq \mu_{\mathrm{ND}})$ is the corresponding eigenvalue. Therefore, MNF transformation can sequence the transformation image as per the quality. The transformation can be expressed as the following matrix form:

$$\boldsymbol{Y}(x) = \boldsymbol{A}^{\mathrm{T}}\mathbf{Z}(x) \tag{3.4}$$

Here, $\boldsymbol{Y}(x) = [\boldsymbol{Y}_1(x), \boldsymbol{Y}_2(x), \ldots, \boldsymbol{Y}_{\mathrm{ND}}(x)]$, $\boldsymbol{A} = [a_1, a_2, \ldots, a_{\mathrm{ND}}]$.

One important feature of MNF transformation is the scale invariance on any band (only depending on the signal to noise ratio). The other important feature is the transformation orthogonal to $\boldsymbol{S}(x), \boldsymbol{N}(x)$ and $\mathbf{Z}(x)$.

In order to obtain the transformation matrix, we need obtaining $\boldsymbol{\Sigma}$ and $\boldsymbol{\Sigma}_N$. $\boldsymbol{\Sigma}$ can be obtained by the sample covariance matrix $\mathbf{Z}(x)$. The introduction to evaluate $\boldsymbol{\Sigma}_N$ is as follows.

In most remote-sensing data, the pixel spectrum has the stronger spatial correlation, while the spatial correlation of the noise is relatively weak. Given Δ is a smaller spatial delay, the selection of the vector $\boldsymbol{a}_i$ should minimize the absolute value of $\mathrm{Corr}(\boldsymbol{Y}_i(x), \boldsymbol{Y}_i(x+\Delta)) - \mathrm{Corr}(\boldsymbol{a}_i^{\mathrm{T}}\mathbf{Z}(x), \boldsymbol{a}_i^{\mathrm{T}}\mathbf{Z}(x+\Delta))$ in the case of meeting the conditions of being orthogonal to $\boldsymbol{Y}_k(x), k < i$. Generally, Δ is valued at (1,0) or (0,1). According to MAF (Minimum/Maximum Autocor-relation Factors, for estimating the covariance matrix of some noise) transformation thought, $\boldsymbol{\Sigma}_\Delta = \mathrm{Cov}\{\mathbf{Z}(x) - \mathbf{Z}(x+\Delta)\}$ can be regarded as the measurement of the noise. The thought assumes that the signal $\boldsymbol{S}(x)$ is irrelevant to the noise $\boldsymbol{N}(x)$ and meets

$$\begin{cases} \mathrm{Cov}\{\boldsymbol{S}(x), \boldsymbol{S}(x+\Delta)\} = b_\Delta \boldsymbol{\Sigma}_S \\ \mathrm{Cov}\{\boldsymbol{N}(x), \boldsymbol{N}(x+\Delta)\} = c_\Delta \boldsymbol{\Sigma}_N \end{cases} \tag{3.5}$$

Here, b_Δ and c_Δ are constant, and b_Δ is more than c_Δ. Thus, it can be inferred:

$$\boldsymbol{\Sigma}_\Delta/2 = (1 - b_\Delta)\boldsymbol{\Sigma} + (b_\Delta - c_\Delta)\boldsymbol{\Sigma}_N \tag{3.6}$$

When the signal space height has self-correlation, the value of $\mathrm{Cov}\{\boldsymbol{S}(x), \boldsymbol{S}(x+\Delta)\}$ is close to the value of $\boldsymbol{\Sigma}_S$, and then it should have $b_\Delta \approx 1$. On the other hand, when the value of $\mathrm{Cov}\{\boldsymbol{N}(x), \boldsymbol{N}(x+\Delta)\}$ is close to the value of $\boldsymbol{\Sigma}_N$, and then it should have $c_\Delta \approx 0$. In these cases, $\boldsymbol{\Sigma}_N \approx \boldsymbol{\Sigma}_\Delta/2$ is correct. Even if b_Δ and c_Δ cannot reach the above limit, the following conclusion can be drawn as well. $\boldsymbol{\Sigma}_\Delta\boldsymbol{\Sigma}^{-1}$ and $\boldsymbol{\Sigma}_N\boldsymbol{\Sigma}^{-1}$ have the similar characteristic vector, and this characteristic is independent of b_Δ and c_Δ. The characteristic vector of $\boldsymbol{\Sigma}_\Delta\boldsymbol{\Sigma}^{-1}(\lambda_i)$ and $\boldsymbol{\Sigma}_N\boldsymbol{\Sigma}^{-1}(\mu_i)$ has the following relation:

$$\mu_i = \frac{\lambda_i/2 - (1 - b_\Delta)}{b_\Delta - c_\Delta} \tag{3.7}$$

From $0 \le \mu_i \le 1$, it gets $c_\Delta \le 1 - \lambda_i/2 \le b_\Delta$. Therefore, we can obtain the upper bound of c_Δ and the lower bound of b_Δ through the maximum and minimum λ_i. When $b_\Delta \approx 1$ and $c_\Delta \approx 0$, $1 - \lambda_i/2$ becomes the relevance of the i MAF component and the neighborhood component:

$$1 - \lambda_i/2 = \mathrm{Corr}\{\boldsymbol{Y}_i(x), \boldsymbol{Y}_i(x+\Delta)\} \tag{3.8}$$

Thus, the relation between $\boldsymbol{\Sigma}_N$ and $\boldsymbol{\Sigma}_\Delta$ is established, and MNF transformation is finished.

2. **PCA transformation**

See Sect. 7.1.2 of this book for the basic theory of PCA. It can be proved that, when various bands possess the irrelevant equal covariance σ_N^2 noise, MNF transformation is equivalent to PCA transformation.

3. **PPI (Pixel Purity Index) method**

As shown in Fig. 3.1, the main feature of PPI algorithm (Boardman et al. 1995) lies in the supervision. Firstly, carry out the dimension reduction processing by utilizing MNF (or PCA) transformation. Then, in Nd dimension data after MNF transformation, for each point in the image cube, it randomly generates the L straight lines in the random direction, so as to calculate the pixel purity. All points in the data space are projected to these straight lines, and the points falling into the end point of each straight line are counted. Upon vast statistics, those pixel points with the count value exceeding the presetting quantity are affirmed as the spectral endmember. Furthermore, those spectral endmembers are determined by the corresponding class in the mode of manual, supervised or unsupervised clustering modes.

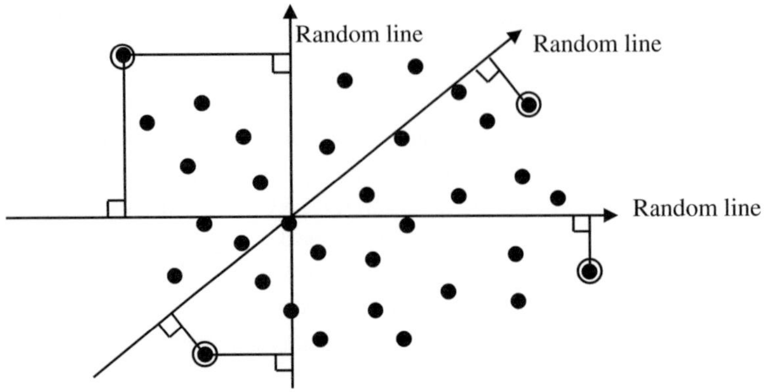

Fig. 3.1 Schematic diagram of PPI algorithm

3.1.2 N-FINDR Algorithm

Suppose that the land object classes in the hyperspectral image data have the corresponding spectral endmembers, while each pixel is mixed by one or more spectral endmembers among them. Thus, according to the convex geometric theory, all pixels form into a convex polyhedron in the hyperspectral data space, and each spectral endmember corresponds to a vertex of the convex polyhedron, as shown in Fig. 3.2. In this case, the task of the spectral endmember extraction becomes the extraction of the vertex of the convex polyhedron formed by the data space. Because the convex polyhedron with the vertex of all spectral endmembers possesses the maximum volume, the task turns into seeking the pixel of assigned numbers, to make convex polyhedron with the vertex of them possess the maximum volume. It is emphasized that, the implementation of the algorithm is made in the transformation space upon MNF, rather than in the original data space. One of the purposes for the dimension reduction processing is to enable the volume calculation of the convex polyhedron to implement. As a matter of fact, in the practical

Fig. 3.2 Schematic diagram of N-FINDR algorithm

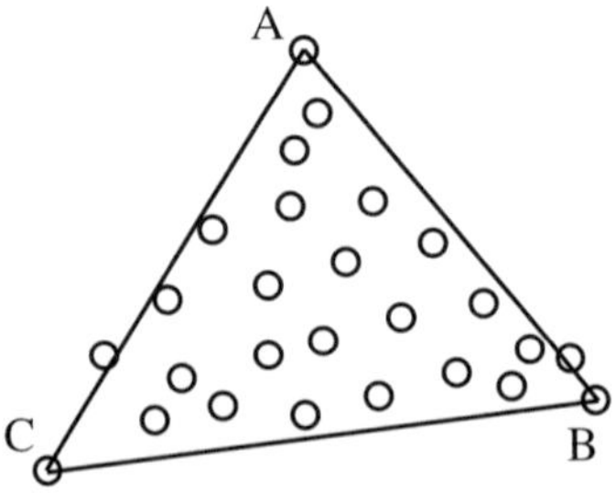

application, MNF transformation is often replaced by PCA transformation, because the latter one possesses more intuitive physical significance and smaller calculated amount and have the effect of approaching the former one upon a large number of the experimental demonstrations. We will further describe this process below.

Given $\boldsymbol{s}_t(t = 1, 2, \ldots, \mathrm{Np})$ is the pixel data in the transformation domain after MNF transformation of the original data. Nd is the corresponding spectral dimension. If $\boldsymbol{e}_i(i = 1, 2, \ldots, \mathrm{Nd} + 1)$ is all spectral endmembers of the space, each datum in $\boldsymbol{s}_t(t = 1, 2, \ldots, \mathrm{Np})$ can be expressed as the linear synthesis of all spectral endmembers as follows:

$$
\begin{aligned}
\boldsymbol{s}_t &= \sum_{i=1}^{\mathrm{Nd}+1} \boldsymbol{\alpha}_i^t \boldsymbol{e}_i + \varepsilon \\
\text{s.t.} & \sum_{i=1}^{\mathrm{Nd}+1} \boldsymbol{\alpha}_i^t = 1, \quad 0 \le \boldsymbol{\alpha}_i^t \le 1. \quad t = 1, 2, \ldots, \mathrm{Np}
\end{aligned}
\tag{3.9}
$$

Here, $\boldsymbol{\alpha}_i^t$ is the mixing ratio of the i spectral endmember in the t pixel spectrum, and ε is error term. As previously mentioned, the task of the spectral endmember extraction is to seek the pixel of assigned numbers, and maximize the volume of the convex polyhedron expanded by them. For this purpose, the algorithm randomly selects $(\mathrm{Nd} + 1)$ pixels as the initial spectral endmember, and correspondingly calculates the volume of the convex polyhedron expanded by them. Then, each pixel is used for successively substituting each currently selected spectral endmember. If some substitute can gain the convex polyhedron with larger volume, such substitute can be retained as the effective substitute, or, it will be eliminated as the invalid substitute. Repeat such basic processes till the substitute cannot result in the volume increase of the convex polyhedron. At this time, the currently selected result will be selected as the final spectral endmember.

The volume calculation formula of the convex polyhedron expanded by the $(\mathrm{Nd} + 1)$ pixel $\boldsymbol{p}_1, \boldsymbol{p}_2, \ldots, \boldsymbol{p}_{\mathrm{Nd}+1}$ is as follows:

$$
V(\boldsymbol{E}) = \frac{1}{(\mathrm{Nd} + 1)!} abs(|\boldsymbol{E}|) \tag{3.10}
$$

where,

$$
\boldsymbol{E} = \begin{bmatrix} 1 & 1 & \ldots & 1 \\ \boldsymbol{p}_1 & \boldsymbol{p}_2 & \ldots & \boldsymbol{p}_{\mathrm{Nd}+1} \end{bmatrix} \tag{3.11}
$$

Here, $\mathrm{abs}(\boldsymbol{E})$ and $|\boldsymbol{E}|$ are respectively the absolute value and determinant operator.

3.2 Distance Measure-Based Fast N-FINDR Algorithm

As the automatic selection of the spectral endmember, N-FINDR algorithm has been widely applied due to the automaticity and high efficiency. However, the following two aspects greatly restrict the efficiency of this method, involving a large number of the volume calculations, and the random selection of the spectral endmember initial value and the blind search of the spectral endmember update point. This book will improve on account of these two aspects.

3.2.1 Substituting Distance Measure for Volume One

It can be seen that the algorithm includes a large number of the volume calculations, which is the most time consuming section. Moreover, the complexity of the volume calculation (i.e., mainly determinant calculation) will show the cubic increase with the increasing number of the selected spectral endmember, and accordingly reduce the operating speed of the algorithm greatly. For this purpose, we put forward an improvement method of converting the volume calculation in the algorithm into the distance calculation from the point to the hyperplane.

In order to be convenient for describing and gaining the visualization effect, we firstly consider the situation of the two-dimensional space. In Fig. 3.3, $\boldsymbol{A}$, $\boldsymbol{B}$ and $\boldsymbol{C}$ as the vertex form a triangle (two-dimensional convex polyhedron), noting its volume as V_{old}. Given $\boldsymbol{A}_0$ is the point differing from $\boldsymbol{A}$, $\boldsymbol{B}$ and $\boldsymbol{C}$, $\boldsymbol{A}_0$, $\boldsymbol{B}$ and $\boldsymbol{C}$ form a new triangle, noting its volume as V_{new}. Then the line segments $\boldsymbol{AD}$ and $\boldsymbol{A}_0\boldsymbol{D}_0$ are respectively noted as the distance from $\boldsymbol{A}$ and $\boldsymbol{A}_0$ to the line segment BC. So, whether replacing $\boldsymbol{A}_0$ by $\boldsymbol{A}$ is effective only needs comparing the size of V_{old} and V_{new} of the area (two-dimensional volume). For this purpose, the original N-FINDR

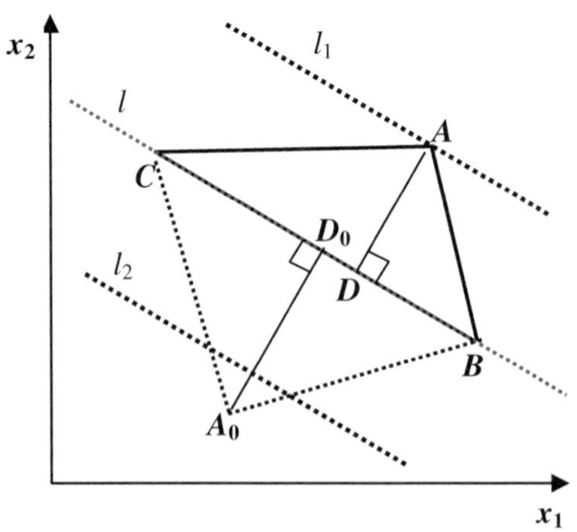

Fig. 3.3 Schematic diagram of measure substitute in N-FINDR algorithm

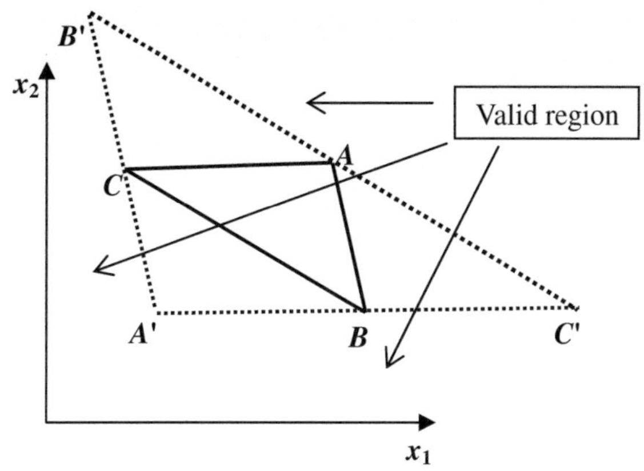

Fig. 3.4 Schematic diagram of effective area of spectral endmember update

algorithm needs specifically calculating V_{old} and V_{new} in accordance with the formula (3.10). However, we see that the size relations between V_{old} and V_{new} is consistent with the size relations between $\boldsymbol{AD}$ and $\boldsymbol{A}_0\boldsymbol{D}_0$. Hence, comparing V_{old} and V_{new} can be finished by comparing $\boldsymbol{AD}$ and $\boldsymbol{A}_0\boldsymbol{D}_0$. Figure 3.3 provides such intuitive descriptions. l is the straight line formed by the points $\boldsymbol{B}$ and $\boldsymbol{C}$. l_1 is the straight line crossing the point $\boldsymbol{A}$ and parallel to l. l_2 is the symmetric straight line of the straight line l_1 with respect to the straight line l. From Fig. 3.3, $\boldsymbol{A}$ can be replaced by $\boldsymbol{A}_0$, if and only if $\boldsymbol{A}_0$ falls outside the area between the parallel lines of l_1 and l_2.

When the vertexes $\boldsymbol{A}$, $\boldsymbol{B}$, and $\boldsymbol{C}$ are considered at the same time, the effective area of the spectral endmember update will be beyond the area defined by the dashed triangle in Fig. 3.4.

Now we consider the situation of the Nd dimension space. Given $\boldsymbol{s}_1, \boldsymbol{s}_2, \ldots, \boldsymbol{s}_{\text{Nd}+1}$ is the (Nd + 1) point in this space, and $\boldsymbol{s}_0$ is the other point differing from them. We apply the distance measure to calculate whether the point $\boldsymbol{s}_0$ can replace the point $\boldsymbol{s}_i (1 \le i \le \text{Nd} + 1)$. In this case, we only need comparing the distance from the point $\boldsymbol{s}_0$ and point $\boldsymbol{s}_i$ to the hyperplane formed by $\boldsymbol{s}_1, \boldsymbol{s}_2, \ldots, \boldsymbol{s}_{i-1}, \boldsymbol{s}_{i+1}, \ldots, \boldsymbol{s}_{\text{Nd}+1}$. According to the high-dimensional geometric theory, the expression of the hyperplane from $\boldsymbol{s}_1, \boldsymbol{s}_2, \ldots, \boldsymbol{s}_{i-1}, \boldsymbol{s}_{i+1}, \ldots, \boldsymbol{s}_{\text{Nd}+1}$ is provided by the following formula:

$$\sum_{i=1}^{N} \boldsymbol{\alpha}^{\text{T}} \boldsymbol{x}_i + b = 0 \tag{3.12}$$

where, $\boldsymbol{\alpha} = (\alpha_1, \alpha_2, \ldots, \alpha_{\text{Nd}})^{\text{T}}$ is the solution of the following equation:

$$[\boldsymbol{s}_1, \boldsymbol{s}_2, \ldots, \boldsymbol{s}_{i-1}, \boldsymbol{s}_{i+1}, \ldots, \boldsymbol{s}_{\text{Nd}+1}]^{\text{T}} \cdot \boldsymbol{\alpha} + b \cdot \boldsymbol{I}_{\text{Nd}} = 0 \tag{3.13}$$

In the formula, $\boldsymbol{I}_d$ is the column vector of $\text{Nd} \times 1$. When $\boldsymbol{s}_1, \boldsymbol{s}_2, \ldots, \boldsymbol{s}_{i-1}, \boldsymbol{s}_{i+1}, \ldots, \boldsymbol{s}_{\text{Nd}+1}$ has the linear correlation with the original point, b values at 0, or b is valued at 1. In fact, the opportunity of b valuing at 0 is the minimum. Thus, the distance from s_0 to the hyperplane (3.12) can be provided by the following formula:

$$D(s_0) = \alpha^{\mathrm{T}} s_0 + b \tag{3.14}$$

3.2.2 PPI Concept-Based Pixel Indexing

In order to gain the spectral endmember iterative search with rapid convergence, each pixel point should be evaluated and sequenced in advance in accordance with the potential purity. According to the basic thought of PPI algorithm, the spectral endmember should be in the margin of all data spectral space. Specifically, when we project each data spectrum into numerous test vectors (skewers) in the random direction, the spectral endmember will fall the larger probability over the terminal of the test vector projection. We can make the spectral endmember extraction in this way. While compromising and considering the complexity of the calculation and the accuracy of the selection, we only select each dimension coordinate of the spectral space as the test vector. So, all projection results can be exempted from any calculation, and can be gained directly by the coordinate value of the data. Therefore, we can sequence in accordance with the following way.

1. Select and arrange the dot pair corresponding to the maximum value and minimum value coordinates successively from the first dimension to the final dimension of the original data space.
2. Make the first step operation in the remaining data space.
3. Continue such process, till all data points are selected and arranged.

After all data points are sequenced, $(\mathrm{Nd}+1)$ data point in the headmost has the reason to be selected as the initial spectral endmember due to the maximum purity, and the iteration update process is also made successively in accordance with the sequencing result. In this way, selecting the initial spectral endmember and selecting the spectral endmember update search order accordingly will quicken the convergence rate of the algorithm, and meanwhile reduce the local convergence opportunity of the algorithm.

Two acceleration ways proposed above can be used independently and also can be used in combination by any check mode. In the experiment, adopt various acceleration modes and accordingly reflect the acceleration effect of each way exactly. For the convenience of quotation, the algorithm from independently using the first way (replacing the volume measure by the distance measure), the second way (PPI concept-based rearrangement) and the combination use of two ways shall be respectively noted as D-N-FINDR, P-N-FINDR and PD-N-FINDR.

3.2.3 Complexity Analysis and Efficiency Assessment

In the implementation of N-FINDR algorithm, the frequency of the hyperplane creation is equal to the frequency of the spectral endmember update and the product of the number of spectral endmembers, which will be less than the frequency of the distance calculation. Therefore, in the improved N-FINDR algorithm, the complexity of the calculation can ignore the hyperplane creation and can be approximately equal to the distance calculation. From the formula (3.14), a distance calculation is only a dot product calculation, and the complexity of the calculation will have the linear increase with the increase of the number of the spectral endmembers. By contrast, the volume calculation needs calculating the determinant of a $(\mathrm{Nd}+1)\times(\mathrm{Nd}+1)$ matrix, and the complexity will have the cubic increase with the increase of the number of the spectral endmembers. Except for the differences of the calculated amount, the algorithm before and after the improvement will be fully equivalent for the final spectral endmember extraction result.

3.3 Linear LSSVM-Based Distance Calculation

The distance formula employed in the last section has the simple calculation, but the implementation needs the assistance of the dimension reduction method, i.e., the distance formula can only be used in the corresponding low-dimensional space. The assistance of the dimension reduction method increases the calculated amount, and also results in the relevant endmember extraction algorithm failing to implement in the original data space, with some influence on the reliability. For this purpose, this section puts forward a distance calculation method of continuously keeping the linear complexity which can be implemented in any space.

The optimization problem expression of the linear LSSVM:

$$\begin{aligned}&\min_{w,b,e} J(\boldsymbol{w},\boldsymbol{e})=\frac{1}{2}\|\boldsymbol{w}\|^2+\frac{\gamma}{2}\sum_{i=1}^{\mathrm{Ntr}}e_i^2\\&\text{s.t.}\quad \boldsymbol{y}_i=\langle\boldsymbol{w},\boldsymbol{x}_i\rangle+b+e_i,\\&i=1,2,\ldots,\mathrm{Ntr},\quad \gamma>0.\end{aligned} \tag{3.15}$$

The corresponding discrimination function is:

$$f(\boldsymbol{x})=\sum_{i=1}^{\mathrm{Ntr}}\alpha_i K(\boldsymbol{x}_i,\boldsymbol{x})+b=\langle\boldsymbol{w}^*,\boldsymbol{x}\rangle+\boldsymbol{b}^* \tag{3.16}$$

where, the vector $\boldsymbol{w}^*=\sum_{i=1}^{\mathrm{Ntr}}\alpha_i\boldsymbol{x}_i\boldsymbol{y}_i$, $\boldsymbol{b}^*=-\frac{1}{2}\left(\max_{y_i=0}(\langle\boldsymbol{w}^*,\boldsymbol{x}_i\rangle)+\min_{y_i=+1}(\langle\boldsymbol{w}^*,\boldsymbol{x}_i\rangle)\right)$.

The basic SVM classification theory is used for handling two types of the classification problems. For the multi-class classification problem, it will be transformed into multiple two-class classification problems for solving. Different transformation modes correspond to different multi-class classifier structures. At present, one of the most common multi-class classifier structures is 1-*a*-*r* (1-against-rest) type multi-class classifier. Take *N* class problem as an example. 1-*a*-*r* type classifier is to construct *N* two-class target subclassifiers. The *k* subclassifier regards the training sample in the *k* class as one class training sample, and the rest is regarded as the other class training sample. For some input sample, the classification result will output the corresponding class with the maximum value for various subclassifiers.

Take a three-class classification problem as an example. In one of three subclassifiers of the linear LSSVM, the spectral endmember $\boldsymbol{A}$ in Fig. 3.5 is stipulated as one class, and $\boldsymbol{B}$ and $\boldsymbol{C}$ are stipulated as the other class:

$$f(\boldsymbol{A}) = 1, \quad f(\boldsymbol{B}) = 1, \quad f(\boldsymbol{C}) = 0 \tag{3.17}$$

Given $\boldsymbol{P}$ is any point differing from $\boldsymbol{A}, \boldsymbol{B}$ and $\boldsymbol{C}, \boldsymbol{D}$ expresses the intersection point of the extension cord for the line segment $L_{\boldsymbol{BC}}$ and the line segment $L_{\boldsymbol{AP}}$. Now given $\boldsymbol{P} = \lambda \boldsymbol{A} + (1 - \lambda)\boldsymbol{D}$, here $0 < \lambda < 1$. From (3.16), the following relational expression is correct:

$$\begin{aligned} f(\boldsymbol{P}) &= f[\lambda \boldsymbol{A} + (1 - \lambda)\boldsymbol{D}] = f(\lambda \boldsymbol{A}) + f[(1 - \lambda)\boldsymbol{D}] - b \\ &= [\lambda f(\boldsymbol{A}) + (1 - \lambda)b] + [(1 - \lambda)f(\boldsymbol{D}) + \lambda b] - b \\ &= \lambda f(\boldsymbol{A}) + (1 - \lambda)f(\boldsymbol{D}) = \lambda \end{aligned} \tag{3.18}$$

From the relational expression (3.18), it can be inferred that the discrimination function $f(\boldsymbol{P})$ defines a directed distance in direct proportion to Euclidean distance from the point $\boldsymbol{P}$ to the straight line $\boldsymbol{BC}$. Similarly, the other two subclassifiers of the three-class linear LSSVM also, respectively, define the directed distance from the point $\boldsymbol{P}$ to the straight line $\boldsymbol{AB}$ and the straight line $\boldsymbol{AC}$. By a logical extension of

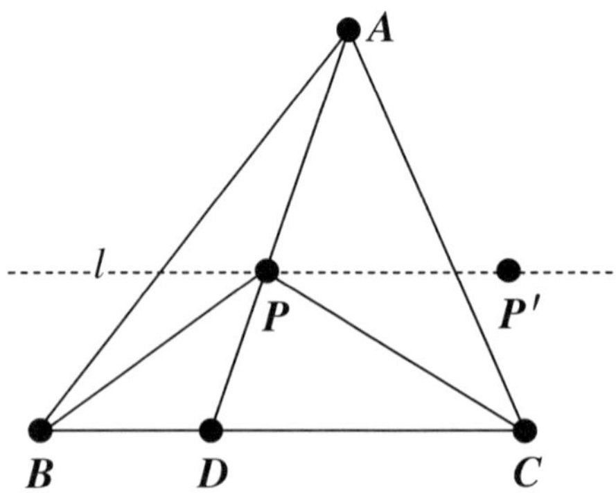

Fig. 3.5 Distance calculation function of linear LSSVM

this point, for N class classification problem, N subclassifiers of the linear LSSVM define the directed distance function of N directions. In fact, the above distance calculation function relevant to the linear LSSVM shall be valid for other linear support vector machines (LSVM).

With the above distance calculation function, we can achieve the fast N-FINDR algorithm free from dimension reduction pre-processing rapidly by the linear LLSVM.

3.4 Robust Method in Endmember Extraction

As we mentioned above, the outlier points exist extensively in the remote-sensing image, and generally are relatively far away from the data center. Therefore, in the N-FINDR algorithm, they will be selected as the spectral endmember by the higher probability. In order to overcome this adverse effect, this section will carry out the robustness control on it.

It will research on the robustness measures respectively in the data pre-processing stage and spectral endmember extraction stage below.

3.4.1 *In the Pre-processing Stage: Obtaining of Robust Covariance Matrix*

While making the data pre-processing of N-FINDR algorithm, whether using PCA transformation or MNF transformation, they need solving the covariance matrix of the data set. In the case of the outlier points in the data, the calculation of the covariance matrix will be influenced and will not be accurate. For this purpose, we weaken the influence of the outlier points by the covariance matrix calculation method with the robust characteristic (Hubert et al. 2005).

Still given the total pixels of Np, and the dimension of ND. All pixels can be stored in the matrix $\boldsymbol{X} = \boldsymbol{X}_{\mathrm{Np,ND}}$ of $\mathrm{Np} \times \mathrm{ND}$. The main steps of obtaining the robust covariance matrix can be simply summarized as follows. Firstly, transform the original data into the subspace with the dimension of at most $\mathrm{Np} - 1$. Then, create the covariance matrix $\boldsymbol{S}_0$ for selecting the k dimension subspace to be retained finally, and accordingly generate the k dimension subspace of the data gained from fitting of the previous step. Finally, the data gained in the first step are projected to the k dimension subspace. Make the robustness estimation for the mean vector (also known as robust center, location) and covariance matrix (scatter matrix) once again. Now make the specific description for this process.

Stage I: After make the centralization processing of the original data $\boldsymbol{X}_{\mathrm{Np,ND}}$, finish the first step by the singular value decomposition:

$$\boldsymbol{X}_{\mathrm{Np,ND}} - 1_{\mathrm{Np}}\hat{\boldsymbol{\mu}}'_X = \boldsymbol{U}_{\mathrm{Np},r_0}\boldsymbol{D}_{r_0,r_0}\boldsymbol{V}'_{r_0,\mathrm{ND}} \tag{3.19}$$

where, $\hat{\boldsymbol{\mu}}_X$ is the mean vector, r_0 is the rank of the matrix at the left end of the equation, $\boldsymbol{D}$ is the diagonal matrix of $r_0 \times r_0$, and $\boldsymbol{U}$ and $\boldsymbol{V}$ are the orthogonal matrix. Actually, this step is just the general PCA full-dimensional transformation of the original data.

Stage II: In this stage, n_o samples with the minimum outlier index will be selected from all data points. Their covariance matrixes will be used for obtaining the subspace of the dimension k_o . n_o is generally assigned to be about 75 % of all data.

The determination method of these points is as follows.

1. Define the outlier measurement outl_A as follows:

$$\mathrm{outl}_A(\boldsymbol{x}_i) = \max_{v \in B} \frac{\left| \boldsymbol{x}'_i\boldsymbol{v} - \mathrm{med}_j(\boldsymbol{x}'_j\boldsymbol{v}) \right|}{\mathrm{mad}_j(\boldsymbol{x}'_j\boldsymbol{v})} \tag{3.20}$$

where, $\mathrm{mad}(\boldsymbol{x}'_j\boldsymbol{v}) = \mathrm{med}_j\left|\boldsymbol{x}'_j\boldsymbol{v} - med_j(\boldsymbol{x}'_j\boldsymbol{v})\right|$, $\boldsymbol{v}$ is the vector with different random directions, and can be formed by randomly selecting $\min(C_n^2, 250)$ dot pairs in the original data. Thus, according to the measurement index, n_o data points of the corresponding minimum values can be selected.

2. After selecting n_o data points (the corresponding index set is H_0), utilize them to calculate the mean vector and covariance matrix, and then obtain the k_o principal components of the corresponding maximum eigenvalue by the covariance matrix.
3. The data point is projected into the subspace extended by the k_o principal components, and the new data point set $\boldsymbol{X}^*_{\mathrm{Np},k_0}$ is obtained.

Stage III: In this stage, make the robustness estimation for the covariance matrix of $\boldsymbol{X}^*_{\mathrm{Np},k_0}$. For this purpose, we need select n_o data points of the minimum determinant corresponding to the covariance matrix.

1. Define that $\hat{\boldsymbol{\mu}}_0$ and $\boldsymbol{\Sigma}_0$ are respectively the mean vector and covariance matrix of the data point in H_0. Apply the C-step method to seek n_o data points.
   If $\det(\boldsymbol{\Sigma}_0) > 0$, apply the following formula to calculate the robust distance for all data points:

$$d_{\hat{\mu}_0,\Sigma_0}(i) = \sqrt{(\boldsymbol{x}^*_i - \hat{\boldsymbol{\mu}}_0)'\boldsymbol{\Sigma}_0^{-1}(\boldsymbol{x}^*_i - \hat{\boldsymbol{\mu}}_0)}, \quad i = 1, 2, \ldots, \mathrm{Np} \tag{3.21}$$

Select n_o data points corresponding to the minimum distance to form the subset H_1, and accordingly calculate $\hat{\boldsymbol{\mu}}_1$, $\boldsymbol{\Sigma}_1$ and the distance $d_{\hat{\mu}_1,\Sigma_1}(i)$ of all data points. According to such mode, carry on, till the obtained determinant of the covariance matrix is not reduced. Moreover, when the covariance matrix obtained in some step is the singular matrix, the data point will be mapped into the space extended by the nonzero characteristic vector of the matrix, and then the C-step method will be applied here. After the algorithm convergence, we will gain the searched h data points (the corresponding index set is H_1) and the relevant data matrix $\boldsymbol{X}^*_{\mathrm{Np},k_1}$.

2. Now, apply the fast MCD algorithm on $\boldsymbol{X}^*_{\mathrm{Np},k_1}$. Select multiple (about 250) subsets with $(k_1 + 1)$ data in $\boldsymbol{X}^*_{\mathrm{Np},k_1}$ randomly. Calculate the mean vector and covariance matrix for each subset, and the robust distance for all data points, and accordingly select the n_o data points corresponding to the minimum robust distance and form the subset. Implement the C-step method for the subset. The finally obtained data set is defined as $\tilde{\boldsymbol{X}}_{\mathrm{Np},k}$. Given $\hat{\boldsymbol{\mu}}_2$ and $\boldsymbol{S}_1$ are the mean vector and covariance matrix of n_o data points in the first step, while $\hat{\boldsymbol{\mu}}_3$ and $\boldsymbol{S}_2$ are the mean vector and covariance matrix obtained from the MCD fast algorithm. If $\det(\boldsymbol{S}_1) < \det(\boldsymbol{S}_2)$, it will continue to calculate in accordance with $\hat{\boldsymbol{\mu}}_2$ and $\boldsymbol{S}_1$. To this end, given $\hat{\boldsymbol{\mu}}_4 = \hat{\boldsymbol{\mu}}_2, \boldsymbol{S}_3 = \boldsymbol{S}_1$. Otherwise, given $\hat{\boldsymbol{\mu}}_4 = \hat{\boldsymbol{\mu}}_3, \boldsymbol{S}_3 = \boldsymbol{S}_2$.
3. In order to increase the statistical efficiency, make the weighting processing for $\hat{\boldsymbol{\mu}}_4$ and $\boldsymbol{S}_3$ again. Given $\bar{D}$ is the mean value of all data points relevant to the robust distance of $\hat{\boldsymbol{\mu}}_4$ and $\boldsymbol{S}_3$, denoted as D_i $(i = 1, 2, \ldots, \mathrm{Np})$ is the robust distance of all data points relevant to $\hat{\boldsymbol{\mu}}_4$ and $\boldsymbol{S}_3$. Define the weighting function w:

$$w(D_i) = \begin{cases} 1 & \text{if} \quad D_i \le \sqrt{\chi^2_{k,0.975}} \\ 0 & \text{otherwise} \end{cases} \tag{3.22}$$

Thus, the reweighted $\hat{\boldsymbol{\mu}}_4$ and $\boldsymbol{S}_3$ can be calculated as follows:

$$\boldsymbol{\mu}_4 = \frac{\sum_{i=1}^{n} w(D_i)\tilde{\boldsymbol{x}}_l}{\sum_{i=1}^{n} w(D_i)} \tag{3.23}$$

$$\boldsymbol{S}_3 = \frac{\sum_{i=1}^{n} w(D_i)(\tilde{\boldsymbol{x}}_i - \hat{\boldsymbol{\mu}}_4)(\tilde{\boldsymbol{x}}_i - \hat{\boldsymbol{\mu}}_4)'}{\sum_{i=1}^{n} w(D_i) - 1} \tag{3.24}$$

After obtaining $\boldsymbol{S}_3$, it can get the corresponding characteristic root diagonal matrix $\boldsymbol{L}_2 = \boldsymbol{L}_{k,k}$, and the characteristic vector matrix $\boldsymbol{P}_2 = \boldsymbol{P}_{k,k}$. Upon obtaining the robust covariance matrix, it can be applied to the pre-processing of N-FINDR algorithm with the PCA transformation.

3.4.2 In Endmember Extraction Stage: Deletion of Outliers

The volume calculation-based or the distance test-based spectral endmember iterative search processes are easily interfered by the outlier points. The spectral endmember from the N-FINDR algorithm is the pixel point combination of the corresponding maximum volume, so the outlier point by its special spatial position possesses higher probability of being selected to the spectral endmember than the common pixel point, accordingly resulting in the larger influence and even the total failure of the algorithm. Even though it has only one point like this, the potential influence is immeasurable. Moreover, in the hyperspectral image, the outlier points exist extensively. Therefore, constructing the robust algorithm of this process has greater significance than the previous process. If it can confirm the outlier points and eliminate to consider these points during the spectral endmember iterative search process, our purpose can be achieved. The obtaining method of the robust covariance matrix includes the confirmation method of the outlier points, but this method has poor effect on the algorithm in this stage. This method affirms that the outlier points are conducted relative to all pixel points, so that some class may be misjudged as the outlier point completely due to the less total pixels. Such error has smaller adverse impact during the robust covariance matrix solution or robust hyperplane fitting, while the impact by the spectral endmember extraction is tremendous. For this purpose, we confirm and remove the outlier points in the way of the neighborhood analysis.

The outlier point usually exists in more isolating mode. In this way, each pixel point can be centered to establish the neighborhood window with the fixed size, and the isolation degree measurement index of the central point can be made through calculating the pixel points included in the neighborhood window. The larger the isolation degree measurement index is, the greater possibility this point as the outlier point has. In order to save the calculated amount, we replace the circular neighborhood by the square neighborhood (high-dimensional box), so that we can avoid vast distance calculations on the premise of basically not influencing the removal effect of the outlier point.

3.5 Performance Assessment

3.5.1 Distance Measure-Based N-FINDR Fast Algorithm

Here we lay particular emphasis on the efficiency comparison of N-FINDR fast algorithm based on the direct distance calculation in the dimension reduction space and the original N-FINDR.

In the first experiment, as shown in Fig. 3.6, $\boldsymbol{A}$ (−15,0), $\boldsymbol{B}$ (15,0) and $\boldsymbol{C}$ (0,20) as the spectral endmembers are mixed into 1000 points (the Gaussian noise with the additional variance of 1) by the random normalization coefficient. In the

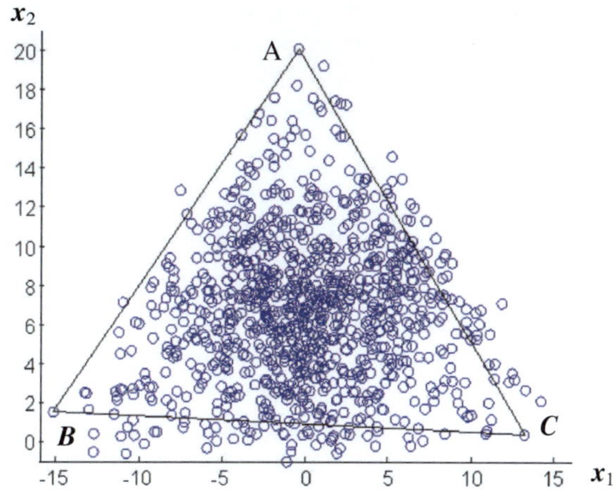

Fig. 3.6 Synthesized data in experiment 1

experiment, two methods (respectively referred to as V-N-FINDR and D-N-FINDR) before and after improvement have gained the same spectral endmember (the triangle vertex in the figure). The comparison of their running time (*E*-time), times of spectral endmember update (EM-times) and times of volume/distance calculation (*V*/*D*-times) is shown in Table 3.1.

In the second experiment, we increase the number of the spectral endmembers to 10, i.e., in nine-dimensional space, nine standard unit vectors and the original points are selected as the spectral endmember, and they are combined into 10,000 data points. The comparison result is shown in Table 3.2.

In regard to the proposed LSSVM-based N-FINDR fast algorithm, due to the same inherent principle of the above direct distance calculation method based on

Table 3.1 Comparison of running time/iterations in experiment 1

EM extraction method	Running time (s)	Times of EM update	Times of volume/distance calculation
V-N-FINDR	2.9380	21	9504
P-N-FINDR	1.0470	2	3725
D-N-FINDR	0.4210	21	9504
PD-N-FINDR	0.1410	2	3725

Table 3.2 Comparison of running time/iterations in experiment 2

EM extraction method	Running time (s)	Times of EM update	Times of volume/distance calculation
V-N-FINDR	961.61	95	689843
P-N-FINDR	132.02	28	105562
D-N-FINDR	22.750	95	689843
PD-N-FINDR	3.3910	28	105562

the dimension reduction space, the endmember extraction process is identical, and only the calculation efficiencies are slightly different. In view of the complexity of the dimension reduction, LSSVM-based fast algorithm can double the execution speed compared with the direct distance calculation.

3.5.2 *Robustness Assessment*

The employed data are some of the first class sample points in iris multispectral data (400 samples, four bands, three types of vegetation) published in the internet. The former two bands of the data are selected for obtaining the two-dimensional image display. Figure 3.7 is the spatial distribution map of these data points.

Figure 3.8 provides the affirmation of the outlier points under different scales. The horizontal axis and vertical axis are respectively Mahalanobis distance and robust distance between the number stronghold and data center. In the figure, the upper left area among four areas indicates the area affirmed as the outlier point in the robust distance sense and affirmed as the regular point in the Mahalanobis distance sense. From the spatial distribution map of the data points, it can be seen that the robust method has more reasonable affirmation effect. From Fig. 3.7, it can be seen that three sample points are relatively far away from the data center. This phenomenon is fully embodied in Fig. 3.8.

The experimental image used in the second group of the experiment is derived from the other AVIRIS military hyperspectral image (San Diego of America). The data have been pre-processed by the atmospheric correction and geometric correction, and stored in the form of the reflectivity data. See Table 3.3 for the specific data parameters. Utilize three-class target spectrum in the data as the spectral endmember. They are randomly combined into 1000 data samples.

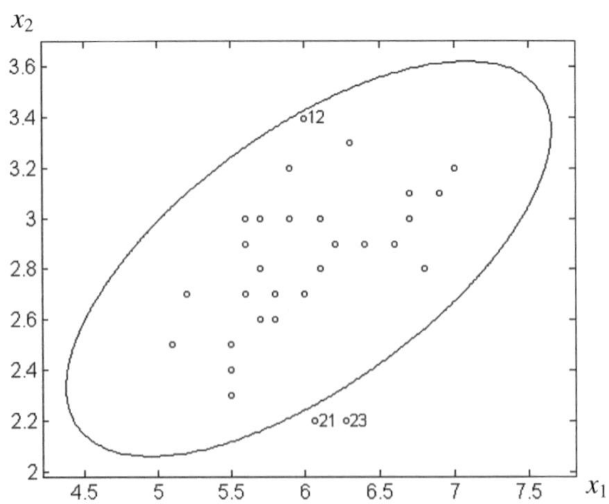

Fig. 3.7 Two-dimensional space display of experimental data

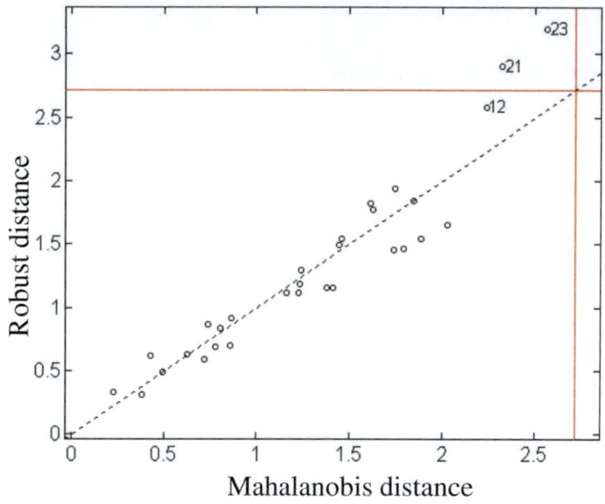

Fig. 3.8 Affirmation of outlier point

Table 3.3 Hyperspectral data parameters of San Diego

Sensor	AVIRIS
Wave length	400–1800 nm
Available spectral segments	126
Ground resolution	3.5 m
Image size	400 × 400
Gray range	0–10,000
Data acquisition location	San Diego of America

Then each spectral endmember adds the random noise respectively, and generates 10 outlier points separately. So, the total samples gained are 1030. Three images provided by Fig. 3.9 are respectively the original spectral endmember, the spectral endmember directly extracted by N-FINDR algorithm, and the spectral endmember extracted by the robustness method.

Thus, it is clear that, the extraction result in (b) is the spectral endmember interfered by the noise, while the extraction result in (c) preferably overcomes the noise interference and the result gained is close to the true spectral endmember.

In the third group of the experiment, apply three classes of the land objects (500 samples respectively for soybean, grass and forest land) of the Indian agriculture and forestry hyperspectral data to make the experiment of the spectral endmember extraction, and make the comparison between the average spectrum of various classes as the true spectrum and the extraction result. It can be seen that, the spectral endmember selected by the robustness method is obviously closer to the true spectrum. Figure 3.10 provides their intuitive comparison.

It is easy to analyze that, each spectral endmember, as the class center of the corresponding class, should fall into the position approximating the vertex rather than the vertex of the data form. Selecting the vertex is just an operation under the

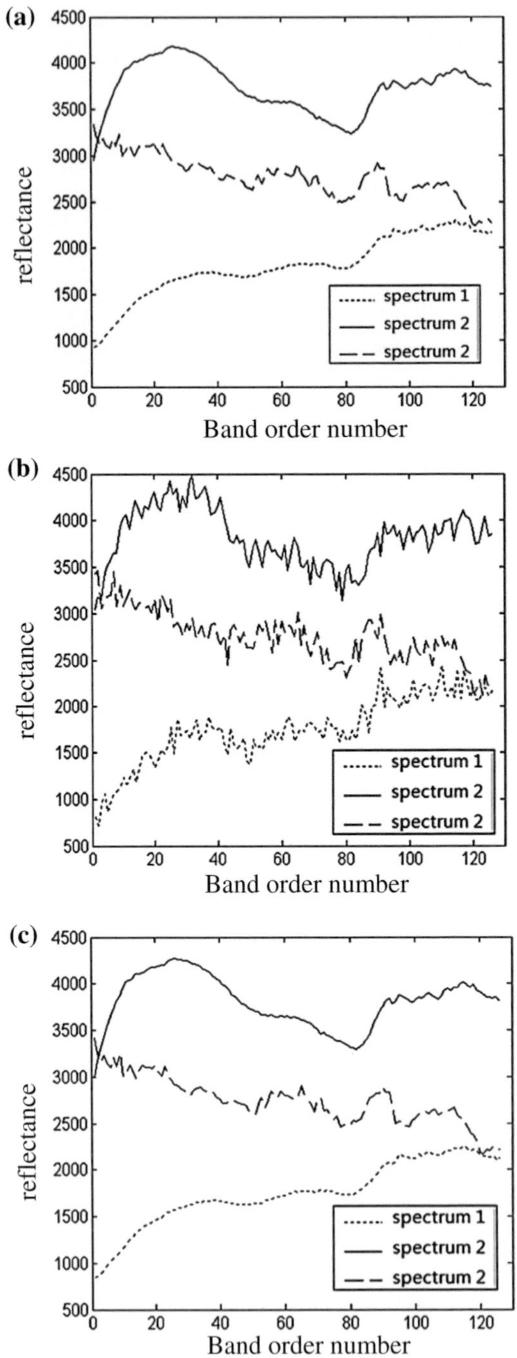

Fig. 3.9 Comparison of spectral endmember extraction result by different methods. **a** True spectral endmember. **b** Original method extraction result. **c** Robust method extraction result

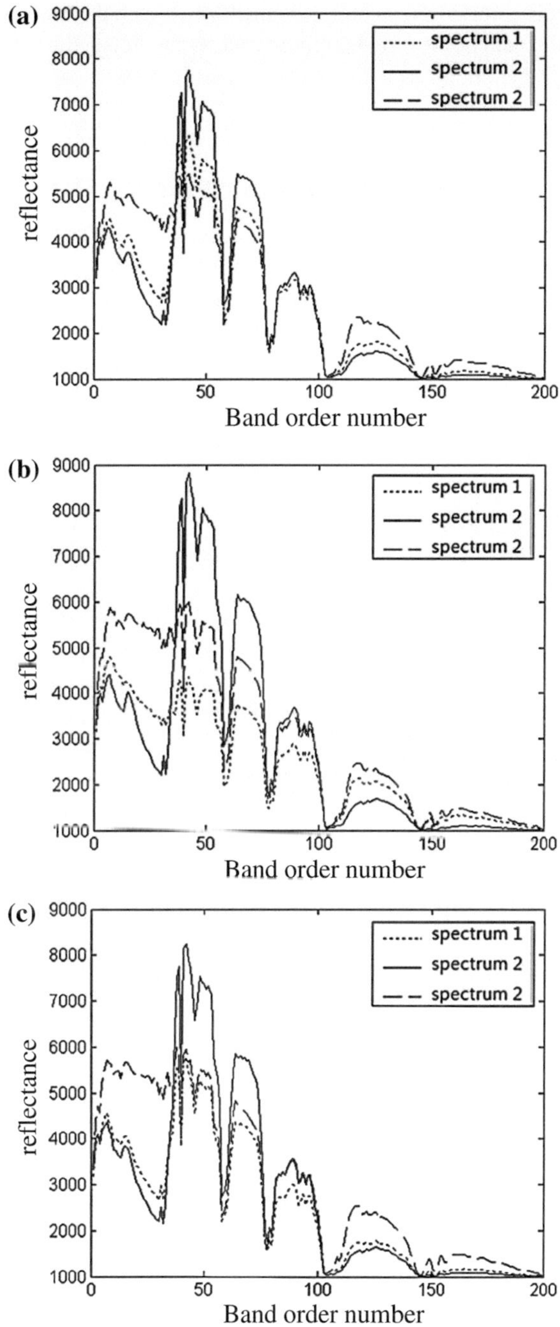

Fig. 3.10 Result comparison of spectral endmember extraction result by different methods. **a** True spectral endmember. **b** Original method extraction result. **c** Robust method extraction result

ideal condition. Thus, whether the hyperspectral data have the outlier point or the data point seriously interfered by the noise, the proposed robustness method will be beneficial to gain more reasonable result.

3.6 Two Applications of Fast N-FINDR Algorithm

Below we utilize the fast N-FINDR algorithm and the core concept to construct the new solving algorithm of LSMM, and construct the fast and unsupervised band selection algorithm for the spectral unmixing.

3.6.1 *Construction of New Solving Algorithm for LSMM*

LSMM generally solves by the least squares method. Without the constraint condition, the least squares method only includes the simple matrix operation. While adding the non-negativity constraint condition, this implementation method includes a large number of complex iterative processes, with unobvious physical significance and relatively low operation efficiency. For this purpose, we utilize the convex geometrical features of the hyperspectral data and the distance measure-based volume comparison to solve LSMM.

Firstly, we describe this method in two-dimensional space. As shown in Fig. 3.11, the pixel s_0 is mixed by the spectral endmembers A, B and C. Given the components of the spectral endmembers A, B and C in the pixel s_0 are P_1, P_2 and P_3, the triangular area formed by A, B and C is S, the triangular area formed by s_0, B and C is S_1, the distance from A and s_0 to the side BC(the location of the straight line) is respectively $D(A)$ and $D(s_0)$. Below it will describe our method with the

Fig. 3.11 Schematic diagram of distance measure-based LSMM solution method

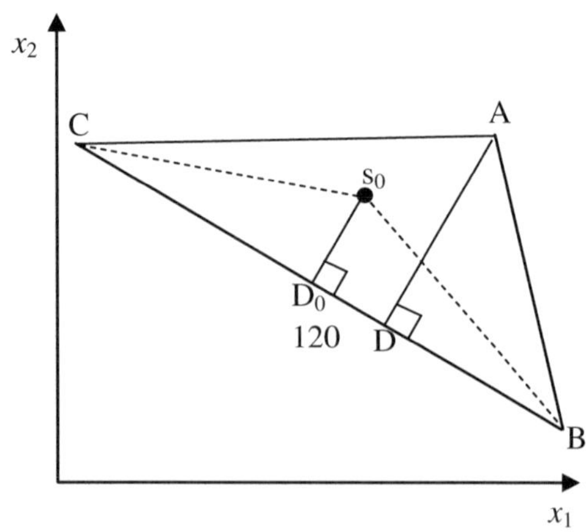

solution of P_1 as an example. It is easily inferred that, P_1, S_1 and S have the following concise relation:

$$P_1 = S_1/S \tag{3.25}$$

Furthermore, this relation can be expressed by the distance from the point to the straight line (side):

$$P_1 = D(s_0)/D(A) \tag{3.26}$$

Thus, P_1 is obtained from a pair of the specific values of the distance. Similarly, P_2 and P_3 can also be obtained simply.

Such relation can be easily expanded into the high-dimensional space. In combination with the distance formula provided in the previous section, the solution formula of the component P_i is:

$$P_i = D(s_0)/D(s_i), \quad i = 1, 2, \ldots, N \tag{3.27}$$

For N hyperplane equations and N distances during the solving process, they can be fully calculated at first, and the calculated amount relative to massive mixed pixel spectral unmixing is negligible. Thus, solving the calculated amount of each component of each mixed pixel only includes an inner product operation and a division operation.

When the mixed pixel s_0 falls into the convex formed by each spectral endmember, the obtained component naturally meets the conditions of the normalization and non-negativity. As a matter of fact, not all pixel points can be combined by the spectral endmembers with the normalization and non-negativity, because all classes included in all pixels are precisely the fuzzy measures. When the mixed pixel s_0 falls outside the convex, the non-negativity condition of the solving component is destroyed. In this case, the negative component can be reset as zero. Then make normalization processing for the other components again. It can be seen that, the distance comparison-based LSMM solution method is equipped with the smaller calculated amount and more intuitive physical significance.

3.6.2 Construction of Fast and Unsupervised Band Selection Algorithm

As previously mentioned, the band selection has the important significance for eliminating the redundant information and reducing the calculated amount. In the case of the priori class information, the supervised band selection method is relatively convenient and effective. However, for lack of the priori class information, such method cannot be implemented. For the existing unsupervised band selection method, most of algorithms possess very great calculated amount. For this purpose, we construct a comparatively effective unsupervised band selection method with the simple calculation by the proposed fast N-FINDR algorithm.

N-FINDR algorithm is initially used for the spectral endmember extraction, but it is noted that, the function of the algorithm is much more than that. In principle, it can also select the representative band, while other bands can be combined linearly by these selected bands. In such sense, according to the determinant property and the solution principle of the linear equation system, applying all bands and only applying the selected bands for the spectral unmixing will gain the similar effect.

Firstly, provide the definitions of two data spaces. Given Ne and ND are respectively the number of pixels and the number of bands of the original data. Note that all pixel points are $\boldsymbol{S}_i = [S_1^i, S_2^i, \cdots, S_{\text{ND}}^i]^{\text{T}}$ $(i = 1, 2, \ldots, \text{Ne})$. For convenience, the space including all pixels is called as the pixel space, while the space including the data $\boldsymbol{s}_i = [S_i^1, S_i^2, \ldots, S_i^{\text{ND}}]^{\text{T}}$ $(i = 1, 2, \cdots, \text{ND})$ is called as the band space. It is easily seen that, the one-to-one correspondence is between the data points in the band space and the bands of the original hyperspectral data. Figure 3.12 provides two-dimensional display of these two spaces (from the true hyperspectral data). In Fig. 3.12a, two coordinate axes are respectively the brightness values of the pixels under two assigned bands, and each point in the figure. corresponds to a two-dimensional pixel vector. In Fig. 3.12b, two coordinate axes are respectively the brightness values of two assigned pixels under various bands, and each point in the figure corresponds to a two-dimensional band vector.

Constructing the band selection method by N-FINDR algorithm is similar to the spectral endmember extraction method in a manner. The difference is the selection method is made in the band space rather than the pixel space, and PCA pre-processing is also made in the band space.

During the PCA transformation process as required by the band selection, it is necessary to make the characteristic unmixing for the matrix $\boldsymbol{C} = \boldsymbol{M} \times \boldsymbol{M}$ of Ne $\times$ Ne. When Ne is larger, the calculated amount of the characteristic unmixing will be very large and even will be difficult to complete. Moreover, for the entire hyperspectral image band selection, the value of Ne is usually larger (generally more than

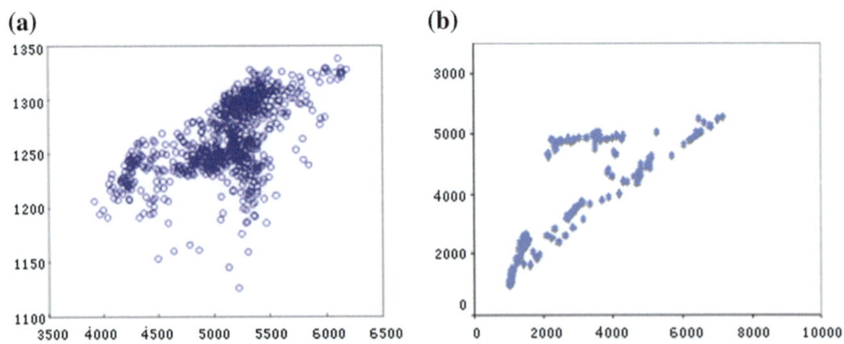

Fig. 3.12 Display of two spaces. **a** Pixel space. **b** Band space

20,000). In order to reduce the calculated amount, the unmixing calculation can be transformed into the characteristic unmixing of ND × ND matrix by the kernel method (Rosipal et al. 2003), to accordingly achieve the purpose of reducing the calculated amount. Now this skill is described as follows.

Given the data $\boldsymbol{s}_i = [S_i^1, S_i^2, \ldots, S_i^{\mathrm{Ne}}]^{\mathrm{T}}$ $(i = 1, 2, \ldots, \mathrm{ND})$ have been centrally processed, i.e., $\sum_{i=1}^{\mathrm{ND}} \boldsymbol{s}_i = 0$. Firstly, utilize the transformation ϕ $(\phi : R^{\mathrm{Ne}} \rightarrow R^{\chi})$. The sample set $\{x_k\}$ is mapped into the eigenspace R^{χ}, then the calculation formula of the covariance matrix $\boldsymbol{\Sigma}_{\varphi}$ in the eigenspace R^{χ} is as follows:

$$\boldsymbol{\Sigma}_{\phi} = \frac{1}{\mathrm{ND}} \sum_{i=1}^{\mathrm{ND}} \phi(x_i)\phi(x_i)^T \tag{3.28}$$

Use $\boldsymbol{V}$ $(\boldsymbol{V} \neq 0)$ to express the characteristic vector of $\boldsymbol{\Sigma}_{\phi}$ corresponding to the eigenvalue λ. Then $\boldsymbol{V}$ can be expanded linearly by the vector in the eigenspace, i.e., $\boldsymbol{V} \in \mathrm{span}\{\phi(x_1), \phi(x_2), \ldots, \phi(x_{\mathrm{ND}})\}$. Given

$$\boldsymbol{V} = \sum_{i=1}^{\mathrm{ND}} \beta_i \phi(x_i) \tag{3.29}$$

$\boldsymbol{\beta} - (\beta_1, \beta_2, \ldots \beta_{\mathrm{ND}})^{\mathrm{T}}$ is called as the dual vector of $\boldsymbol{V}$. From the relational expression of the eigenvalue and characteristic vector,

$$\mathrm{Np}\, \lambda \cdot \boldsymbol{V} = \boldsymbol{\Sigma}_{\phi} \cdot \boldsymbol{V} \tag{3.30}$$

Both sides are multiplied by $\phi(x)$ at the same time, gaining the following relational expression

$$\mathrm{Np}\, \lambda \boldsymbol{\beta} = \boldsymbol{K\beta} \tag{3.31}$$

Here, $\boldsymbol{K}$ is the kernel matrix of ND × ND:

$$\boldsymbol{K}(i,j) = \langle \phi(x_i), \phi(x_j) \rangle \tag{3.32}$$

When we adopt the linear kernel function (i.e., for any $\boldsymbol{x} \in R^{\mathrm{Ne}}$, $\phi(\boldsymbol{x}) = \boldsymbol{x}$), the eigenspace is equal to the original space, and the obtained characteristic vector is also the characteristic vector of the original data, while the matrix unmixing is made for the kernel matrix $\boldsymbol{K}$ of ND × ND.

Below we will verify the property of such band selection method by the experiment. In the experiment, an efficient unsupervised band selection method (BS-FSM method) is used for the effect comparison with our method. Define a similarity measure of maximization information compression index λ_2. Given $\boldsymbol{\Sigma}$ is the covariance matrix between the variables $\boldsymbol{x}$ and $\boldsymbol{y}$, then their maximization information compression index λ_2 is defined as the minimum eigenvalue of the matrix $\boldsymbol{\Sigma}$. The calculation formula is as follows:

$$2\lambda_2(\boldsymbol{x},\boldsymbol{y}) = \mathrm{var}(\boldsymbol{x}) + \mathrm{var}(\boldsymbol{y}) - \sqrt{[\mathrm{var}(\boldsymbol{x}) + \mathrm{var}(\boldsymbol{y})]^2 - 4\mathrm{var}(\boldsymbol{x})\mathrm{var}(\boldsymbol{y})[1 - \rho(\boldsymbol{x}, \boldsymbol{y})^2]} \tag{3.33}$$

where, $\rho(\boldsymbol{x}, \boldsymbol{y}) = \frac{\mathrm{cov}(\boldsymbol{x},\boldsymbol{y})}{\sqrt{\mathrm{var}(\boldsymbol{x})\mathrm{var}(\boldsymbol{y})}}$ is the similarity measure between the random variables $\boldsymbol{x}$ and $\boldsymbol{y}$. When the variables $\boldsymbol{x}$ and $\boldsymbol{y}$ are linearly independent, $\lambda_2(\boldsymbol{x},\boldsymbol{y}) = 0$. Two basic steps of this method are dividing the original characteristic set with the equal interval in accordance with the characteristic similarity measure, and selecting a characteristic representative from each characteristic subset. See also the articles of Pabitra et al. (2002) for the specific process and performance advantages.

The first group of the experiment is to select 10 sample points from San Diego military hyperspectral data (126 bands) as the spectral endmembers, and combine them into 1000 mixed pixels. Select the fitting average error as the assessment index. We make the detailed contrast of the method performance through transforming the number of samples and selecting the number of bands. The experimental result is shown in Table 3.4.

The result shows that, the search process time consuming of the new algorithm is irrelevant to the number of samples, while the pre-processing process time consuming is irrelevant to the number of bands to be selected. When the number of samples is more than the number of bands, the complexity of the band selection algorithm mainly focuses on the fast PCA. In this case, the new method is faster than FS method.

In the second group of the experiment, the experimental image selects 10 classes of samples as the spectral endmembers. They are linearly and normally combined into 1,000 samples for the band selection and unmixing test. Table 3.5 provides the unmixing error by different methods and while selecting different numbers of bands. The contrast result shows the advantages of the new method.

Table 3.4 MSE contrast of two band selection methods

Number of samples	Method	MSE and running time (s) while selecting the number of bands as Nd			
		Nd = 5	Nd = 10	Nd = 15	Nd = 20
Np = 200	FS	0.0141, 0.10	0.0085, 0.13	0.0069, 0.14	0.0058, 0.16
	New method	0.0098, 0.20	0.0056, 0.24	0.0048, 0.44	0.0043, 0.73
Np = 2000	FS	0.0132, 0.23	0.0088, 0.24	0.0072, 0.27	0.0061, 0.28
	New method	0.0101, 0.34	0.0059, 0.41	0.0053, 0.52	0.0045, 0.84
Np = 20,000	FS	0.0250, 3.31	0.0111, 3.35	0.0097, 3.33	0.0095, 3.37
	New method	0.0115, 1.96	0.0065, 2.12	0.0055, 2.44	0.0049, 2.70

Table 3.5 Unmixing accuracy contrast of two band selection methods

Band selection method	Unmixing accuracy (%) while selecting the number of bands as Nd						
	Nd = 3	Nd = 4	Nd = 5	Nd = 6	Nd = 7	Nd = 8	Nd = 9
FS	95.56	95.95	96.48	96.85	97.34	97.92	98.70
New method	95.67	96.00	96.63	96.88	97.42	98.32	98.78

3.7 Chapter Conclusions

This chapter focuses on introducing the fast implementation method of N-FINDR endmember extraction algorithm, and also researches the robustness control and other aspects.

On the basis of setting up a distance formula with the simple calculation, the fast N-FINDR algorithm replaces the volume calculation by applying the distance test, to greatly reduce the complexity of the algorithm. Such replacement does not have the influence on the spectral endmember extraction effect, and the only change is the complexity of the calculation. When the number of the selected spectral endmembers is larger, the advantages of such replacement are more obvious. The rearrangement of the data set and the ordered search of the spectral endmember make the iteration update speed enhance greatly, and reduce the possibility of the local optimum of the algorithm.

Furthermore, utilizing the distance measurement function of the linear LSSVM realizes the distance calculation. This distance has the simple calculation, and can be free from the pre-processing of the dimension reduction. For SVM theory, people mostly understand its classification and regression functions, while this book skillfully utilizes it for the distance test and solves two problems of N-FINDR algorithm. It is noted that the classifier structures other than the nonlinear SVM and 1-*a*-*r* cannot achieve the purpose of this book. Moreover, the least squares type of SVM can be replaced by the common type of SVM in theory, but the efficiency for handling the ultra small sample problem like this book will be influenced greatly.

In addition, this chapter also puts forward the robustness control method in allusion to the weakness of N-FINDR algorithm easily interfered by the outlier points, i.e., in the data pre-processing stage, the general covariance matrix should be replaced by the robust covariance matrix. During the spectral endmember search process, remove the outlier points by the square neighborhood detection method, and increase the robustness control on the outlier point interference during the spectral endmember extraction process. The robustness method and the reduction of the iterative search frequency of the algorithm by the pixel pre-sequencing proposed in this chapter can be borrowed by other endmember extraction method.

It is also noted that, this concept of replacing the volume by the distance absolutely can be applied to the other common endmember extraction algorithm based on the simplex volume maximization of Simplex Growing Algorithms (SGA). See the related articles of the author for the specific method, not to go into details.

References

Boardman JW, Kruse FA, Green RO (1955) Mapping target signatures via partial unmixing of AVIRIS data. In summaries of the V JPL Airborne Earth Science Workshop, Pasadena, CA

Devijver PA, Kittler J (1980) Pattern recognition: a statistical approach. Prentice-Hall, Englewood Cliffs, NJ

Green A, Berman M, Switzer P, Craig M (1988) A transformation for ordering multispectal data in terms of image quality with implications for noise removal. IEEE Trans Geosci Remote Sens 26:65–74

Hubert M, Rousseeuw PJ, Branden KV (2005) ROBPCA: a new approach to robust principal component analysis. Technometrics 47(1):64–79

Jia X, Richards JA (1999) Segmented principal components transformation for efficient hyperspectral remote-sensing image display and classification. IEEE Trans Geosci Remote Sens 37:538–542

Mitra P, Pal SK, Murthy CA (2002) Unsupervised feature selection using feature similarity. IEEE Trans Pattern Anal Mach Intell 24(3):301–312

Pabitra M, Aand Murthy C, Sankar KP (2002) Unsupervised feature selection using feature similarity. IEEE Trans Pattern Anal Mach Intell 3:301–302

Rosipal R, Trejo LJ, Matthews B, Wheeler K (2003) Nonlinear Kernel-based chemometric tools: a machine learning approach. In: Proceedings of 3rd international symposium on PLS and related methods (PLS'03), Lisbon, Portugal, pp 249–260

Winter ME (1999) N-FINDR: an algorithm for fast autonomous spectral end-member determination in hyperspectral data. Proc SPIE Imaging Spectrom 3753:266–275

Chapter 4
Spectral Unmixing Technique of HSI

Relative to the classification technique, the spectral unmixing (Keshava and Mustard 2002) i.e., soft classification technique started late. Although the spectral resolution of the hyperspectral image has been improved greatly, the spatial resolution of the corresponding land object target of the pixel has been relatively low. For example, the spatial resolution of AVIRIS is 20 m × 20 m, so a pixel may include two or more land object targets, i.e., the pixel is mixed. When the interested target is less than one pixel or more pixels, the object to be researched and analyzed will focus on the mixed pixel. If only a mixed pixel is assigned to some class, it will certainly result in some classification error as well as the reduction in the classification accuracy, and accordingly influence the follow-up application of the analysis result. As the leading technique of the mixed pixel processing, the spectral unmixing is to solve the proportion of various mixed elements in the mixed pixel, and is a more accurate classification technique. After introducing the traditional LSMA method, this chapter will focus on the proposed SVM-based unmixing method.

4.1 LSMM-Based LSMA Method

Given the number of the original classes that we are interested in is *Nc*, and the number of the spectral bands is ND. Define a ND × Nc matrix $\boldsymbol{R}$, to make it include the spectral characteristic vector of the spectral endmember, then the mixed model can be expressed as:

$$\boldsymbol{x} = \boldsymbol{RP} + \varepsilon \tag{4.1}$$

In the formula, $\boldsymbol{x}$ is the input spectral vector, $\boldsymbol{P}$ is the mixing ratio vector, and ε is the estimation error term. Generally, $\boldsymbol{R}$ is assumed to be full rank, i.e., all spectral endmembers are linearly independent. Without the constraint condition, the mixing ratio vector $\boldsymbol{P}$ can be directly obtained by both sides of the equation multiplying by $\boldsymbol{R}^{-1}$ at the same time. If $\boldsymbol{R}$ is not the full rank, the least squares solution formula of $\boldsymbol{P}$ is:

L. Wang and C. Zhao, *Hyperspectral Image Processing*,
DOI 10.1007/978-3-662-47456-3_4

$$\boldsymbol{P} = (\boldsymbol{R}^T\boldsymbol{R})^{-1}\boldsymbol{R}^T\boldsymbol{x} \tag{4.2}$$

Define $\boldsymbol{V}$ as the covariance error matrix of the observed value $\hat{\boldsymbol{x}}$, i.e., $\hat{\boldsymbol{x}} = \boldsymbol{x} - \boldsymbol{\varepsilon}$, $\boldsymbol{V} = \boldsymbol{E}(\boldsymbol{\varepsilon\varepsilon}^T)$. If adding the normalization constraint condition $\sum_{j=1}^{Nc}\boldsymbol{P}_j$, then Fletcher (1987), the least square estimation formula of the mixing ratio vector is:

$$\boldsymbol{P} = \boldsymbol{Z}^{-1}\boldsymbol{t} - \frac{\boldsymbol{Z}^{-1}\boldsymbol{1}}{\boldsymbol{1}^T\boldsymbol{Z}^{-1}\boldsymbol{1}}\left(\boldsymbol{1}^T\boldsymbol{Z}^{-1}\boldsymbol{t} - \boldsymbol{1}\right) \tag{4.3}$$

where, $\boldsymbol{Z} = \boldsymbol{R}^T\boldsymbol{V}^{-1}\boldsymbol{R}$ is the weighting autocorrelation matrix, $\boldsymbol{t} = \boldsymbol{R}^T\boldsymbol{V}^{-1}\boldsymbol{x}$ is the weighting cross-correlation vector, and "$\boldsymbol{1}$" is the column vector with the element of 1. So the formula (4.3) can be sorted as:

$$\boldsymbol{P} = \boldsymbol{Z}^{-1}\left(\boldsymbol{I} - \frac{\boldsymbol{1}\boldsymbol{1}^T\boldsymbol{Z}^{-1}}{\boldsymbol{1}^T\boldsymbol{Z}^{-1}\boldsymbol{1}}\right)\boldsymbol{R}^T\boldsymbol{V}^{-1}\boldsymbol{x} + \frac{\boldsymbol{Z}^{-1}\boldsymbol{1}}{\boldsymbol{1}^T\boldsymbol{Z}^{-1}\boldsymbol{1}} \tag{4.4}$$

where, $\boldsymbol{I}$ is the unit matrix.

While adding the non-negativity constraint condition (ANC), the solution optimization expression of the mixing ratio vector is:

$$\min\,(\boldsymbol{x} - \boldsymbol{RP})^T(\boldsymbol{x} - \boldsymbol{RP}),\quad \text{s.t. } \boldsymbol{P} \geq 0 \tag{4.5}$$

Because the constraint condition is the system of inequalities, Lagrange multiplier solution method will not be applicable. For this purpose, introduce the constant positive constrained vector $\boldsymbol{\theta} = [\theta_1, \theta_2, \ldots, \theta_{\mathrm{Nc}}], \theta_i > 0, i = 1, 2, \ldots, \mathrm{Nc}$ as the undetermined formal solution of $\boldsymbol{P}$. So it can get the Lagrange form of the above formula:

$$\boldsymbol{J} = \frac{1}{2}(\boldsymbol{x} - \boldsymbol{RP})^T(\boldsymbol{x} - \boldsymbol{RP}) + \lambda(\boldsymbol{P} - \boldsymbol{\theta}), \boldsymbol{\theta} = \boldsymbol{R} \tag{4.6}$$

Given the derivative with respect to $\boldsymbol{\theta}$ in the above formula is 0, obtaining:

$$\left.\frac{\partial \boldsymbol{J}}{\partial \boldsymbol{\theta}}\right|_{\hat{P}_{\mathrm{NCLS}}} = 0 \Rightarrow \boldsymbol{R}^T\boldsymbol{R}\hat{\boldsymbol{P}}_{\mathrm{NCLSC}} - \boldsymbol{R}^T\boldsymbol{x} + \boldsymbol{\lambda} = 0 \tag{4.7}$$

Thus, obtain two iterative equations of solving the mixing ratio vector:

$$\begin{aligned} \hat{\boldsymbol{P}}_{\mathrm{NCLS}} &= (\boldsymbol{R}^T\boldsymbol{R})^{-1}\boldsymbol{R}^T\boldsymbol{x} - (\boldsymbol{R}^T\boldsymbol{R})^{-1}\boldsymbol{\lambda} = \hat{\boldsymbol{P}}_{\mathrm{LS}} \\ \boldsymbol{\lambda} &= \boldsymbol{R}^T(\boldsymbol{x} - \boldsymbol{R}\hat{\boldsymbol{P}}_{\mathrm{NCLS}}) \end{aligned} \tag{4.8}$$

Below it will provide the iterative solution process with the non-negativity constraint condition (Du and Chein-I 2004):

1. Initialization. Given $\boldsymbol{S}_{+} = \{1, 2, \ldots, p\}$, $\boldsymbol{S}_{-} = \Phi$, $k = 0$.
2. Apply the formula (4.2) to calculate the unrestraint mixing ratio vector $\hat{\boldsymbol{P}}_{\mathrm{LS}}$, given $\hat{\boldsymbol{P}}_{\mathrm{NCLS}}^{(k)} = \hat{\boldsymbol{P}}_{\mathrm{LS}}$.
3. In the k iteration, if the component in $\hat{\boldsymbol{P}}_{\mathrm{NCLS}}^{(k)}$ is the positive value, the algorithm is end, otherwise, it will continue.
4. Given $k = k + 1$.
5. Move the negative component index in $\boldsymbol{S}_{+}^{(k-1)}$ corresponding to $\hat{\boldsymbol{P}}_{\mathrm{NCLS}}^{(k)}$ to $\boldsymbol{S}_{-}^{(k-1)}$. The corresponding index set turns into $\boldsymbol{S}_{+}^{(k)}$ and $\boldsymbol{S}_{-}^{(k)}$. Introduce a new index set $\boldsymbol{S}^{(k)}$ and given it equal to $\boldsymbol{S}_{-}^{(k)}$.
6. Define $\hat{\boldsymbol{P}}_{S^{(k)}}$ as including all components of $\hat{\boldsymbol{P}}_{\mathrm{LS}}$ in $\boldsymbol{S}_{-}^{(k)}$.
7. Form the other matrix $\boldsymbol{\Phi}_{P}^{(k)}$ through deleting all rows and columns of the indexes in the matrix $(\boldsymbol{R}^T\boldsymbol{R})^{-1}$ corresponding to $\boldsymbol{S}_{+}^{(k)}$.
8. Calculate $\boldsymbol{\lambda}^{(k)} = (\boldsymbol{\Phi}_{P}^{(k)})^{-1}\hat{\boldsymbol{P}}_{S^{(k)}}$ if the component of $\boldsymbol{\lambda}^{(k)}$ is negative, the algorithm enters into the 13th step, otherwise, it will continue.
9. Calculate $\boldsymbol{\lambda}_{\mathrm{nax}}^{(k)} = \arg(\max \boldsymbol{\lambda}_{j}^{(k)})$, and move the indexes in $\boldsymbol{S}_{-}^{(k)}$ corresponding to $\boldsymbol{\lambda}_{\mathrm{nax}}^{(k)}$ to $\boldsymbol{S}_{+}^{(k)}$.
10. Form the other matrix $\boldsymbol{\Phi}_{\lambda}^{(k)}$ through deleting all rows and columns of the indexes in the matrix $(\boldsymbol{R}^T\boldsymbol{R})^{-1}$ corresponding to $\boldsymbol{S}_{+}^{(k)}$.
11. Set up $\hat{\boldsymbol{P}}_{S^{(k)}} = \hat{\boldsymbol{P}}_{\mathrm{LS}} - \boldsymbol{\Phi}_{\lambda}^{(k)}\boldsymbol{\lambda}^{(k)}$.
12. Move the negative component in $\hat{\boldsymbol{P}}_{S^{(k)}}$ and belonging to $\boldsymbol{S}^{(k)}$ from $\boldsymbol{S}_{+}^{(k)}$ to $\boldsymbol{S}^{(k)}$. Without the negative component, go to the sixth step.
13. Form the other matrix $\boldsymbol{\Phi}_{\lambda}^{(k)}$ through deleting all rows and columns of the indexes in the matrix $(\boldsymbol{R}^T\boldsymbol{R})^{-1}$ corresponding to $\boldsymbol{S}_{+}^{(k)}$.
14. Set up $\hat{\boldsymbol{P}}_{\mathrm{NCLS}}^{(k)} = \hat{\boldsymbol{P}}_{\mathrm{LS}} - \boldsymbol{\Phi}_{\lambda}^{(k)}\boldsymbol{\lambda}^{(k)}$, and go to the third step.

Make the normative processing for the mixing ratio vector again by the method of setting the threshold value.

While adding the normalization constraint condition, it can introduce the matrix $\tilde{\boldsymbol{R}} = \begin{bmatrix} \delta\boldsymbol{R} \\ \boldsymbol{1}^T \end{bmatrix}$ and $\tilde{\boldsymbol{x}} = \begin{bmatrix} \delta\boldsymbol{x} \\ 1 \end{bmatrix}$, and replace $\boldsymbol{R}$ and $\boldsymbol{x}$ in the above algorithm respectively by them, to obtain the solution algorithm under the full constrained condition. Here, δ is the parameter of controlling the influence of the normalization constraint.

4.2 Two New Solving Methods for Full Constrained LSMA

4.2.1 *Parameter Substitution Method in Iteration Solving Method*

Firstly, given some mixed pixel $\boldsymbol{x}$ is the spectral vector mixed by two types of land objects, the estimation mixing ratios of two types of land objects are respectively p_1 and p_2, and the corresponding spectral endmembers are respectively $\boldsymbol{e}_1$ and $\boldsymbol{e}_2$, ε is the estimation error term, then

$$\boldsymbol{x} = p_1\boldsymbol{e}_1 + p_2\boldsymbol{e}_2 + \varepsilon \tag{4.9}$$

Given $p_1 = \sin^2\theta$ and $p_2 = \cos^2\theta$, then $p_1 + p_2 = 1$ is correct, meanwhile meeting $0 \leq p_1 \leq 1$ and $0 \leq p_2 \leq 1$, i.e., meeting the normalization and non-negativity constraint conditions of the mixing ratio. Moreover, the trigonometric function is the periodic function. When θ traverses the value of $[0, \pi/2]$, it can guarantee p_1 and p_2 taking any value in $[0, 1]$. Now, the error term is:

$$\begin{aligned}
\varepsilon &= \boldsymbol{x} - (p_1\boldsymbol{e}_1 + p_2\boldsymbol{e}_2) \\
&= \boldsymbol{x} - (\sin^2\theta\boldsymbol{e}_1 + \cos^2\theta\boldsymbol{e}_2) \\
&= \boldsymbol{x} - [(1 - \cos^2\theta)\boldsymbol{e}_1 + \cos^2\theta\boldsymbol{e}_2] \\
&= \boldsymbol{x} - [\boldsymbol{e}_1 + \cos^2\theta(\boldsymbol{e}_2 - \boldsymbol{e}_1)] \\
&= \boldsymbol{x} - 0.5[\cos 2\theta(\boldsymbol{e}_2 - \boldsymbol{e}_1) + \boldsymbol{e}_1 + \boldsymbol{e}_2]
\end{aligned} \tag{4.10}$$

i.e., the fitness function with θ ($\theta \in [0, \pi/2]$) as the independent variable and ε as the function value in the optimization algorithm. In this way, the process of solving the minimum value of the minimum mean square error is transformed into the optimizing problem of the function extremum without the constraint condition. The triangle identical transformation in the formula (4.10) can reduce the complexity of the optimization calculation.

By a logical extension of this point, when the number of the mixed classes is $N \geq 3$, the mixing ratios are respectively $p_1, p_2 \ldots p_{N-1}, p_N$, and the spectral endmembers are respectively $\boldsymbol{e}_1, \boldsymbol{e}_2 \ldots \boldsymbol{e}_{N-1}, \boldsymbol{e}_N$. ε is still the estimation error term. The corresponding mixed model is:

$$x = p_1\boldsymbol{e}_1 + p_2\boldsymbol{e}_2 + \cdots + p_{N-1}\boldsymbol{e}_{\mathrm{N}-1} + p_N\boldsymbol{e}_N + \varepsilon \tag{4.11}$$

Then

$$
\begin{aligned}
p_1 &= \sin^2\theta_1 \\
p_2 &= \cos^2\theta_1 \sin^2\theta_2 \\
&\vdots \\
p_{N-1} &= \prod_{i=1}^{N-2} \cos^2\theta_i \sin^2\theta_{N-1} \\
p_N &= \prod_{i=1}^{N-2} \cos^2\theta_i \cos^2\theta_{N-1}
\end{aligned}
\tag{4.12}
$$

$p_1 + p_2 + \cdots + p_{N-1} + p_N = 1$, and $0 \le p_i \le 1 (i = 1 \sim N)$ are correct. As previously mentioned, the value range of $\theta_i (i = 1 \sim N)$ can be defined as $[0, \pi/2]$. Now the error term is:

$$
\begin{aligned}
\boldsymbol{\varepsilon} &= \boldsymbol{x} - (p_1\boldsymbol{e}_1 + p_2\boldsymbol{e}_2 + \cdots + p_{N-1}\boldsymbol{e}_{N-1} + p_N\boldsymbol{e}_N) \\
&= \boldsymbol{x} - (\sin^2\theta_1\boldsymbol{e}_1 + \cos^2\theta_1\sin^2\theta_2\boldsymbol{e}_2 + \cdots \\
&\quad + \prod_{i=1}^{N-2}\cos^2\theta_i\sin^2\theta_{N-1}\boldsymbol{e}_{N-1} + \prod_{i=1}^{N-2}\cos^2\theta_i\cos^2\theta_{N-1}\boldsymbol{e}_N)
\end{aligned}
\tag{4.13}
$$

Similarly it becomes the optimizing problem without the constraint condition. The formula (4.13) can also carry out the identical transformation to reduce the complexity of the calculation.

Further, the model can solve by the Taguchi algorithm. Please refer to the related articles published by the author for details.

4.2.2 Geometric Solving Method

In the above Sect. 3.1.6, we introduce the geometric-solving method of LSMM in the simple case. Here we further introduce the solving method under the fully constrained condition. Noting

$$
\begin{aligned}
\boldsymbol{v}_i &= \boldsymbol{e}_i - \boldsymbol{e}_{i-1}, i = 1, 2, \ldots, d-1 \\
\boldsymbol{V} &= [\boldsymbol{v}_1, \boldsymbol{v}_2, \ldots, \boldsymbol{v}_{d+1}]
\end{aligned}
\tag{4.14}
$$

The corresponding space of all vectors linearly and normally combined by the endmember pixel $\boldsymbol{e}_1, \boldsymbol{e}_2, \ldots, \boldsymbol{e}_d$ is the $d-1$ dimension space Γ extended by various columns of the matrix $\boldsymbol{V}$. The corresponding space of all vectors linearly combined with full constraint by the endmember pixel $\boldsymbol{e}_1, \boldsymbol{e}_2, \ldots, \boldsymbol{e}_d$ is the space of $d - 1$ dimension convex polyhedron Γ_B extended by d vertexes of the endmember pixel $\boldsymbol{e}_1, \boldsymbol{e}_2, \ldots, \boldsymbol{e}_d$. Γ_B is a bounded domain in Γ. According to the basic theory of the least squares estimation, the essence of the mixed pixel is the normalization constraint least squares estimation of s_0 is to search the pixel point s_0^{Γ} nearest to s_0 in Γ of d -1 dimension space. This point is the projection of s_0 in the space Γ. The essence of full constraint least squares estimation of s_0 is to search the pixel point $s_0^{\Gamma_B}$ nearest to s_0 in the convex polyhedron Γ_B. Thus, it can be seen that, the solution of FCLS-LSMM is the constrained least squares estimation problem. based on this recognition, we set up the following geometric-solving method of FCLS-LSMM.

1. **Situation of 3-endmember**

2-endmember geometric unmixing can be completed only once, and is very simple. So we will begin the analysis with 3-endmember. Under the assistance of the above dimension reduction method, the following discussion contents assume that the pixel to be unmixed is located in the $d-1$ dimension space Γ extended by d vertexes of the endmember pixel $e_1, e_2, \ldots, e_d$.

When the mixed pixel s_0 falls into the convex formed by various spectral endmembers, the mixing ratio obtained by the method in this book naturally meets the normalization and non-negativity condition, and then the obtained result is the optimal solution in the full constraint least squares sense. However, when the mixed pixel s_0 falls outside the convex, the positive and negative measures will have the situation that some mixing ratios are negative, while the normalization condition is destroyed for the positive type measure. The pixel points outside the convex actually exist in quantity, so it is necessary to make the reasonable treatment. It indicates that, for the geometric-solving method of the spectral unmixing, an unmixing usually can not gain the optimal full constraint least squares solution. Below it will put forward a geometric method of the progressive solution.

As shown in Fig. 4.1, s_1, s_2 and s_3 are located outside the triangle $\boldsymbol{ABC}$. When the unmixing component corresponding to only one endmember is negative, such as s_1 and s_2, the other two endmembers should be used for the second geometric unmixing. When the unmixing components corresponding to two endmembers are

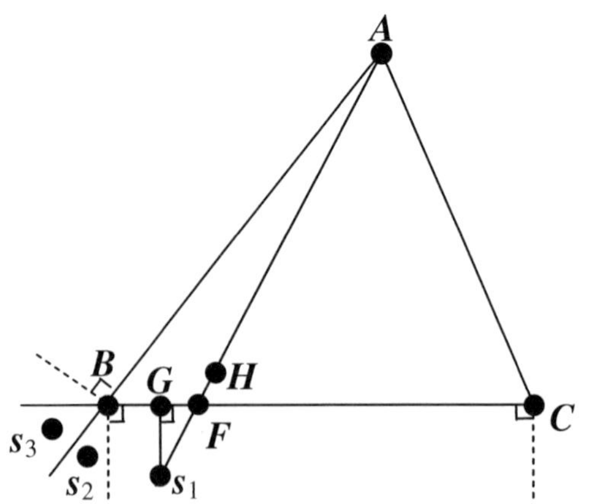

Fig. 4.1 Schematic diagram of full constraint least squares optimal estimation

negative, given $\boldsymbol{A}$ and $\boldsymbol{C}$, such s_3, two 2-endmember combinations $(\boldsymbol{A},\boldsymbol{B})$ and $(\boldsymbol{B},\boldsymbol{C})$ should be used for the second geometric unmixing. If the geometric unmixing of one of 2-endmember combinations gains the full constraint result, the unmixing must be optimal. If two 2-endmember combinations $(\boldsymbol{A},\boldsymbol{B})$ and $(\boldsymbol{B},\boldsymbol{C})$ fails to gain the full constraint result, such as s_2 and s_3, the unmixing component corresponding to some endmember must be positive, and the endmember is the optimal estimation of the corresponding pixel.

In Fig. 4.1, for the first geometric unmixing based on positive and negative measures of the pixel $\boldsymbol{s}_1$, the endmember $\boldsymbol{A}$ corresponds to the negative unmixing component, and the corresponding unmixing component is zero, but the non-negativity condition is destroyed. If the negative unmixing component is directly set as zero, and then the new component combination is processed normally. At this time, the full constraint condition is met, but the corresponding unmixing error is increased to $|\boldsymbol{F}\boldsymbol{s}_1|$. If using the positive measure in the first geometric unmixing, the corresponding unmixing error upon the normalization processing of the unmixing result is $|\boldsymbol{H}\boldsymbol{s}_1|$. Here $\boldsymbol{H}$ meets $|\boldsymbol{F}\boldsymbol{H}| = |\boldsymbol{F}\boldsymbol{s}_1| \times |\boldsymbol{A}\boldsymbol{F}|/(|\boldsymbol{A}\boldsymbol{F}| + 2|\boldsymbol{F}\boldsymbol{s}_1|)$. Moreover, for the proposed geometric unmixing of the positive and negative measures, the error upon unmixing twice by the endmembers $\boldsymbol{B}$ and $\boldsymbol{C}$ is only $|\boldsymbol{G}\boldsymbol{s}_1|$. Obviously, $\boldsymbol{G}$ is FCLS optimal estimation of the pixel $\boldsymbol{s}_1$.

Furthermore, the 3-endmember geometric unmixing process of the above pixel $\boldsymbol{s}$ can be summarized as follows.

1. Firstly, all three endmembers are used for the geometric unmixing. If the unmixing result meets the full constraint condition, the result will be optimal. If the unmixing components corresponding to two endmembers are negative, such as $\boldsymbol{A}$ and $\boldsymbol{B}$, and the interior angle $\boldsymbol{ACB}$ of the triangle is the acute angle, the optimal full constraint estimation of the pixel $\boldsymbol{s}$ is $\boldsymbol{C}$, and it only needs an unmixing.
2. If the unmixing component corresponding to one endmember in the geometric unmixing of all three endmembers is negative, such as $\boldsymbol{A}$, the other two endmembers are used for the secondary unmixing. If the secondary unmixing result is full constraint, the result is optimal. Otherwise, the optimal full constraint estimation of the pixel $\boldsymbol{s}$ is the one corresponding to the positive unmixing component in $\boldsymbol{A}$ and $\boldsymbol{B}$, and it needs unmixing twice.
3. If the unmixing components corresponding to two endmembers in the geometric unmixing of all three endmembers are negative, such as $\boldsymbol{A}$ and $\boldsymbol{B}$, and the interior angle $\boldsymbol{ACB}$ of the triangle is the obtuse angle, two-endmember combinations $(\boldsymbol{A},\boldsymbol{C})$ and $(\boldsymbol{B},\boldsymbol{C})$ are successively used for the geometric unmixing of s, and one and only one group of the result is full constraint as well as optimal. And it needs unmixing at most for three times.

In fact, most interior angles are acute angles (the obtuse angle probability for the interior angle of the triangle is 1/6), and accordingly the above unmixing process has small calculated amount. The calculation of the interior angle is one-off, so the complexity can be ignored.

2. **Situation of more than 3-endmember**

Before establishing the geometric FCLS-LSMA method, firstly it emphasizes that the following fact is correct:

1. For d endmember unmixing, in the case of only one negative endmember (i.e., the endmember corresponding to the negative unmixing component), the negative endmember must not be included in the optimal endmember combination.
2. If multiple m endmember combinations can produce the full constraint result, and any $m+1$ endmember combinations cannot produce the full constraint result, one and only one m endmember combination is the optimal endmember combination.
3. For d endmember unmixing, in the case of multiple negative endmembers, some negative endmembers may be included in the optimal endmember combination.
4. In the case of d_0 negative endmembers in d endmember unmixing, remove a negative endmember successively and form $d_0(d-1)$ endmember combination. At most only one $(d-1)$ endmember combination is optimal.
5. In multiple m endmember combinations derived from different $(m+1)$ endmember combinations, multiple m endmember combinations may produce the full constraint result.

The understanding of the contents in Article 4 is slightly complex. Now explain this fact by Fig. 4.2. In 4-endmember unmixing of the pixel $\boldsymbol{s}$, if the endmembers $\boldsymbol{A}$ and $\boldsymbol{C}$ are negative, $\boldsymbol{s}$ will certainly fall into the space formed by $\boldsymbol{BF}$, $\boldsymbol{BE}$, $\boldsymbol{BD}$, $\boldsymbol{DG}$ and $\boldsymbol{DH}$. When the included angle θ between the plane $\boldsymbol{ABD}$ and the plane $\boldsymbol{CBD}$ is the acute angle, two 3-endmember combinations $(\boldsymbol{A}, \boldsymbol{B}, \boldsymbol{D})$ and $(\boldsymbol{C}, \boldsymbol{B}, \boldsymbol{D})$ can not produce the full constraint result. If θ is the obtuse angle, only one of $(\boldsymbol{A}, \boldsymbol{B}, \boldsymbol{D})$ and $(\boldsymbol{C}, \boldsymbol{B}, \boldsymbol{D})$ can produce the full constraint result. Furthermore, if $\boldsymbol{D}$ is also the

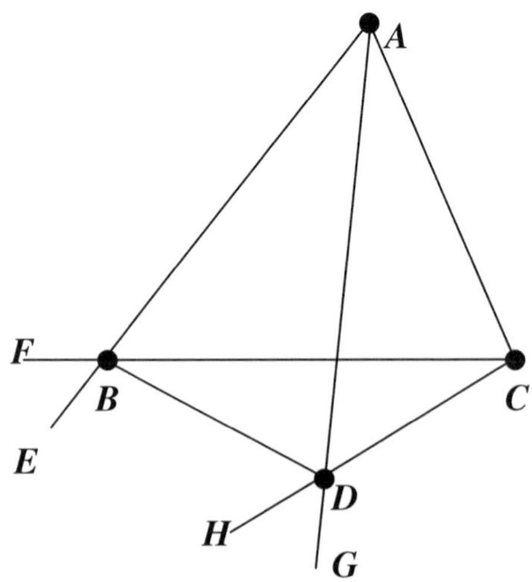

Fig. 4.2 Schematic diagram of searching only optimal 3-endmember combination

negative endmember, according to the above analysis, at most one of any two 3-endmember combinations can produce the full constraint result. Thus, at most one of all 3-endmember combinations can produce the full constraint result.

By utilizing the above fact, for d endmember unmixing problem of the pixel $\boldsymbol{s}$, the geometric FCLS-LSMA process can be described as follows:

Step 1: Firstly, unmix the pixel $\boldsymbol{s}$ by all d endmembers. If the gained component naturally meets the full constraint condition, the gained result is optimal, and the unmixing is finished. Otherwise, the unmixing processing goes into the next step.

Step 2: In the case of d_1 negative endmembers in the previous step, it can remove a negative endmember in d endmember combination every time, and form d_1 $(d - 1)$ endmember combinations. Then, unmix the pixel $\boldsymbol{s}$ by successively utilizing d_1 $(d - 1)$ endmember combinations. If some unmixing result naturally meets the full constraint condition, the gained result is optimal, and the unmixing is finished. If d_1 unmixing fails to gain the full constraint result, the unmixing processing goes into the next step.

Step 3: If removing a negative endmember in all $(d - 1)$ endmember combinations in the previous step successively every time, it can gain d_2 $(d - 2)$ endmember combinations differing from each other. Unmix the pixel $\boldsymbol{s}$ by successively utilizing d_2 $(d - 2)$ endmember combinations. The optimal unmixing result surely exists in the unmixing result naturally meeting the full constraint. For multiple groups of the full constraint unmixing results, it can determine the optimal result by the unmixing error comparison. Without the unmixing result naturally meeting the full constraint, the unmixing processing goes into the next step.

Step 4: Continue, till 2-endmember combination analysis. If all 2-endmember combination unmixing fails to gain the full constraint result, the pixel $\boldsymbol{s}$ is incorporated into some endmember by 100 % through comparing the unmixing error (it can replace the unmixing error comparison by the spectral angle matching). Finally, for the endmember excluding in the optimal endmember combination, the corresponding unmixing component is set as 0, and the unmixing is finished.

Through the above low complexity process, the optimal FCLS-LSMA result will be generated. The above process also includes several unmixing, but the essence is totally different from the traditional iterative solution mode.

Starting from the least squares criterion, this full constraint adjustment way is different from the traditional full constraint adjustment way proposed in the section of Foreword. In Fig. 4.2, s_0 is a mixed pixel outside the triangle formed by the endmembers $\boldsymbol{A}$, $\boldsymbol{B}$ and $\boldsymbol{C}$. In theory, it can be inferred that the line segment $\boldsymbol{G}s_0$, line segment $\boldsymbol{F}s_0$ and line segment $s_0\boldsymbol{s}^0$ respectively correspond to the traditional adjustment mode of the positive measure method, the traditional adjustment mode of the positive and negative measure method, and the corresponding error vectors by the adjustment way in this book. With regard to the position of $\boldsymbol{s}^0$, here it meets $|\boldsymbol{A}\boldsymbol{s}^0| = |\boldsymbol{A}s_0| \times |\boldsymbol{AF}|/2|\boldsymbol{s}^0 s_0|$. $|\cdot|$ shows the length of the line segment. It will not provide the specific derivation process due to the limited space. The reader is requested to verify voluntarily. It is thus clear that, the mode in this book possesses the theoretically minimum unmixing error. In addition, it can be seen that the positive and negative measures have smaller unmixing error than the positive measure, and also can provide the spatial positional information of the pixel to be analyzed to a certain extent. In fact, the positive and negative measures can be transformed into the positive measure upon the absolute value, but this transformation does not have the benefit.

4.3 The Principle of LSVM-Based Spectral Unmixing

This section firstly describes the feasibility of applying LSVM into the spectral unmixing by the equality proof of two spectral unmixing models of LSVM and LSMM, and further demonstrates the unique advantages of the unmixing application.

4.3.1 *Equality Proof of LSVM and LSMM for Spectral Unmixing*

By virtue of Fig. 4.3, for the visualization, it considers the situation of three spectral endmembers. In the sense of the least squares, the closed formula of LSMM of the point $\boldsymbol{s}$ unmixing is:

$$\begin{aligned} &\boldsymbol{F}_{\mathrm{LSMM}}^{A,B,C}(\boldsymbol{P}) = \boldsymbol{\Theta s} \\ &\boldsymbol{\Theta} = \left([\boldsymbol{A},\boldsymbol{B},\boldsymbol{C}]^{T}[\boldsymbol{A},\boldsymbol{B},\boldsymbol{C}]\right)^{-1}[\boldsymbol{A},\boldsymbol{B},\boldsymbol{C}]^{T} \end{aligned} \tag{4.15}$$

Define Δ_{xyz} as the triangle composed by $\boldsymbol{x}$, $\boldsymbol{y}$ and z, the area is expressed as $S(\Delta_{xyz})$. L_{xy} is the straight line formed by $\boldsymbol{x}$ and $\boldsymbol{y}$, and the length is $\mathrm{le}(L_{xy})$. Suppose that the area $S(\Delta_{ABC})$ is 1, from the connotation of LSMM, it can be inferred

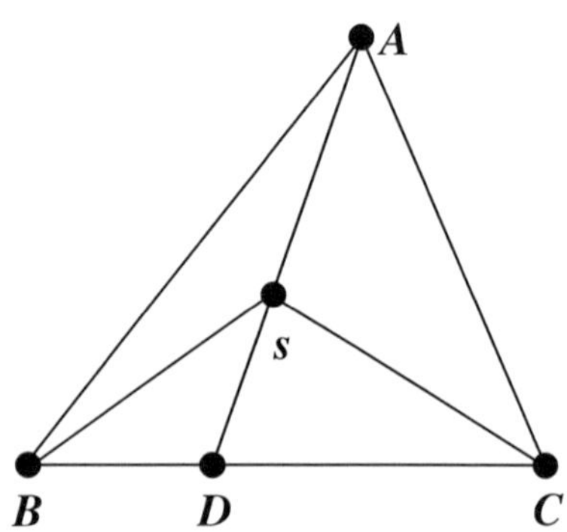

Fig. 4.3 Distance measure function of linear LSSVM

that, the pixel $\boldsymbol{s}$ is under LSMA, and the proportion $F^{\boldsymbol{A}}_{\text{LSMM}}(\boldsymbol{s})$ of the spectral endmember $\boldsymbol{A}$ is equal to $S(\Delta_{\boldsymbol{PBC}})$, then

$$F^{\boldsymbol{A}}_{\text{LSMM}}(\boldsymbol{P}) = S(\Delta_{\boldsymbol{PBC}}) = \frac{\text{le}(L_{\boldsymbol{PD}})}{\text{le}(L_{\boldsymbol{AD}})} \tag{4.16}$$

Now we prove that the unmixed component $F^{\boldsymbol{A}}_{\text{LSMM}}(\boldsymbol{s})$ of LSVM is equal to (4.16). The spectral endmember $\boldsymbol{A}$ is stipulated as the class of "1", $\boldsymbol{B}$ and $\boldsymbol{C}$ (so $\boldsymbol{D}$) is the class of "0", i.e.,:

$$f(\boldsymbol{A}) = 1,\ f(\boldsymbol{B}) = f(\boldsymbol{C}) = f(\boldsymbol{D}) = 0 \tag{4.17}$$

According to the discriminant form of LSVM, the component $F^{\boldsymbol{A}}_{\text{LSVM}}(\boldsymbol{s})$ can be calculated as follows.

$$F^{\boldsymbol{A}}_{\text{LSVM}}(\boldsymbol{s}) = f(\boldsymbol{s}) = f\left(\frac{\boldsymbol{sD}}{\boldsymbol{AD}}\boldsymbol{A} + \frac{\boldsymbol{As}}{\boldsymbol{AD}}\boldsymbol{D}\right) = \frac{\boldsymbol{sD}}{\boldsymbol{AD}} \tag{4.18}$$

From (4.16) and (4.18), it can be seen that two results are identical. This equal conclusion can be promoted to the component corresponding to the endmembers $\boldsymbol{B}$ and $\boldsymbol{C}$.

Given that l expresses the parallel lines of the line segment $L_{\boldsymbol{BC}}$ via the point $\boldsymbol{s}$. Given that $\boldsymbol{s}'$ is the other point differing from $\boldsymbol{s}$ in the straight line l, for whether $\boldsymbol{s}'$ falls in the Δ_{ABC}, it can obtain that the unmixing results of $\boldsymbol{s}'$ and $\boldsymbol{s}$ are the same in two models.

In order to promote the formula (4.18), given that $F^j(\cdot)$ expresses the j class of the unconstraint unmixing component of $\cdot$, and $f_j(\cdot)$ is the j output of 1-a-r type SVM, the following formula is correct:

$$F^j(\boldsymbol{x}) = f_j(\boldsymbol{x}) \tag{4.19}$$

The above analysis shows that, in the case of only utilizing the spectral endmember information and unconstraint condition, LSVM and LSMA are identical in the unmixing effect. As a matter of fact, both are the distance scale-based method, with different directions of the distance measure. It is observed that, the complexity of LSVM and LSMA is identical. In conformity with the physical significance, the spectral unmixing generally applies the constraint condition in the following mode. The negative value in the result is set as 0, while the value of more than 1 is set as 1. Then the sum of the unmixing component of each pixel is set as 1 through zooming. It can be seen that, under the constraint method, LSVM and LSMA still keep the same effect.

4.3.2 The Unique Superiority of LSVM-Based Unmixing

The further discussion will illustrate that LSVM possesses the incomparable advantages of LSMA. Firstly, it is the expansion and utilization of the information. Each class of the land object in LSMA can be expressed by an endmember information, and this expression is inaccurate when the intra-class spectral changes are greater, accordingly resulting in the decline of the unmixing accuracy. LSVM can conveniently utilize multiple training samples to depict a class, and accordingly gain more reasonable analysis model. In two classes of the true hyperspectral data in Fig. 4.4, obviously LSVM gains more reasonable partition model.

Secondly, it is the nonlinear promotion of the model. The actual hyperspectral image spectral unmixing problem is often the nonlinear problem, rather than the linear problem, while LSMA is just proposed for solving the linear problem. At present, there is not the effective method to promote to the nonlinear problem. LSVM can easily do it by introducing the nonlinear mapping. Fig. 4.5 is a schematic diagram of the nonlinear unmixing of two-dimensional data. The straight line l_4 is the corresponding decision function of LSVM, while l_2 is the corresponding decision function of nonlinear SVM (NLSVM).

Through the analysis, it can be found that, the data points in the shadow area will be wrongly decided as the pure pixel by LSVM, while NLSVM will make more reasonable mixed decision on it. Although NLSVM may be weaker than LSVM in disposing some data points (such as point $\boldsymbol{Q}$), but for the total effect, the unmixing advantages of NLSVM will be very obvious.

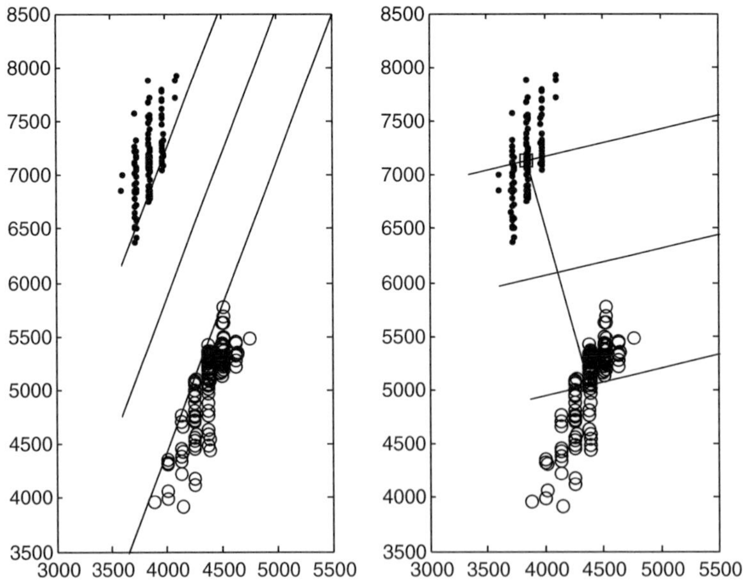

Fig. 4.4 Partition model comparison of LSVM and LSMA (*Left* LSVM; *Right* LSMA)

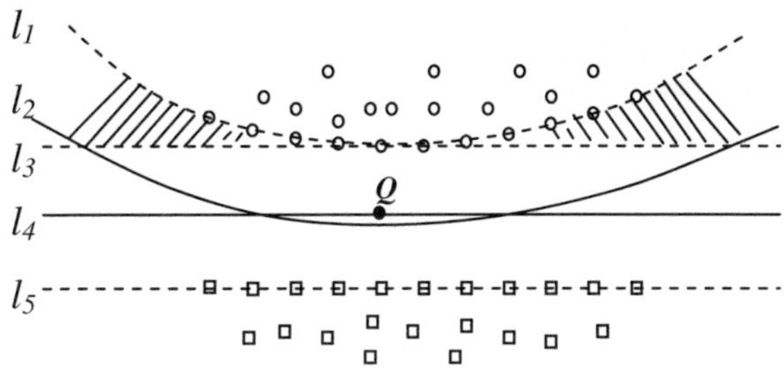

Fig. 4.5 Partition model comparison of LSVM and NLSVM

4.4 Spatial–Spectral Information-Based Unmixing Method

Due to the participation of the unrelated classes during unmixing and the insufficiency of the spectral unmixing model, the traditional unmixing mode is unsatisfactory in the unmixing effect. In allusion to this problem, some scholars make the correlation research focusing on the selection of the related classes and the establishment of the new unmixing model. The step-by-step unmixing concept in the article of Winter et al. (2003) was to make the selection of the related classes by the spectral information. However, this method had not been applied extensively due to the comparatively complex implementation. Qing and Zhen (1999) and Junwu et al. (2002) proposed making the selection of the related classes by the spatial information. Such selection had stronger reasonableness and relatively simple implementation. Before selecting the related classes, these two methods respectively used the simple spectral matching method and subspace projection method, with relatively low unmixing accuracy. This chapter puts forward a linear least squares SVM (LSSVM) model-based unmixing method in combination with the spatial information, for gaining better unmixing effect. Now, this method is described and assessed.

A large number of mixed pixels exist in the hyperspectral image, but each mixed pixel often includes a small quantity of the related classes, and these classes have close relations with the neighborhood pixel classes, as shown in Fig. 4.6. If using these related classes properly during unmixing, it will undoubtedly gain the best unmixing effect. Moreover, introducing the unrelated classes in unmixing will have the adverse effects on the unmixing accuracy. Therefore, judging the classes related to the mixed pixel to be unmixed correctly is especially important.

The spatial information is relatively effective for determining the related classes. For this purpose, firstly apply all classes to all pixels for the initial unmixing, and then distinguish the mixed pixels and unmixed pixels. According to the spatial correlation of the land object classes, admit that the unmixing process has some error at the same time. If some pixel is unmixed into the approximate spectral endmember (i.e., some class of the mixing ratio has the absolute advantage, such as more than 90 %), and the eight- neighborhood pixels are unmixed into the spectral endmember approximating such class, we have the reason to affirm the pixel as pure

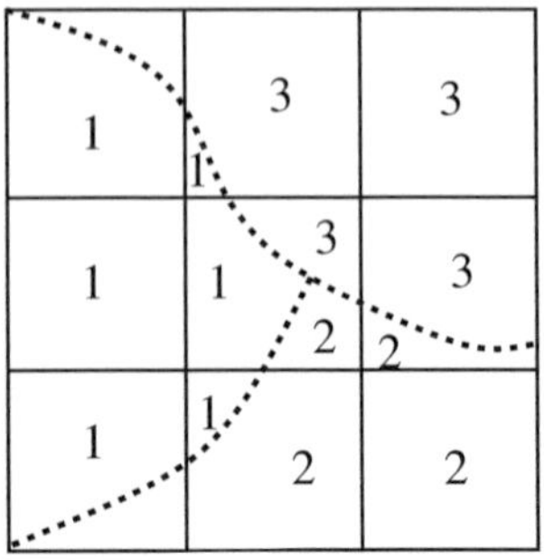

Fig. 4.6 Class relation between mixed pixel and neighborhood pixels

pixel of such class and correspondingly adjust the unmixing value. Other pixels are deemed as the mixed pixel processing. For these mixed pixels, decide the related classes by the unmixing class information provided by the eight- neighborhood pixels, i.e., accumulate various classes of the component values unmixed by the eight- neighborhood pixels, abandon the classes with smaller accumulated value (for example, the accumulated value is less than 5 %), and meanwhile ensure no excessive residual number of classes (such as less than 4 classes). After determining the related classes of all mixed pixels, we can make more accurate secondary unmixing for the mixed pixel by using the related classes. Figure 4.7 for the overall process.

4.5 SVM-Based Spectral Unmixing Model with Unmixing Residue Constraints

The existing SVM model is to incorporate the hard classification error constraint condition into the optimization function, while the general assessment principle of the spectral unmixing is the unmixing error, i.e., soft classification error. Both still have some differences. For this purpose, this book will establish the least squares SVM new model with unmixing residue constraints. At the same time, this book will also infer the corresponding closed form solution, and the method of substituting the variation endmembers for fixed one in the new model.

The least squares SVM (LSSVM) is widely popular due to the convenience of solution and the assessment method of conforming to the mean square error minimum principle. This model replaces the inequality constraint in the standard SVM model by the equality constraint in this model, to make the model solution transform from the original complex quadratic programming to the concise closed form. Therefore, we select LSSVM as the unmixing model. But the reader should note that, the following research contents are still valid for other types of SVM except for the derivation of the closed form solution in Sect. 4.5.2.

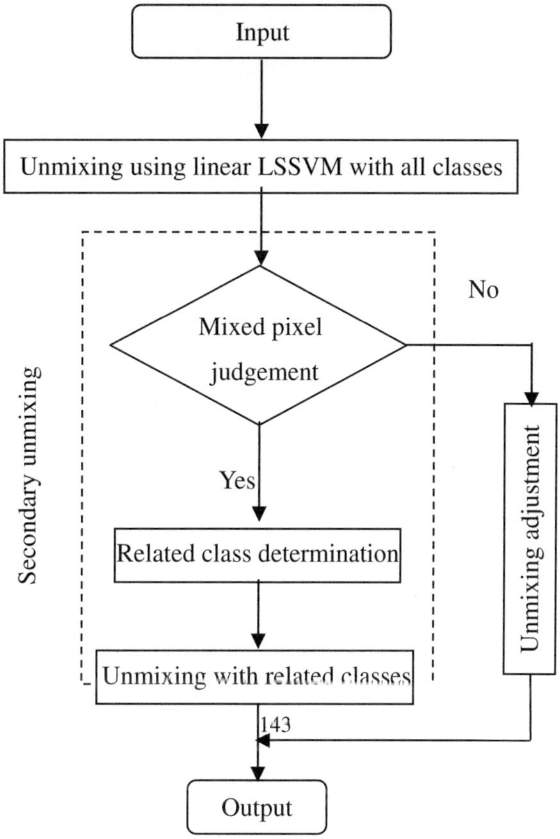

Fig. 4.7 Block diagram of spectral unmixing method in combination with spatial information

4.5.1 Original LSSVM-Based Spectral Unmixing

In order to better understand LSSVM spectral unmixing model with unmixing residue constraints to be proposed, we firstly introduce the original hard constraint LSSVM unmixing model. In this model, the famous "class kernel" concept (Brown et al. 2000) have been embodied well. "Class kernel" means a specific zone in the eigenspace, and the pixels in the zone are the pure pixels belonging to one class. This is a very reasonable affirmation, because the pixels in the same class have larger differences of the spectrum due to different regions and different interference degrees of the air and noise during the imaging process.

Fig. 4.8 provides a two-class (ω_A and ω_B), linearly separable SVM spectral unmixing model schematic diagram. The support vectors are the points in two classification boundaries. These pixels are called as "positive pure" pixels, and the points outside two classification boundaries are called as "pure" pixels. The point between two classification boundaries is affirmed as the mixed pixel. At the moment, it can establish the following unmixing decision rule.

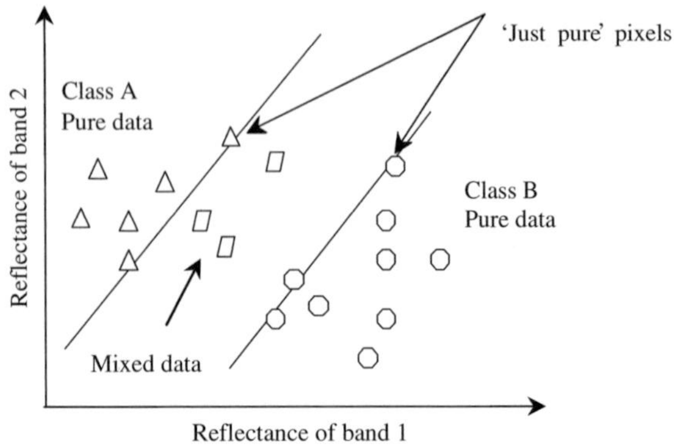

Fig. 4.8 LSVM-based spectral unmixing

Given n point training data set $\{\boldsymbol{x}_i, y_i\}_{i=1}^{\mathrm{Ntr}}$, including the input spectral data $\boldsymbol{x}_i \in \boldsymbol{R}^d$, and output class mark $y_i = 1, 0$ and $i = 1, 2, \ldots, \mathrm{Ntr}$. The corresponding discrimination function can be expressed as:

$$f(\boldsymbol{x}) = \sum_{i=1}^{\mathrm{Ntr}} \alpha_i^* \langle \phi(\boldsymbol{x}_i), \phi(\boldsymbol{x}) \rangle + b^* \tag{4.20}$$

Then the spectral unmixing decision rule is:

$$\begin{aligned} & \text{if}\, f(\boldsymbol{x}) \geq 1 \\ & \boldsymbol{x} \in \omega_A,\ \text{i.e. } \beta_x(\omega_A) = 1, \beta_x(\omega_B) = 0; \\ & \text{if}\, f(\boldsymbol{x}) \leq 0 \\ & \boldsymbol{x} \in \omega_B,\ \text{i.e. } \beta_x(\omega_A) = 0, \beta_x(\omega_B) = 1; \\ & \text{if } 0 < f(\boldsymbol{x}) < 1 \\ & \boldsymbol{x} \in \{\omega_A, \omega_B\}, \beta_x(\omega_A) = f(\boldsymbol{x}), \beta_x(\omega_B) = 1 - f(\boldsymbol{x}) \end{aligned} \tag{4.21}$$

Here, $\beta_x(\omega_A)$ and $\beta_x(\omega_B)$ are the component values of the classes ω_A and ω_B.

For the multi-class problem, it can be transformed into multiple two-class problems by one-against-rest (1-a-r) type multi-class classifier structure. For the optimal component value $\beta_{\boldsymbol{x}}^r(\omega_i)$ gained by each SVM submodel, make the normalization processing by the following formula:

$$\beta_{\boldsymbol{x}}(\omega_i) = \beta_{\boldsymbol{x}}^r(\omega_i) / \sum_{i=1}^{Nc} \beta_{\boldsymbol{x}}^r(\omega_i) \tag{4.22}$$

Make $\sum_{i=1}^{Nc} \beta_{\boldsymbol{x}}(\omega_i) = 1$.

In essence, SVM unmixing is the supervised method. Each supervised sample is regarded as the endmember spectrum with different importance, and SVM automatically gives different weights to reflect their different effects on the class division. This mode is obviously superior to LSMA method of the traditional single endmember. SVM is also superior to the practice of equally treating each endmember spectrum in the multi-endmember spectral unmixing method of Asner and Lobell (2000) and Bateson et al. (2000). In addition, the automatic endmember weighting mode of SVM enables the mixed supervised data to be used for training SVM model just like the pure supervised data, which the multi-endmember spectral unmixing method of Asner and Lobell (2000) and Bateson et al. (2000) can not do.

4.5.2 Construction of Spectral Unmixing Model Based on Unmixing Residue Constrained LSSVM and Derivation of Its Closed form Solution

Without loss of generality, for *Nc* class of the spectral unmixing problem, and meanwhile considering Nc submodels, we request that the unmixing error is as small as possible, i.e.,:

$$\begin{aligned} &\min_{w,b,e} J(\boldsymbol{w},\boldsymbol{e}) = \tfrac{1}{2}\sum_{c=1}^{\mathrm{Nc}} \|\boldsymbol{w}_c\|^2 + \tfrac{\gamma}{2}\sum_{i=1}^{\mathrm{Ntr}} \boldsymbol{e}_i^T \boldsymbol{e}_i \\ &\text{s.t.}\quad \boldsymbol{e}_i = \sum_{\iota-1}^{\mathrm{Nc}} \boldsymbol{E}_c(\langle \boldsymbol{w}_c, \boldsymbol{x}_i\rangle + b_c) - \boldsymbol{x}_i, \\ &\qquad i = 1,2,\ldots,\mathrm{Ntr} \end{aligned} \tag{4.23}$$

where, Ntr is the number of training samples, and Nc is the number of classes (meanwhile setting the number of bands as ND), and $\boldsymbol{E}_c$ is the spectral endmember of the *c* class. The above constraint condition is Ntr × ND equations, which will be more obvious in the following equivalent forms:

$$\begin{aligned} &\min_{w,b,e} J(\boldsymbol{w},\boldsymbol{e}) = \frac{1}{2}\sum_{c=1}^{\mathrm{Nc}} \|\boldsymbol{w}_c\|^2 + \frac{\gamma}{2}\sum_{i=1}^{\mathrm{Ntr}}\sum_{j=1}^{\mathrm{ND}} e_{i,j}^2 \\ &\text{s.t.}\quad e_{i,j} = \sum_{c=1}^{\mathrm{Nc}} \boldsymbol{E}_{c,j}(\langle \boldsymbol{w}_c, \phi(\boldsymbol{x}_i)\rangle + b_c) - x_{i,j} \\ &\qquad i = 1,2,\ldots,\mathrm{Ntr},\ \ j = 1,2,\ldots,\mathrm{ND} \end{aligned} \tag{4.24}$$

where, $\boldsymbol{E}_{c,j}$ is the component value of the *j* band in the spectral endmember of the *c* class land object.

Take two linear problems as the example. Given that two subclassifiers are interrelated by the normalization condition in two problems, i.e., $\sum_{c=1,2} (\langle \boldsymbol{w}_c, \varphi(\boldsymbol{x}_i)\rangle + b_c) = 1$, the above formula becomes:

$$\begin{aligned} &\min_{\boldsymbol{w},\boldsymbol{b},\boldsymbol{e}} J(\boldsymbol{w},\boldsymbol{e}) = \frac{1}{2}\|\boldsymbol{w}\|^2 + \frac{\gamma}{2}\sum_{i=1}^{\mathrm{Ntr}}\sum_{j=1}^{\mathrm{ND}} e_{i,j}^2 \\ &\text{s.t.}\quad e_{i,j} = \boldsymbol{E}_{1,j}(\langle \boldsymbol{w},\phi(\boldsymbol{x}_i)\rangle + \boldsymbol{b}) + \boldsymbol{E}_{2,j}(1 - \langle \boldsymbol{w},\phi(\boldsymbol{x}_i)\rangle - \boldsymbol{b}) - \boldsymbol{x}_i \\ &\qquad\qquad i = 1,2,\ldots,\mathrm{Ntr};\quad j = 1,2,\ldots,\mathrm{ND} \end{aligned} \tag{4.25}$$

The corresponding Lagrange equation is:

$$\begin{aligned} L = &J(\boldsymbol{w},\boldsymbol{e}) \\ &+ \sum_{i=1}^{\mathrm{Ntr}}\sum_{j=1}^{\mathrm{ND}} \alpha_{i,j}\left[\boldsymbol{E}_{1,j}(\langle \boldsymbol{w},\phi(\boldsymbol{x}_i)\rangle + \boldsymbol{b}) + \boldsymbol{E}_{2,j}(1 - \langle \boldsymbol{w},\phi(\boldsymbol{x}_i)\rangle - \boldsymbol{b}) - \boldsymbol{x}_i - e_{i,j}\right] \end{aligned} \tag{4.26}$$

If the mixing ratio of each sample data is known and meets the normalization condition, i.e., when the positive class mixing ratio of the sample $\boldsymbol{x}_i$ is y_i, the mixing ratio of the other class is $1 - y_i$, then:

$$\begin{aligned} e_{i,j} &= \left([\langle \boldsymbol{w}_c,\phi(\boldsymbol{x}_i)\rangle + b_c]\boldsymbol{E}_{1,j} + [1 - \langle \boldsymbol{w}_c,\phi(x_i)\rangle - b_c]\boldsymbol{E}_{2,j}\right) - \boldsymbol{x}_{i,j} \\ &= \left([\langle \boldsymbol{w}_c,\phi(\boldsymbol{x}_i)\rangle + b_c]\boldsymbol{E}_{1,j} + [1 - \langle \boldsymbol{w}_c,\phi(x_i)\rangle - b_c]\boldsymbol{E}_{2,j}\right) - \left(y_i\boldsymbol{E}_{1,j} + [1 - y_i]\boldsymbol{E}_{2,j}\right) \\ &= [\langle \boldsymbol{w}_c,\phi(\boldsymbol{x}_i)\rangle + b_c - y_i]\boldsymbol{E}_{1,j} + [1 - \langle \boldsymbol{w}_c,\phi(x_i)\rangle - b_c - 1 + y_i]\boldsymbol{E}_{2,j} \\ &= [\langle \boldsymbol{w}_c,\phi(\boldsymbol{x}_i)\rangle + b_c - y_i]\boldsymbol{E}_{1,j} - [\langle \boldsymbol{w}_c,\phi(x_i)\rangle + b_c - y_i]\boldsymbol{E}_{2,j} \\ &= [\langle \boldsymbol{w}_c,\phi(\boldsymbol{x}_i)\rangle + b_c - y_i][\boldsymbol{E}_{1,j} - \boldsymbol{E}_{2,j}] \end{aligned} \tag{4.27}$$

The dual problem corresponding to the formula (4.26) is:

$$\min_{w,b,e,\alpha} L(\boldsymbol{w},b,\boldsymbol{e},\alpha) = J(\boldsymbol{w},\boldsymbol{e}) - \sum_{j=1}^{\mathrm{ND}}\sum_{i=1}^{\mathrm{Ntr}} \alpha_{i,j}\left\{e_{i,j} - [y_i - \langle \boldsymbol{w},\phi(\boldsymbol{x}_i)\rangle - b][\boldsymbol{E}_{1,j} - \boldsymbol{E}_{2,j}]\right\} \tag{4.28}$$

The optimal KKT condition is:

$$\begin{cases} \frac{\partial L}{\partial w} = 0 \Rightarrow \boldsymbol{w} = \sum_{j=1}^{\mathrm{ND}}\sum_{i=1}^{\mathrm{Ntr}} \alpha_{i,j}\phi(\boldsymbol{x}_i)[\boldsymbol{E}_{2,j} - \boldsymbol{E}_{1,j}] \\ \frac{\partial L}{\partial b} = 0 \Rightarrow \sum_{j=1}^{\mathrm{ND}}\sum_{i=1}^{\mathrm{Ntr}} \alpha_{i,j}[\boldsymbol{E}_{1,j} - \boldsymbol{E}_{2,j}] = 0 \\ \frac{\partial L}{\partial e_{i,j}} = 0 \Rightarrow \alpha_{i,j} = \gamma e_{i,j} \quad i = 1,2,\ldots,\mathrm{Ntr};\ j = 1,2,\ldots,\mathrm{ND} \\ \frac{\partial L}{\partial \alpha_{i,j}} = 0 \Rightarrow e_{i,j} - [y_i - \langle \boldsymbol{w},\phi(\boldsymbol{x}_i)\rangle - b][\boldsymbol{E}_{1,j} - \boldsymbol{E}_{2,j}] = 0, \quad i = 1,2,\ldots,\mathrm{Ntr};\ j = 1,2,\ldots,\mathrm{ND} \end{cases} \tag{4.29}$$

After eliminating $\boldsymbol{w}$ and $\boldsymbol{e}$ by the method of elimination, it can get the formula:

$$\sum_{j=1}^{\mathrm{ND}}\sum_{i=1}^{\mathrm{Ntr}} \alpha_{i,j} K(\boldsymbol{x}_i, \boldsymbol{x}_{i_0})(\boldsymbol{E}_{1,j} - \boldsymbol{E}_{2,j})(\boldsymbol{E}_{1,j_0} - \boldsymbol{E}_{2,j_0}) + b(\boldsymbol{E}_{1,j_0} - \boldsymbol{E}_{2,j_0})$$
$$- \alpha_{i_0,j_0}/\gamma = y_{i_0}(\boldsymbol{E}_{1,j_0} - \boldsymbol{E}_{2,j_0})$$
$$i_0 = 1, 2, \ldots, \mathrm{Ntr}; \quad j_0 = 1, 2, \ldots, \mathrm{ND} \tag{4.30}$$

The corresponding matrix equation can be expressed as:

$$\begin{bmatrix} 0 & \vec{\boldsymbol{S}}_{1,\mathrm{Ntr}\times\mathrm{ND}} \\ \underset{\rightarrow}{\boldsymbol{S}}_{\mathrm{Ntr}\times\mathrm{ND},1} & (\boldsymbol{K}_{\mathrm{Ntr}\times\mathrm{Ntr}})_{\mathrm{ND}\times\mathrm{ND}} \otimes \left(\vec{\boldsymbol{S}}_{\mathrm{Ntr}\times\mathrm{ND},1} \times \underset{\rightarrow}{\boldsymbol{S}}_{1,\mathrm{Ntr}\times\mathrm{ND}}\right) + \boldsymbol{I}/\gamma \end{bmatrix} \begin{bmatrix} b \\ \vec{\boldsymbol{\alpha}}_{\mathrm{Ntr}\times\mathrm{ND},1} \end{bmatrix} = \begin{bmatrix} 0 \\ \left(\vec{\boldsymbol{y}}_{\mathrm{Ntr}\times 1}\right)_{\mathrm{ND}\times 1} \otimes \vec{\boldsymbol{S}}_{\mathrm{Ntr}\times\mathrm{ND},1} \end{bmatrix} \tag{4.31}$$

where, the vectors $\vec{\boldsymbol{S}}_{1,\mathrm{Ntr}\times\mathrm{ND}}$ and $\vec{\boldsymbol{\alpha}}_{\mathrm{Ntr}\times\mathrm{ND},1}$ are respectively shown as the formulas (4.32) and (4.33). The vector $\boldsymbol{1}_{1\times\mathrm{Ntr}}$ is the 1 × Ntr row vector with the element of 1. $\vec{\boldsymbol{S}}_{\mathrm{Ntr}\times\mathrm{ND},1}$ is the transposition of $\vec{\boldsymbol{S}}_{1,\mathrm{Ntr}\times\mathrm{ND}}$. $\boldsymbol{I}$ is the unit matrix with the order of Ntr × ND. $\left(\vec{\boldsymbol{y}}_{\mathrm{Ntr}\times 1}\right)_{\mathrm{ND}\times 1}$ is (Ntr × ND) × 1 column vector connected by ND column vectors $\vec{\boldsymbol{y}}_{\mathrm{Ntr}\times 1} = [y_1, y_2, \ldots, y_{\mathrm{Ntr}}]^T$ in order. The operator ⊗ shows that the corresponding position elements of two vectors or matrixes with the same dimension are multiplied to gain the vectors or matrixes with the invariable dimension. $\boldsymbol{K}_{\mathrm{Ntr}\times\mathrm{ND}}$ is the kernel matrix of Ntr × ND, i.e., $K(\boldsymbol{x}_m, \boldsymbol{x}_n) = \langle\phi(\boldsymbol{x}_m), \phi(\boldsymbol{x}_n)\rangle$. $(\boldsymbol{K}_{\mathrm{Ntr}\times\mathrm{Ntr}})_{\mathrm{ND}\times\mathrm{ND}}$ is Ntr × ND order square matrix gained by ND × ND repetition with $\boldsymbol{K}_{\mathrm{Ntr}\times\mathrm{Ntr}}$ as the subblock.

$$\vec{\boldsymbol{S}}_{1,N\mathrm{tr}\times\mathrm{ND}} = \left[(\boldsymbol{E}_{1,1} - \boldsymbol{E}_{2,1})\boldsymbol{1}_{1\times\mathrm{Ntr}}, (\boldsymbol{E}_{1,2} - \boldsymbol{E}_{2,2})\boldsymbol{1}_{1\times\mathrm{Ntr}}, \ldots, (\boldsymbol{E}_{1,\mathrm{ND}} - \boldsymbol{E}_{2,\mathrm{ND}})\boldsymbol{1}_{1\times\mathrm{Ntr}}\right] \tag{4.32}$$

$$\vec{\boldsymbol{\alpha}}_{\mathrm{Ntr}\times\mathrm{ND},1} = \left[\alpha_{1,1}, \alpha_{2,1} \ldots, \alpha_{\mathrm{Ntr},1}, \alpha_{1,2}, \alpha_{2,2} \ldots, \alpha_{\mathrm{Ntr},2}, \ldots, \alpha_{1,\mathrm{ND}}, \alpha_{2,\mathrm{ND}} \ldots, \alpha_{\mathrm{Ntr},\mathrm{ND}}\right]^T \tag{4.33}$$

Define (α^*, b^*) is the solution of Eq. (4.31), then the corresponding soft classification discrimination function is:

$$f(x) = \sum_{i=1}^{\mathrm{Ntr}} \alpha_i^* K(\boldsymbol{x}_i, \boldsymbol{x}) + b^* \tag{4.34}$$

$\phi(\boldsymbol{x}) = \boldsymbol{x}$ corresponds to the linear situation. Then the above discrimination function can gain the following form:

$$f(x) = \langle \boldsymbol{w}^*, \boldsymbol{x} \rangle + b^* \tag{4.35}$$

Here, $\boldsymbol{w}^*$ can be obtained by the solution (4.29). The complexity of solving the unmixing component value by the function (4.35) is obviously smaller than the traditional LSMA.

4.5.3 Substituting Multiple Endmembers for Single One in the New Model

We know that, the reason that the unmixing error of the traditional LSMA is less than satisfactory is mainly because LSMM use the single endmember spectrum as the representative of one class, while this mode is inaccurate for the hyperspectral image with greater intra-class spectral changes. The original SVM model uses multiple pure samples to depict a class, and overcomes the defects of LSMM model. The above SVM new model is improved in the constraint condition, but the formula (4.27) introduces such rigid expression once again. For this purpose, it needs utilizing the pure training sample again to substitute for the fixed endmember.

Because each mixed sample is composed by us upon utilizing the known pure pixel and known ratio, relative to the fixed endmember in the traditional LSMM model, these pure pixels can be regarded as the variation endmembers, for fully depicting the intra-class changes of the hyperspectral data. Thus, for each mixed training sample $\boldsymbol{x}_i$, there is always the relevant specific endmember combination $\{\boldsymbol{E}_1^i, \boldsymbol{E}_2^i\}$, to introduce them into SVM new model and fully inherit the advantages of SVM unmixing model and avoiding the problems arising out of the endmember rigidity.

For multi-class unmixing problem, in fact, we cannot make the joint consideration of Nc subclassification models of Nc class problem like the formula (4.23). The theory and experiment manifest that, the above process will result in the underdetermined matrix in Eq. (4.31) without the accurate solution. Thus, we must decompose it into Nc two-class problems for solving in accordance with the classic 1-a-r mode. At the moment, we need determining the specific endmember combination $\{\boldsymbol{E}_1^i, \boldsymbol{E}_2^i\}$ for each sample in each sub unmixing model. During the synthetic process of the mixed data, the pure pixels (i.e., endmembers) relevant to the sample $\boldsymbol{x}_i$ is Nc: $\{\boldsymbol{E}_1^i, \boldsymbol{E}_2^i, \ldots, \boldsymbol{E}_{Nc}^i\}$, and the corresponding synthesis ratio is $\{\lambda_1^i, \lambda_2^i, \ldots, \lambda_{Nc}^i\}$, so $x_i = \sum\limits_{j=1}^{\mathrm{Nc}} \lambda_j^i \boldsymbol{E}_j^i$. For the $k(1 \leq k \leq \mathrm{Nc})$ sub unmixing models, the k class is the positive class, so $\boldsymbol{E}_k^i$ is the first corresponding endmember, with the mixing ratio of λ_k^i. If the rest ingredients are deemed as a whole, the mixing ratio is

$\lambda_{\bar{k}}^i = 1 - \lambda_k^i$. Thus, it can obtain the other corresponding endmember $\boldsymbol{E}_{\bar{k}}^i$: $\boldsymbol{E}_{\bar{k}}^i = \left(x_i - \lambda_k^i \boldsymbol{E}_k^i\right) / \left(1 - \lambda_k^i\right)$ of x_i, meeting the relational expression $x_i = \lambda_k^i \boldsymbol{E}_k^i + \lambda_{\bar{k}}^i \boldsymbol{E}_{\bar{k}}^i$.

4.6 Performance Assessment

4.6.1 Performance Assessment for Original SVM-Based Spectral Unmixing

The assessment of the spectral unmixing result generally includes the unmixing accuracy, spectral unmixing component diagram, and reliability curve, etc. Before applying, firstly it is described.

Unmixing accuracy: Given the total number of pixel points as Ne, and the total number of mixed classes as Nc. The true ratio and unmixing ratio of the j class component in the i pixel are respectively P_i^j and $\hat{P}_i^j$, and the unmixing accuracy Acc can be obtained by the following formula: $\text{Acc} = \sum_{i=1}^{\text{Ne}} \sum_{j=1}^{\text{Nc}} (P_i^j - \hat{P}_i^j)^2$. It is an objective assessment index, and also the most convenient and common assessment index.

Spectral unmixing component diagram: It is two-dimensional gray display for the ratio of some class component in each pixel of the image, and each class corresponds to an unmixing component diagram.

Reliability curve: It is two-dimensional plane curve, indicating the accuracy (Y-axis) of the corresponding sample unmixing under different allowable unmixing errors (X-axis).

Select 500 two-class data (the corresponding class marks are respectively 14 and 11) respectively in Indian agriculture and forestry hyperspectral data, and gain two types of the representative spectrums upon averaging, as shown in Fig. 4.9. Divide

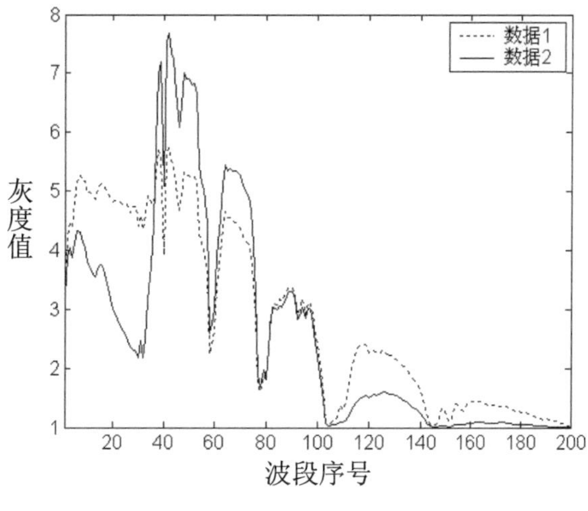

Fig. 4.9 Two types of representative spectrums

each class data into 10 groups equally, and give the corresponding coefficient increasingly. In the k group, the spectrum of the first class data multiples by the weight coefficient $0.05 + (k - 1) \times 0.1$, and the spectrum of the second class data multiples by the weight coefficient $0.95 - (k - 1) \times 0.1$. Then obtain the k group synthesized data by the corresponding summation. Two-class data are synthesized into 500 mixed data successively in such mode. Such data mixed state can be expressed (the first class corresponds to white, and the second class corresponds to black) intuitively in Fig. 4.10a. In the experiment, select the remaining 500 data points respectively in the above two classes as the training samples of SVM learning, and meanwhile apply them to obtain two types of the representative spectrums as the spectral endmember of LSMA unmixing process. Fig. 4.10b, c, and Table 4.1 respectively provide the comparison of unmixing result component diagram and the data comparison. It is obvious that the SVM-based spectral unmixing method has better unmixing effect. The spectral unmixing reliability curve in Fig. 4.11 also intuitively displays this comparison effect.

Also, the other group of the experiment convincingly states the advantages of SVM unmixing. As shown in Fig. 4.12, the upper row is the subimage of Indian agriculture and forestry hyperspectral image and the spectral unmixing component diagram under two methods, with the image size of 15 × 15. Select 40 training samples (20 samples for each class respectively) manually. It is observed that, two types of the land objects are embodied better in SVM unmixing result. The experimental image in the lower row is derived from the subimage of the San Diego military hyperspectral image and the related unmixing component diagram, with the image size of 40 × 50. In the experiment, the target spectrum and background spectrum are selected manually (100 units for each class). In the LSMA unmixing result, the background unmixing error in the top left corner is larger, while in the SVM unmixing result, the target and background are displayed accurately.

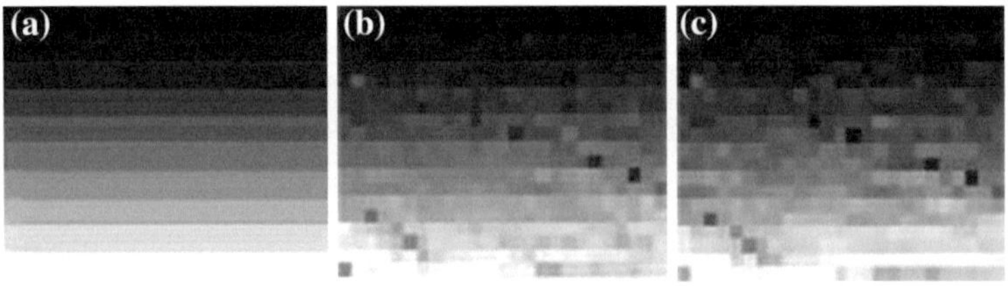

Fig. 4.10 Comparison of spectral unmixing component diagram of two methods. **a** Original image. **b** SVM unmixing image. **c** LSMA unmixing image

Table 4.1 Comparison of unmixing result of two methods (%)

Assessment method	Unmixing method	
	SVM	LSMA
Unmixing accuracy	93.12	90.92
Absolute error mean	0.0329	0.0471

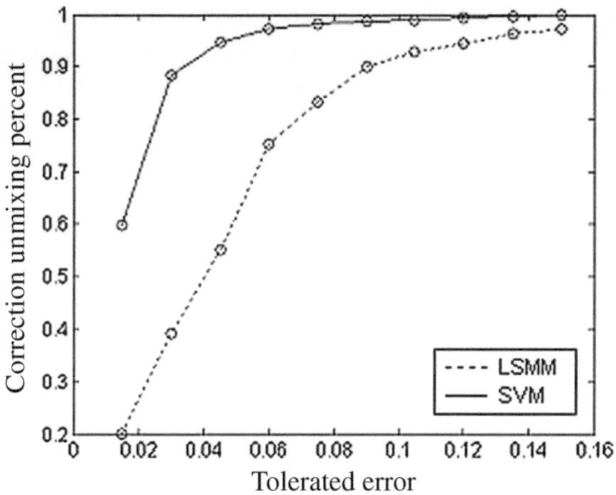

Fig. 4.11 Comparison of unmixing reliability curve of two methods

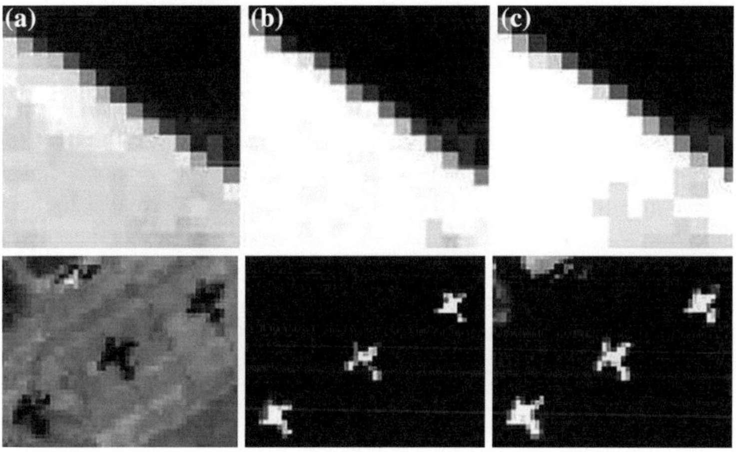

Fig. 4.12 Comparison of spectral unmixing component diagram of two methods. a Original image. **b** SVM unmixing image. **c** LSMA unmixing image

4.6.2 Assessment on Robust Weighted SVM-Based Unmixing

This chapter does not provide the detailed robust weighted unmixing method here, which means the sample weighting method by SVM comprehensive weighting classification of the last chapter, so as to obtain the robustness of the atypical samples during the process of SVM unmixing. Below, it will indicate the efficiency of the proposed distance-based weighting method (hereinafter referred to as

D-WLSSVM) by the comparison with the traditional training error-based least squares SVM weighting (hereinafter referred to as E-WLSSVM) method. It continues to adopt the agriculture and forestry hyperspectral data. Three groups of different class pairs, SMO algorithm and MSE (Mean Square Error) assessment criteria are applied to this experiment. MSE criteria are defined as follows:

$$\mathrm{MSE} = \frac{1}{\mathrm{Nte}} \sum_{k=1}^{\mathrm{Nte}} [y(k) - f(\boldsymbol{x}_k)]^2 \tag{4.36}$$

Here, Nte is the total number of test samples, and $f(\boldsymbol{x}_k)$ is the fuzzy decision function corresponding to the unmixing method.

Firstly, in Experiment 1, make the comparison for the generalization ability of two methods by two classes (500 training samples respectively and 500 test samples respectively) of soybean and forest land in Indian agriculture and forestry hyperspectral data and the linear kernel function. The comparison result is shown in Fig. 4.13. The value of the parameter γ is shown by 2^k, and the integer k is continuously changed to 10 from -10. The X-axis is the value of k, and the Y-axis is the corresponding MSE value. From Fig. 4.13, D-WLSSVM has better generalization ability than E-WLSSVM.

Experiment 2 is aimed at the comparison test for the robust characteristics of two methods. For this purpose, we increase Gaussian noise (mean value of 0, and the variance σ^2 of 0.1, 0.2 and 0.3 respectively) in partial data (100 training samples respectively and 100 test samples respectively) in Experiment 1 data, and synthesize the test data into 100 supervised mixed data. Table 4.2 provides the unmixing result under this condition, and the result shows that the new method possesses better robust characteristics.

The weighting mean calculation time of these two experiments is given in Table 4.3. When the number of the training samples is less (less than 200), the time

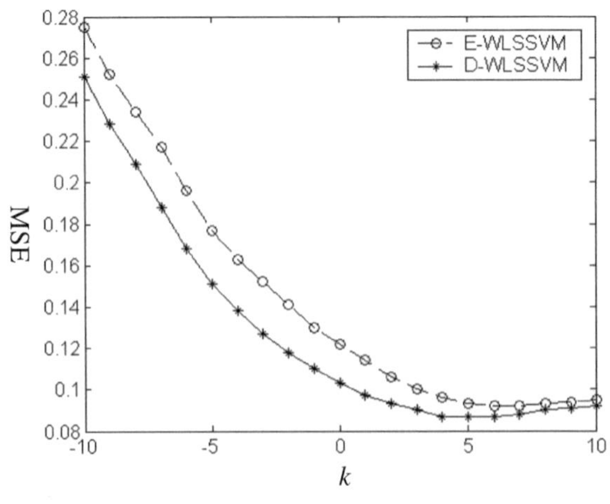

Fig. 4.13 Comparison of generalization ability (Linear Kernel)

Table 4.2 Comparison of robust performance

Unmixing method	Unmixing accuracy under different parameters σ^2(%)		
	$\sigma^2 = 0.1$	$\sigma^2 = 0.2$	$\sigma^2 = 0.3$
E-WLSSVM	81.71	81.69	81.43
D-WLSSVM	84.61	84.37	84.47

Table 4.3 Comparison of running time (s)

Unmixing method	Running time	
	Experiment 1	Experiment 2
E-WLSSVM	156.0	6.960
D-WLSSVM	10.32	2.460

for D-WLSSVM algorithm is about one third of E-WLSSVM algorithm. While the number of training samples is increased to 1,000, the time for D-WLSSVM algorithm is less than one-fifteenth of E-WLSSVM algorithm. From the previous weighting calculation formula, it is not hard to see that, the complexity of calculating the weight in D-WLSSVM algorithm linearly increases only with the increasing of the number of training samples, while E-WLSSVM shows the secondary increase trend with the increasing of the number of training samples. Hence, for the weighting problem of the mass training samples, D-WLSSVM has higher efficiency. For the situation of the nonlinear kernel function, the relative efficiency of D-WLSSVM algorithm is higher. Here we will not specify the experimental result.

In the final experiment, we make comparisons for the sparse approximation ability of two methods by applying the meadow—forest land classes. Because the support vectors from LSSVM algorithm possess the denseness, we cannot make the comparison for the number of support vectors under different parameters γ, and we make the comparison for the unmixing effect under the same number of support vectors. Upon adopting the previously mentioned sparse approximation method, 95 % of the current training samples in the iteration are selected as the support vectors. The experiment selects the number of the support vectors corresponding to the $0, 5, 10, \ldots, 60$ iterations. Fig. 4.14 shows that D-WLSSVM has better sparse approximation ability. When the number of support vectors is reduced to 10, D-WLSSVM unmixing effect is still preferable, while E-WLSSVM unmixing effect is seriously deteriorated.

4.6.3 Assessment on Spatial–Spectral Unmixing Method

The hyperspectral image is difficult to obtain the ground real mixed supervised data, so the assessment problem of the spectral unmixing is very difficult. In view of this problem, Hassan Emami came up with a hard supervision index-based assessment method, and provided an assessment criterion for our experimental analysis. Now the assessment method is described as follows.

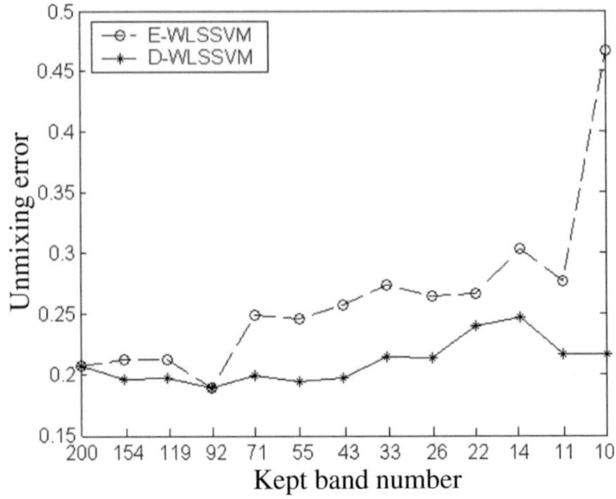

Fig. 4.14 Comparison of sparse approximation ability

For the correction coefficient CC, firstly it is necessary to gain the binary image of each land object in the ground real image. If expressing the binary image of the i land object by B_i,

$$B_i(j,k) = \begin{cases} 1 & \text{if point } (j,k) \text{ in truth image belongs } i\text{th class} \\ 0 & \text{otherwise} \end{cases} \quad i = 1, 2, \ldots, \mathrm{Nc} \tag{4.37}$$

Then noting AM_i and F_i respectively as the unmixing accuracy diagram and unmixing component diagram of the i class of the land object, then:

$$\mathrm{AM}_i(i,j) = B_i(j,k) \cdot F_i(j,k), i = 1, 2, \ldots, N \tag{4.38}$$

Noting

$$S_i = \sum_{j=1}^{r} \sum_{k=1}^{c} \mathrm{Am}_i(j,k) \tag{4.39}$$

So, the correction coefficient CC can be obtained by the following formula:

$$\mathrm{CC} = (\sum_{i=1}^{N} S_i)/\mathrm{Ng} \tag{4.40}$$

The correction coefficient CC can also be obtained class by class:

$$\mathrm{CC}_i = S_i/\mathrm{Ng}_i \tag{4.41}$$

where, Ng_i shows the total number of pixels in the i class of the land object.

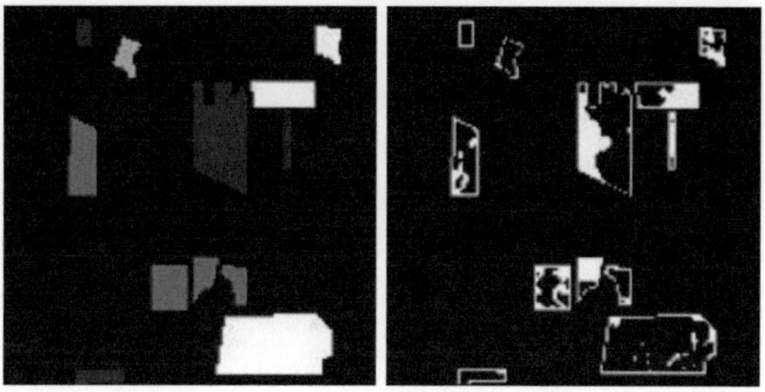

Fig. 4.15 Extraction of original image and mixed area

Table 4.4 Comparison of unmixing accuracy (%)

Unmixing method	Various unmixing accuracy				Average unmixing accuracy
Traditional method	78.26	68.50	91.96	83.69	80.62
Proposed method	90.03	83.69	97.61	88.68	90.00

The experiment still utilizes the Indian agriculture and forestry hyperspectral image. Select four classes of land objects (soybean, forest land, grass, and alfalfa) in the real image as the experimental data (as shown in Fig. 4.15, four different brightness on behalf of four different land object classes). In Fig. 4.15, the bright area is the mixed pixel area extracted under the proposed method. Table 4.4 provides the comparison of unmixing accuracy between the proposed method and traditional LSMA method. The spatial–spectral unmixing method proposed in this book has improved for about 10 % than the traditional spectral unmixing method in the unmixing accuracy. Fig. 4.16 is the comparison of spectral unmixing component diagram in two methods. Obviously, the proposed method has the significant advantages than the traditional unmixing method.

4.6.4 Performance Assessment on New SVM Unmixing Model with Unmixing Residue Constraints

For the sake of convenience, various unmixing models are denoted as follows. The linear SVM with original hard constraints is referred to as LSVM. The endmember linear SVM with unmixing residue constraints is referred to as FLSVM. The variation endmember linear SVM with unmixing residue constraints is referred to as VLSVM. The fixed endmember nonlinear SVM with unmixing residue constraints is referred to as FNLSVM. The variation endmember nonlinear SVM with unmixing residue constraints is referred to as VNLSVM. It is noted again that, “F” and “V” are unmixing residue constraints methods proposed in this book, and “V” also increases the improvement means of “variation endmember”. Separating them is to better reflect the unmixing performance enhancement situation of different improved points.

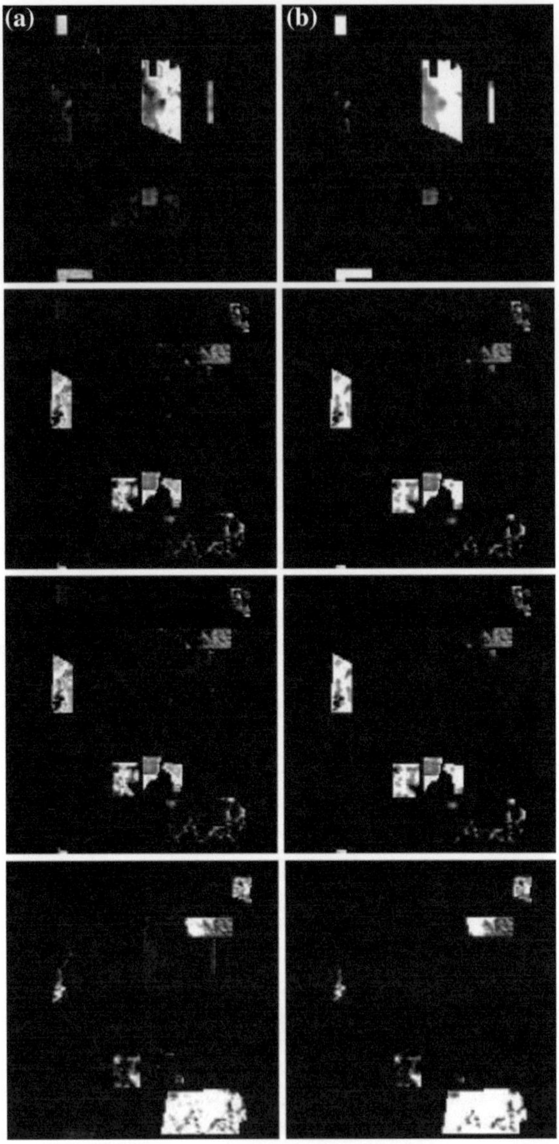

Fig. 4.16 Comparison of unmixing component diagram for four classes of land objects. **a** Traditional method. **b** Proposed method

It still adopts the agriculture and forestry remote sensing image to verify our model. We apply the band selection method of Wang and Jia (2009), and select 5 bands for the comparative experiment. These five bands are 17, 29, 41, 97 and 200.

In the first group of the experiment, centrally extract the former 100 pure pixels $\boldsymbol{a}(1)$, $\boldsymbol{a}(2)$, $\boldsymbol{a}(3), \ldots, \boldsymbol{a}(100)$ and $\boldsymbol{b}(1), \boldsymbol{b}(2), \boldsymbol{b}(3), \ldots, \boldsymbol{b}(100)$ respectively in two classes, i.e., corn and forest land. The mixing ratio is set as $\alpha_k = k \times 0.01 - 0.005$ and $k = 1, 2, \ldots, 100$. Then 100 mixed pixels are gained by the following formula:

Table 4.5 Comparison of training error and test error (Centrally selecting by training samples)

Method	VLSVM	FLSVM	LSVM	LSMA
Training error	0.0730	0.0737	0.0832	0.0879
Test error	0.0875	0.0894	0.0935	0.1276

$$\boldsymbol{m}(\mathrm{k}) = \alpha_k \boldsymbol{a}(\mathrm{k}) + (1 - \alpha_k)\boldsymbol{b}(\mathrm{k}), k = 1,2,\ldots,100 \tag{4.42}$$

Similarly, selecting 1000 pure pixels respectively in each class can get 1000 test mixed data. The unmixing error of various unmixing models is shown in Table 4.5. The unmixing error formula is:

$$e = \sqrt{\frac{1}{\mathrm{Ntr} \times \mathrm{ND}} \sum_{i=1}^{\mathrm{Ntr}} \sum_{j=1}^{\mathrm{ND}} e_{i,j}^2} \tag{4.43}$$

Here, $e_{i,j}$ is the single unmixing error. The training error of LSMA means the unmixing error of the model implementation in the training sample set. From Table 4.5, it can be seen that the effect of various methods is successively VLSVM, FLSVM, LSVM and LSMA from good to poor.

In the second group of the experiment, uniformly extract 100 pure pixels respectively in two classes, and then synthesize the training samples and test samples of the mixed data in accordance with the above mode. The unmixing error obtained from four methods is shown in Table 4.6. Like Experiment 1, VLSVM gains the best unmixing effect once again, and LSMA is still the worst. By comparing the training error variation situation in Tables 4.5 and 4.6, it can be found that different pure pixel extraction modes have smaller impact on the training error of LSVM, but have larger impact on LSMA, because the latter cannot embody the intra-class pixel distribution. When the training data are uniformly extracted, the distribution is more dispersed, resulting in poor training effect of LSMA. Then, by comparing the test expression of three SVM models in Tables 4.5 and 4.6, it can be found that, selecting the training samples well will fully express the entire class distribution, and can enhance the generalization ability of the model.

In the third group of the experiment, the total number of the unmixing classes is also increased to three classes (corn, forest land and newborn soybean), and nonlinear SVM with Gaussian kernel function also participates in the comparative experiment. The result in Table 4.7 shows that, the effect of nonlinear SVM model is better than that of the linear SVM model, which is consistent with the comparison situation in the traditional hard classification. Although the soft classification is not identical with the hard classification, both have the close connections. Upon the

Table 4.6 Comparison of training error and test error (Uniformly selecting by training samples)

Method	VLSVM	FLSVM	LSVM	LSMA
Training error	0.0781	0.0852	0.0896	0.1335
Test error	0.0839	0.0858	0.0909	0.1276

Table 4.7 Comparison of training error and test error (Uniformly selecting by training samples)

Method	VNSVM	FNSVM	VLSVM	FLSVM	LSMA
Test error	0.0677	0.0716	0.0784	0.0791	0.1858

comparison between this experiment and the above two experiments, it is easy to see that, with the increase in the number of classes, SVM has more significant advantages than LSMM.

Finally, for two unmixing problems in Experiment 2 and Experiment 3, it will make the effect comparison between the proposed method and three efficient multiple endmember spectral unmixing methods, i.e., Monte Carlo, Bundle and ViperTools methods. The result in Fig. 4.17 shows that, the proposed method has better effect than other comparative methods. Upon analyzing, it is known that, Bundle and ViperTools superior to the traditional LSMA method lies in automatically expanding the traditional fixed endmember of each class to a reasonable range. However, such method will have slightly poor ability of considering the class distribution due to selecting less training samples. Although Monte Carlo considers a large number of training samples, it is only the statistical average of the simple model and lack of the reasonableness and intelligence. SVM method considers more training samples, while the intelligence overcomes the deficiencies of Monte Carlo method and considers the class distribution most sufficiently, without the problem that the theoretical result is not unique.

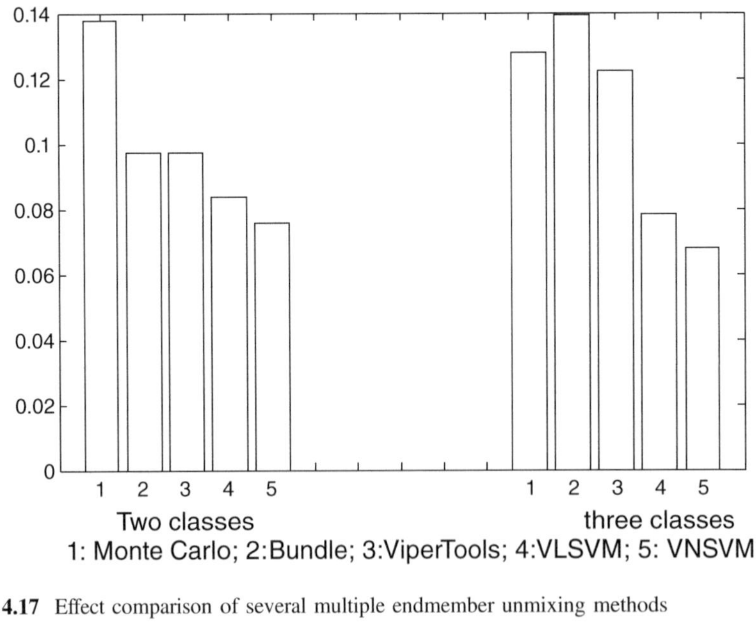

Fig. 4.17 Effect comparison of several multiple endmember unmixing methods

4.7 Fuzzy Method of Accuracy Assessment of Spectral Unmixing

4.7.1 Fuzzy Method of Accuracy Assessment

The unmixing performance assessment generally adopts the root mean square error:

$$\text{error} = \frac{1}{IK}\sum_{i=1}^{\text{Nc}}\sum_{k=1}^{\text{Ne}}(\beta_k(\omega_i)^* - \beta_k(\omega_i)) \tag{4.44}$$

Here, $\beta_k(\omega_i)^*, \beta_k(\omega_i)$ are the unmixing result and real supervised result of the k pixel. Nc is the total number of classes, and Ne the total number of pixels.

The crisp/deterministic assessment assumes that the spectral data is absolutely right. But the assumption is unrealistic due to the artificial error, registration error and intra-class changes. The development and evolution of the spectral unmixing model allow for this problem, and similarly the corresponding assessment criteria should also consider it. For this purpose, this section puts forward a fuzzy assessment mode in view of various indeterminacies.

For the hard classification problem of the continuously changed classes, the scholar establishes the fuzzy assessment method, which is finished by expanding the main diagonal elements of the corresponding confusion matrix (Strahler et al. 2006). In the traditional deterministic assessment, only the elements in the main diagonal of the confusion matrix are deemed as the correct classification. Moreover, in the fuzzy assessment criteria, the classification result of the minor diagonal elements is also considered to be acceptable. We promote and apply this mode to the soft classification assessment, and set up the fuzzy accuracy of the pixel level and the fuzzy mean error of the unmixing component level.

Express the allowable threshold value of the unmixing estimation error by T_u. The corresponding fuzzy mean error criterion is established as follows:

$$\begin{aligned}&\text{if}(\beta_k(\omega_i)^* - \beta_k(\omega_i)) < T_u\\&\text{then}\, e_k^{(f)}(\omega_i) = 0\\&\text{else}\ e_k^{(f)}(\omega_i) = (\beta_k(\omega_i)^* - \beta_k(\omega_i)) - T_u\\&\text{end.}\end{aligned} \tag{4.45}$$

The total fuzzy mean error is

$$e^{(f)} = \frac{1}{\text{Cn}}\sum_{i=1}^{C}\sum_{k=1}^{n} e_k^{(f)}(\omega_i). \tag{4.46}$$

Below we will establish the unmixing error matrix to set up the fuzzy accuracy criteria of the pixel level. Given the ground true resolution of the unmixing component is r_g%, there are $(100/r_g) + 1(=M)$ continuous change levels for each class.

Table 4.8 Description of error matrix ($r_u = r_g$)

Error matrix		Ground true value (%)					
		0	20	40	60	80	100
Thematic mapping value (%)	0						
	20						
	40						
	60						
	80						
	100						

These change levels are similar to each class of the continuous change in the above hard classification, while "class" continuity here is stronger. Thus, it is more applicable to establish the above fuzzy rules. Then, given the calculation resolution of the unmixing component is $r_u\%$, there are $(100/r_u) + 1 = (= N)$ continuous change levels. Thus, it can obtain an unmixing error matrix of N row M column. When $r_u = r_g$, the unmixing error matrix is the square matrix. Table 4.8 is the situation of $M = 6$ and $N = 6$, i.e., $r_u = r_g = 20$. When $r_u < r_g$, the unmixing error matrix is the rectangle matrix. Table 4.9 is the situation of $M = 6$ and $N = 11$, i.e., $r_u = 10$ and $r_g = 20(r_g = 2r_u)$.

The deterministic assessment criteria only deems that the blue black elements in Tables 4.8 and 4.9 are the pixels with the correct classification. Moreover, in the fuzzy criteria, the "classification" result of the light blue elements is also considered to be acceptable.

Ideally, the ground true resolution and the calculation resolution of the unmixing component are infinitely high, i.e., r_u and r_g are zero. At the moment, the unmixing error matrix turns into the square space, as shown in Table 4.10. The unmixing situation of some pixel determines the spatial position of the pixel falling in the unmixing error matrix. For example, the calculation component of the pixel relevant to some class is a, and the true component is b, and then the space coordinate of the pixel in the unmixing error matrix is (a, b). If the pixel strictly falling into the diagonal with "zero area" only has the correct "classification" in accordance with the deterministic assessment criteria, while the accuracy is generally zero in the probability. Moreover, in the fuzzy rule, the pixel falling into the diagonal centered banded area expanded by the upper and lower definite threshold values shall be deemed as the correct "classification".

Like the fuzzy assessment of the hard classification, it can establish the producer's accuracy and user's accuracy here.

Table 4.9 Description of error matrix ($r_u = 2r_g$)

Error matrix		Ground true value (%)					
		0	20	40	60	80	100
Thematic mapping value (%)	0						
	10						
	20						
	30						
	40						
	50						
	60						
	70						
	80						
	90						
	100						

Table 4.10 Description of error matrix ($r_u = r_g = 1$)

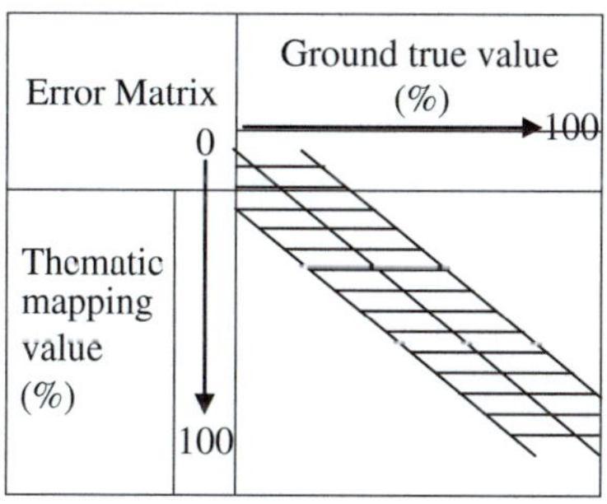

When the number of classes is 2, under the normalization constraints, the unmixing absolute error of each pixel relevant to Class 1 is exactly equal to the unmixing absolute error relevant to Class 2, so it only needs calculating a group of the fuzzy accuracy. When the number of classes is more than 2, it can calculate a group of the fuzzy accuracy for each class.

4.7.2 Application of Fuzzy Method of Accuracy Assessment in Experiments

Select 500 pure data in Sect. 4.5.1 once again, and use them to synthesize the mixed data.

1. Experiment 1

The first group of the experiment uses the synthesized hyperspectral data, and the resolution of the supervised component is 10 %.

The original 1000 units of 200-dimensional hyperspectral pure pixels are noted as $\boldsymbol{a}(1), \boldsymbol{a}(2), \boldsymbol{a}(3), \ldots, \boldsymbol{a}(500)$ and $\boldsymbol{b}(1), \boldsymbol{b}(2), \boldsymbol{b}(3), \ldots, \boldsymbol{b}(500)$. They are originated from Indian agriculture and forestry remote sensing data. The mixed data are synthesized as follows.

Given the mixed component is $\alpha_j = 0.1 \times j$ and $j = 1, 2, \ldots, 10$. For each α_j, generate 100 pairs of the random integers p_k and q_k between 1 and 500, and $k = 1, 2, \ldots, 100$. The mixed spectrum is synthesized as per the following formula:

$$\begin{aligned} \boldsymbol{m}_j(k) &= \alpha_j \boldsymbol{a}(p_k) + (1 - \alpha_j) \boldsymbol{b}(q_k), \\ k &= 1, 2, \ldots, 100, j = 1, 2, \ldots, 10. \end{aligned} \tag{4.47}$$

The random combination of the pure pixel simulates the spectral indeterminacy changes under the noise environment.

Among 500 pure pixels, 300 pixels are used for training, and other 200 pixels are used for testing with the mixed data. Upon comparison of 6 unmixing methods, they are the mean-based linear spectral unmixing model with fully constrained least squares (u_FCLS), Monte Carlo method (u_MC), cluster unmixing method (u_Bundle), ViperTool-based multiple endmember spectral mixture analysis (u_VT), linear SVM unmixing method (u_LSVM), and nonlinear SVM unmixing method (u_NSVM). The nonlinear SVM adopts the Gaussian kernel function. u_FCLS adopts the mean of the pure training pixel as the endmember. Other methods utilize the entire pure training pixel set as the reference spectrum library.

For the sake of visualization, the test data and unmixing result are displayed in the image form of 56 × 25, as shown in Fig. 4.19. Figure 4.19a is the true component diagram. The black pixels in the front eight rows are Class 2 pure class pixels, and the white pixels in the following eight rows are Class 1 pure class pixels. The others are the mixed pixels. The mixing ratio relevant to Class 1 is reduced to 0 from 90 %, and the decreasing step length is 10 % with decreasing one every four rows. The unmixing components of different methods relevant to Class 1 are given by Fig. 4.18b–g.

The absolute component error of the unmixing result is shown in Table 4.11. From Table 4.11, u_NLSVM method gains the best result.

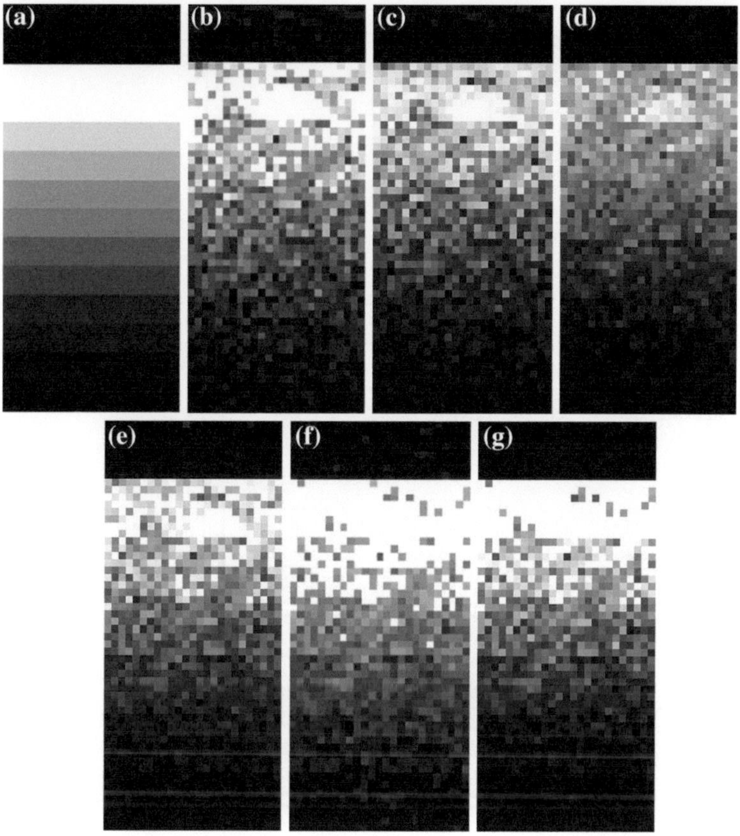

Fig. 4.18 **a** Test data; **b** u_FCLS test result; **c** u_MC; **d** u_Bundle test result; **e** u_VT test result; **f** u_LSVM test result; **g** u_NLSVM test result

Table 4.11 Unmixing error of synthesized data

Method	Deterministic component error	Fuzzy (10 %) component error
u_FCLS	0.0942	0.0286
u_MC	0.0883	0.0217
u_Bundle	0.0796	0.0178
u_VT	0.0712	0.0135
u_LSVM	0.0690	0.0129
u_NLSVM	0.0673	0.0121

In order to gain the assessment of the pixel level, we provide the error matrix diagrams of various methods, as shown in Fig. 4.19. The deterministic classification accuracy of the pixel level is calculated by the total number of the pixels in the main diagonal cell. In the fuzzy assessment criteria, the pixel in the minor diagonal cell is

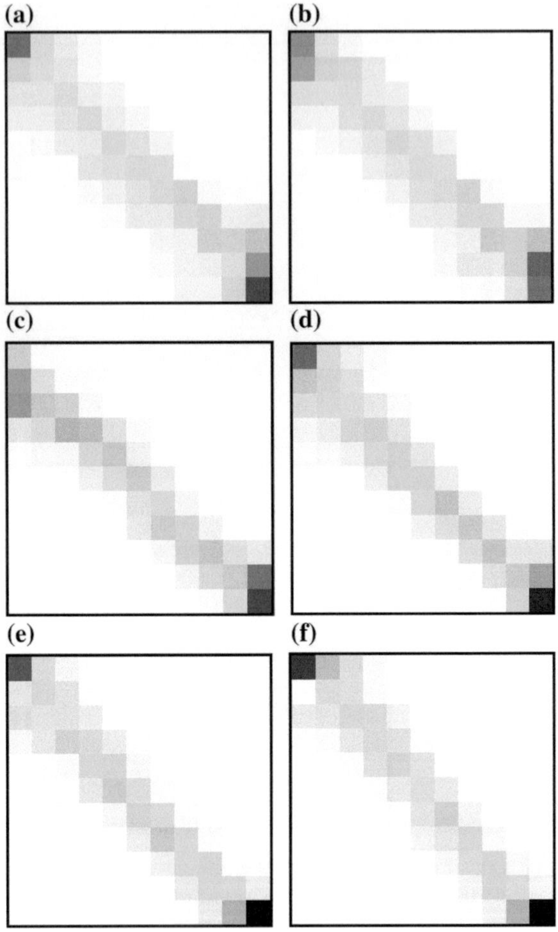

Fig. 4.19 Comparison of fuzzy error matrix of various methods under 10 % resolution: **a** u_FCLS; **b** u_MC; **c** u_Bundle; **d** u_VT; **e** u_LSVM; **f** u_NLSVM

also deemed as the acceptable pixel with the correct classification, so the fuzzy accuracy is certainly higher than the deterministic accuracy, as shown in Fig. 4.20.

2. Experiment 2

The second group of the experiment still adopts the synthesized data, and then selects the ground component resolution as 1 %. The synthesis mode of the mixed supervised data is the same as the previous group of the experiment.

Fig. 4.21 is the comparison of the unmixing reliability curve of various methods. The vertical axis is the unmixing absolute error, and the horizontal axis is the allowable error of the changes from 0 % (deterministic assessment) to 10 %. It can be seen that, u_LSVM and u_NSVM gain the best result.

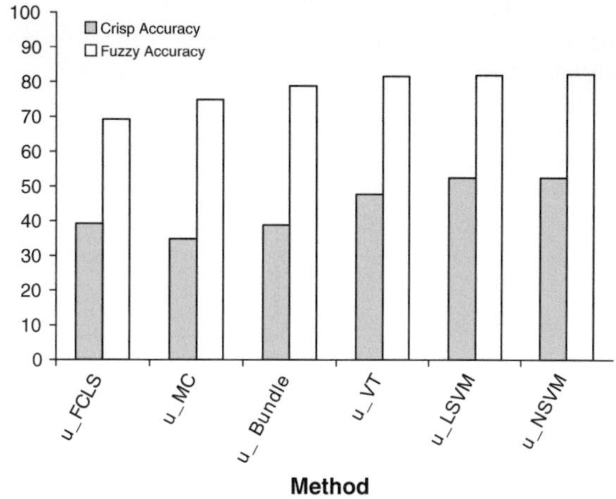

Fig. 4.20 Pixel accuracy assessment based on deterministic method and fuzzy method

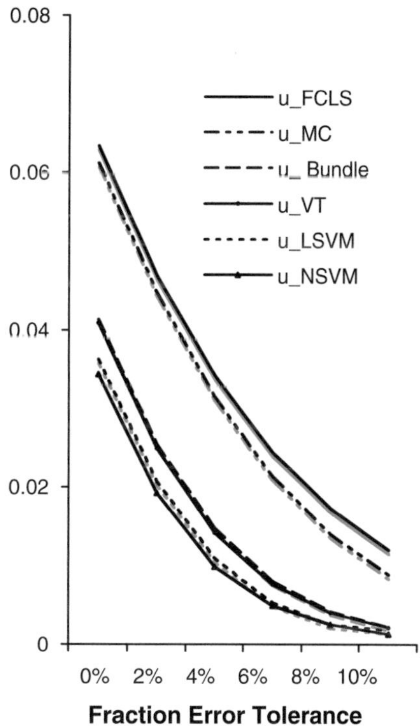

Fig. 4.21 Comparison of unmixing component error under given allowable error

Fig. 4.22 provides the fuzzy matrix when the calculation resolution is 1 %. Most pixels are distributed in the accessory of the diagonal, indicating that various methods gain the acceptable effects as a whole. According to these error matrixes, the fuzzy accuracy of the pixel level can be calculated, and the corresponding result

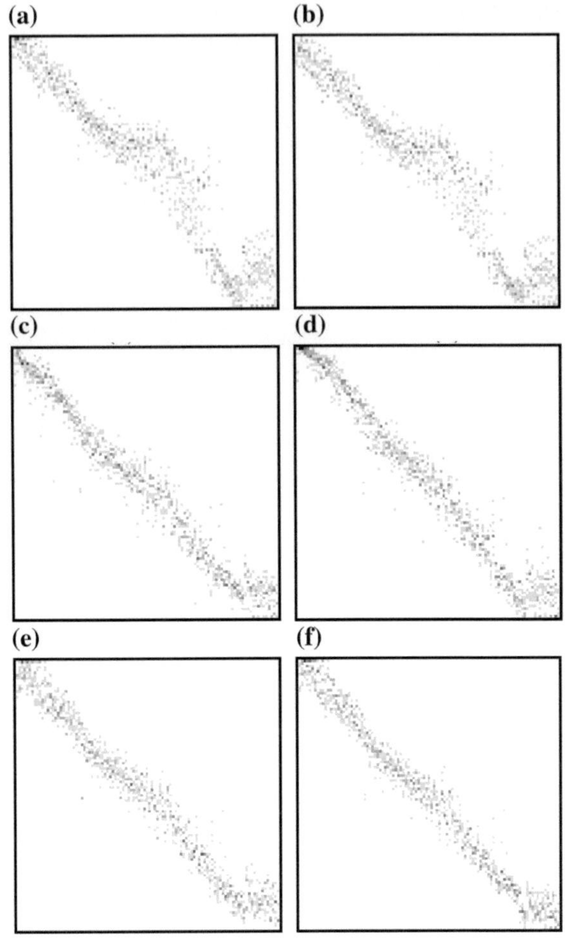

Fig. 4.22 Comparison of fuzzy error matrix of various methods under 1 % resolution: **a** u_FCLS; **b** u_MC; **c** u_Bundle; **d** u_VT; **e** u_LSVM; **f** u_NLSVM (1 % resolution)

is shown in Fig. 4.23. When the calculation resolution is relatively high, the deterministic classification accuracy is very low of less than 10 %. Moreover, the fuzzy accuracy is improved significantly. When the allowable error is 10 %, the classification accuracy of u_Bundle, u_VT, u_LSVM and u_NSVM is more than 90 %, and the accuracy of u_NSVM is the highest.

3. Experiment 3

In the third group of the experiment, Soybeans-min also participates in the experimental classes, i.e., the number of classes is increased to 3 classes. The reliability curve in Fig. 4.24 provides the comparison of the fuzzy accuracy of

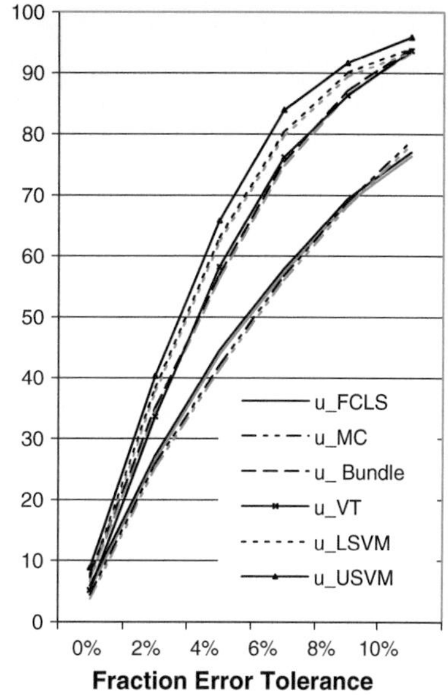

Fig. 4.23 Comparison of pixel level unmixing component error under given allowable error

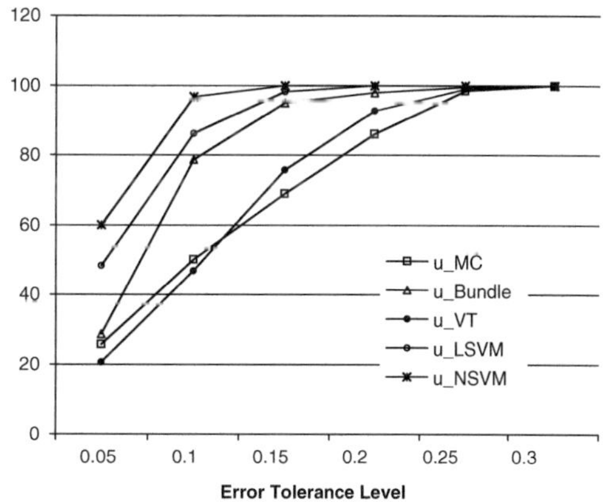

Fig. 4.24 Comparison of reliability curve of various unmixing methods

various methods when the allowable error is respectively 5, 10, 15, 20, 25 and 30 %. Similarly, SVM method gains the best effect.

For the running time, u_LSVM is the fastest, and the time for u_FCLS and u_NLSVM is equal, and they are relatively fast, with dozens of times than u_LSVM. The time for other methods is above thousand times than u_LSVM. Besides unmixing, the endmember extraction process of ViperTools is also time-consuming.

4.8 Chapter Conclusions

SVM-based spectral unmixing method is equipped with the advantages, such as the automatic extraction of the spectral endmember, flexible processing of the unmixing problems of the linear inseparable spectral endmember, and high unmixing accuracy.

One of the main contents in this chapter is the proposed spatial–spectral unmixing method, which breaks through the defects of the traditional method of unmixing purely by the spectral information and all priori classes, and gains favorable unmixing effects. The effective utilization of the spectral information is conducive to the related class selection, and accordingly decreases the unmixing error arising out of introducing the unrelated classes.

The other main content in this chapter is the proposed LSSVM unmixing new model with unmixing error constraints and the derivation of its closed form solution. Furthermore, the fixed endmember in the new model is replaced with the variable endmember for enhancing the performance. Generally speaking, SVM unmixing model is superior to LSMA, the unmixing error constraints are superior to the class information error constraints, the variable endmember is superior to the fixed endmember, and the nonlinear model is superior to the linear model. It is noted that, the constraint condition in the new model is increased to the times of bands of the original model and the dominant solution formula will be faced with the problems of the excessive matrix dimension, so it needs using in combination with the effective dimension reduction method.

In addition, in view of the indeterminacy of the spectrum acquisition and quantitative analysis, or in view of the necessary existence of the equipment error and human error, this chapter also researches two fuzzy accuracy assessment methods of the spectral unmixing, involving the fuzzy error in the unmixing component level and the fuzzy accuracy in the pixel level, aiming at providing a more reasonable unmixing assessment criteria for various unmixing methods and analysts.

References

Asner GP, Lobell DB (2000) A biogeophysical approach for automated SWIR unmixing of soils and vegetation. Remote Sens Environ 74:99–112

Bateson CA, Asner GP, Wessman CA (2000) Endmember bundles: a new approach to incorporating endmember variability into spectral mixture analysis. IEEE Trans Geosci Remote Sens 38:1083–1094

Brown M, Lewis H, Gunn S (2000) Linear spectral mixture models and support vector machine for remote sensing. IEEE Trans Geosci Remote Sens 38:2346–2360

Du Q, Chein-I Chang (2004) Linear mixture analysis-based compression for hyperspectal image analysis. IEEE Trans Geosci Remote Sens 42(4):875–891

Emami H (2005) Introducing correctness coefficient as an accuracy measure for sub pixel classification results. http://www.ncc.org.ir/articles/poster83/H.Emami.pdf

Fletcher R (1987) Practical Methods of Optimization. Chichester. Wiley, UK

Junwu L, Roger LK, Nicolas Y (2002) An unmixing algorithm based on vicinal information. Geosci Remote Sens Symp 3:1453–1455

Keshava N, Mustard JF (2002) Spectral unmixing. IEEE Trans Sig Process Mag 19(1):44–57

Qing H, Zhen X (1999) Neighbor field-based mixed pixel interpretation. J Northern Jiaotong Univ 23(4):118–121

Strahler AH, Boschetti L, Foody GM et al (2006) Global land cover validation: recommendations for evaluation and accuracy assessment of global land cover maps. http://nofc.cfs.nrcan.gc.ca/gofc-gold/Report%20Series/GOLD_25.pdf

Wang LG, Jia XP (2009) Integration of soft and hard classification using extended support vector machine. IEEE Trans Geosci Remote Sens Lett 6(3):544–547

Winter ME, Lucey PG, Steuter D (2003) Examining hyperspectral unmixing error reduction due to stepwise unmixing. Proc SPIE-Int Soc Opt Eng 5093:380–389

Chapter 5
Subpixel Mapping Technique of HSI

Spatial resolution means the minimum target that the sensor can distinguish, or the ground area expressed by a pixel point in the image. It is one of the important indexes of assessing sensor performance and remote-sensing information, and also the important basis of identifying the land object shape and size. The hyperspectral image has very high spectral resolution, but the spatial resolution is relatively low. The mixed pixels exist commonly, and can only show the fuzzy land object distribution information. It leads to greater difficulties for the corresponding application. The accurate land object distribution information has great significance with regard to military intelligence, industrial mining, agricultural planning, and environmental management. The increase of hyperspectral image spatial resolution can be realized by decreasing the size of the photosensitive elements and increasing the pixel density and sampling frequency using hardware, but it is very difficult to realize due to the high cost. Research on enhancing hyperspectral image spatial resolution by adopting image-processing techniques has been a very active project in the remote-sensing field.

In fact, the common aim of enhancing spatial resolution is to obtain the specific distribution of various land objects under higher spatial resolution, and subpixel mapping is exactly a technique for this purpose. The spectral unmixing (or soft classification) technique acquires the proportions of various land objects in the mixed pixel, but their respective distribution is still unknown. The subpixel mapping technique divides the mixed pixels as per the scale S: that is, a pixel is divided into S^2 subpixels, and then the subpixels are assigned to various land objects, to display the land object distribution in the higher resolution. For example, the mixed pixels include three land objects, such as A, B, and C, and the proportions are, respectively, 24, 36, and 40 %. If carrying out the division of $S = 5$, then 6, 9, and 10 subpixels are assigned to Class A, Class B, and Class C, respectively. The ultimate aim of subpixel mapping is to determine which subpixels are assigned to them. In essence, subpixel mapping is a hard classification (HC) technique at a higher scale. It has been successfully applied in various aspects, such as land cover drawing (Tatem et al. 2003), lake boundary extraction, and change detection.

In recent years, many scholars have been dedicated to subpixel mapping. Atkinson (1997) came up with the theoretical basis of subpixel mapping based on

L. Wang and C. Zhao, *Hyperspectral Image Processing*,
DOI 10.1007/978-3-662-47456-3_5

spatial correlation. It is assumed that the earth surface class distribution possessed the spatial correlation in the pixel and between the pixels; that is, the pixels with shorter distance are more likely to belong to the same type compared to pixels with longer distance. The basic principle is to maximize the spatial correlation between the pixels and in the pixel.

For the sake of understanding, the spatial correlation of pixel distribution is described here by a simple example. Figure 5.1 is a simple schematic diagram including two different land objects of A and B, shown by the black and white pixels, respectively, where (a) is an original low-resolution raster image with a 3 × 3 pixel, which can be acquired from the component diagram upon spectral unmixing. The digit marked on each pixel shows the percentage composition of Class A in the pixel, that is, its component value. If dividing each original low-resolution pixel into 2 × 2 (i.e., the amplification ratio of 2) subpixels, it can calculate the number of subpixels, respectively, in two classes by the percentage composition of Classes A and B of (a) in the original pixel. For example, after 75 % of the original pixel in A is decomposed into 2 × 2 subpixels, three subpixels should be assigned to A, and the remaining one subpixel should be assigned to B. (b) and (c) indicate the possible spatial distribution of two land objects after making the super-resolution partition for the original pixel. Moreover, we know that land object distribution in nature has continuity, and the land objects in the same class are more likely to get together. Compared with (b), the intrapixel and interpixel spatial distribution correlation under the distribution of (c) is greater. Thus, it can be considered that, in spatial correlation theory, the possibility of the subpixel distribution as shown in (c) is obviously greater than that in Fig. 5.1b, which is a more reasonable subpixel mapping result.

With the spatial correlation theory as the footstone, many subpixel mapping techniques have been developed. Verhoeye and De Wulf (2002) extracted the mathematical model of spatial correlation theory, transferred subpixel mapping into a linear optimization problem, and maximized spatial correlation among various subpixels in the neighborhood pixel and the central mixed pixel. Mertens et al. (2006) utilized the subpixel/pixel spatial attraction model (SPSAM), and calculated the attraction among various subpixels in the neighborhood pixel and the central mixed pixel, so as to determine the class of each subpixel. This method made the simple and valid solution for the mathematical model proposed by Verhoeye and De Wulf (2002). Ge et al. (2009) drew up the boundary line of various land objects directly in the central mixed pixel in accordance with the land object component in

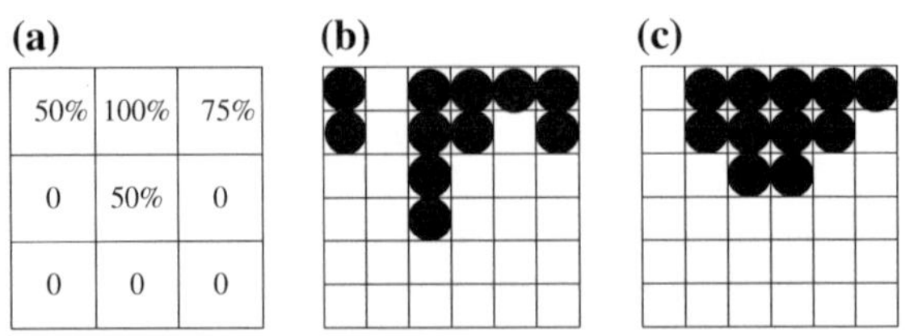

Fig. 5.1 Spatial correlation principle: **a** component value of Class A; **b** possible distribution 1 of Class A; **c** possible distribution 2 of Class A

the neighborhood pixel. Mertens et al. (2003a), Wang et al. (2006), and Zhang et al. (2008) learned the corresponding relation between the class value of each subpixel in the mixed pixel and various component values in the neighborhood pixel by prior information, that is, a high-resolution image, and an accordingly trained BP neural network (BPNN). The trained BPNN could be used for realizing the subpixel mapping of a low-resolution remote-sensing image. Tatem et al. (2001, 2002) and Nguyen et al. (2005) regarded the Hopfield neural network (HNN) as the energy tool, with each subpixel as a nerve cell, and various land object components in the mixed pixel as the bound term, and adopted the constraint energy minimum principle for the output nerve cell to solve the result of the subpixel mapping. Mertens et al. (2003b) put forward an objective function of embodying spatial correlation, which counted the total number of neighborhood subpixels with the same class to the subpixel in the mixed pixel, solved the objective function in virtue of the genetic algorithm, and searched the most probable spatial distribution of various classes in the mixed pixel. Later, Atkinson (2005) proposed the pixel swapping algorithm (PSA): during each iterative process, select two subpixels in great need of swapping class in the mixed pixel, and iterate until a stable mapping result is gained. Teerasit et al. (2005) and Tolpekin and Hamm (2008), Tolpekin and Stein (2009) adopted the Markov random field (MRF) model, and solved the probability of the subpixel assigned to each class.

This chapter introduces several new subpixel mapping techniques, including a subpixel mapping technique for a land class with linear features using the least square support vector machine (LSSVM), a subpixel mapping technique based on a modified spatial attraction model and a mixed spatial attraction model (MSAM), and a subpixel mapping technique using MRF with subpixel shifted remote-sensing images (SSRSI).

5.1 Subpixel Mapping for a Land Class with Linear Features Using a Least Square Support Vector Machine (LSSVM)

Some existing subpixel mapping methods, such as the Hopfield neural network model, PSA, genetic algorithm, and MRF model, need solving by definite iterations, with a larger calculated amount and longer times. Mertens et al. (2003) proposed utilizing the BPNN method and learning the corresponding relation between the class value of each subpixel in the mixed pixel and various component values in the neighborhood pixel by a priori information, and the trained BPNN could be used for realizing the subpixel mapping of a low-resolution remote-sensing image. However, the greatest deficiency of BPNN was, first, the prediction accuracy of the BPNN model was related to the initialization network weights, and the random initialization gave the output indeterminacy. Second, the BPNN had a relatively small rate of convergence, and was easy to get into the local optimum. Third, the BPNN training effect depended on a large number of training samples, and these samples were difficult to acquire sufficiently in the practical application. Such a series of defects resulted in taking away from the application values of the BPNN in the subpixel mapping technique.

It is undeniable that, for large-scale land objects such as buildings, agriculture and forestry land objects, and roads, they have regular spatial distribution characteristics, and show the obvious linear distribution state. For example, the land object boundary of the building is parallel to or perpendicular to the principal axis. For the sake of management and growth, the plantation of agriculture and forestry land objects has geometric laws, focusing on the square. Road construction is generally straight line or cross type. Therefore, according to these characteristics, the subpixel mapping for the land class with linear features can artificially synthesize the training samples, and accordingly get rid of complete dependence on a priori information. In addition, in view of the defects of BPNN in learning, this section puts forward the LSSVM-based subpixel mapping method. LSSVM can effectively handle high-dimension pattern recognition problems, small samples, and nonlinearity, and in addition, the training process time is very short. These advantages make the application of LSSVM in subpixel mapping possible.

5.1.1 Subpixel Mapping Based on the Least Square Support Vector Machine (LSSVM)

During the general SVM learning process, it does not include knowledge relevant to the geometric construction of the problem. However, as a matter of fact, the works of SVM theory have indicated that the learning process could reflect the cognition for the geometric property of the problem; that is, the important features could be formed by the interconnected pixels, and could greatly reduce the dimension of the eigenspace. The class value of each subpixel in the mixed pixel has close corresponding relations with the neighborhood pixel in the pixel. If the land object distribution under high resolution is known, these corresponding relations can be used for training an intelligent learning machine, and accordingly gain a subpixel mapping method based on the training mode.

First, here is a brief review of LSSVM. Given the data set $(\boldsymbol{x}_i, y_i), i = 1, \ldots, n$ composed of n points, $\boldsymbol{x}_i \in R^d$ is the input sample data, and $y_i \in \{1, 0\}$ is the corresponding class mark, that is, output. d is the input data dimension. The training process of LSSVM is aimed at obtaining the optimal classification hyperplane, and getting the decision function upon completion of training:

$$f(\boldsymbol{x}) = \sum_{i=1}^{n} \alpha_i K(\boldsymbol{x}_i, \boldsymbol{x}) + b \tag{5.1}$$

where $\boldsymbol{\alpha} = [\alpha_1, \alpha_2, \ldots, \alpha_n]^{\mathrm{T}}$, b is a constant, x is the test sample, and K is the kernel function. In the LSSVM model, $\boldsymbol{\alpha}$ and b are obtained by solving an equation, thus the overall training process is very fast.

Given are the number of classes C in the local window is 2, and a pixel $p_{i,j}$ is divided into 2 × 2 subpixels. In this case, construct four subclassifiers. In the single element subpixel mapping expression:

$$p_{i,j} \rightarrow \begin{bmatrix} p_{i,j}^1 & p_{i,j}^2 \\ p_{i,j}^3 & p_{i,j}^4 \end{bmatrix} \tag{5.2}$$

The present central task is the problem relevant to the construction of training samples. Given $\boldsymbol{x}_{i,j}$ and $y_{i,j}^k$ are, respectively, the input vector and output value of the k subclassifier, they can be given by the formula:

$$y_{i,j}^k = \begin{cases} 1, & \text{if } p_{i,j}^k \text{ belongs to target class;} \\ 0, & \text{otherwise} \end{cases}, \quad k = 1, 2, 3, 4. \tag{5.3}$$

$$\boldsymbol{x}_{i,j} = \frac{\left[F_{i-1,j-1}, F_{i-1,j}, F_{i-1,j+1}, F_{i,j-1}, F_{i,j}, F_{i,j+1}, F_{i+1,j-1}, F_{i+1,j}, F_{i+1,j+1}\right]^{\mathrm{T}}}{\left\| \left[F_{i-1,j-1}, F_{i-1,j}, F_{i-1,j+1}, F_{i,j-1}, F_{i,j}, F_{i,j+1}, F_{i+1,j-1}, F_{i+1,j}, F_{i+1,j+1}\right]^{\mathrm{T}} \right\|} \tag{5.4}$$

where $F_{i,j}$ is the component value of the target class in $p_{i,j}$. Each land object is successively deemed the target class, to enable the spatial distribution of each land object to be expressed as the training sample, for the purpose of LSSVM learning. Thus, each subclassifier obtains a pair of training samples, with the same input sample, and the 2 × 2 unit space upon outputting their corresponding partitions possesses the subpixels with the same spatial position. Once enough training samples have been obtained, they can be used for training each LSSVM. Two special training sample creation modes involving the local analysis method and normalization processing in formula (5.4) will make different images correspond to similar training samples. Thus, the training algorithm will have universality to a certain extent. If $Z = S \times S$ ($S > 2$, S is the amplification ratio), it needs construction of $S \times S$ LSSVM. We set forth the construction process of the training sample by a simple example.

In the case of a high-resolution image, the number of classes is Nc. It can acquire Nc binary images thereof. In the k (k = 1,2, …, Nc) image, according to each pixel assigned to the k class and non-k class, the grey values are marked 1 and 0. Do the resampling fuzzy processing for Nc binary images. As shown in Fig. 5.2, each 2 × 2 pixel in the known high-resolution image is fuzzy to a low-resolution pixel, and the grey value is the mean grey value of these four pixels. In this way, upon down-sampling, Nc low-resolution images are acquired, and the grey value of each pixel in these images is between [0, 1]. They can be regarded as the component diagrams, and the spatial distribution of four subpixels in each pixel of the image is known. It can acquire the input of training samples and four outputs by these synthesized component diagrams, and then train the LSSVM.

Upon completion of training, $\boldsymbol{\alpha}$ and b in formula (5.1) can be obtained. Then, according to the input of various test samples extracted from the low-resolution image, it estimates the class value of the subpixel, that is, output. It is noted that, for

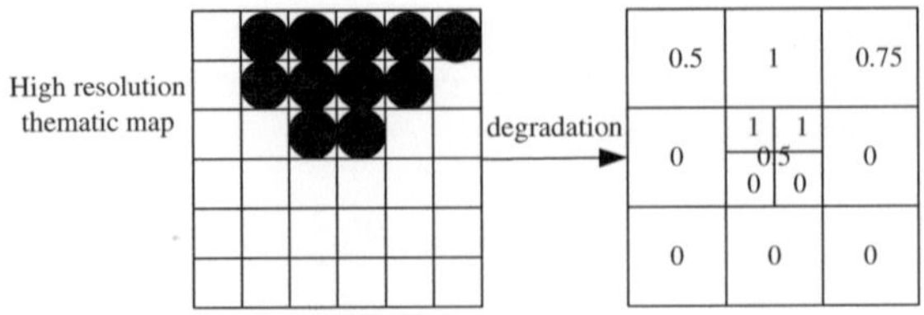

Fig. 5.2 Acquisition method of training samples

each class, the output in this process is not always the integer 1 or 0, but it represents the possibility of being assigned to such class. We can decide each subpixel as the class of the maximum corresponding output.

5.1.2 Artificially Synthesized Training Samples

As mentioned above, the LSSVM-based subpixel mapping method needs training sample information. In practical application, most often a priori information is unknown or difficult to acquire. However, for large-scale land objects involving buildings, agriculture and forestry land objects, and roads, they have regular spatial distribution characteristics, and show the obvious and regular linear distribution state in the overall vision. Below we analyze and verify the spatial distribution of two hyperspectral remote-sensing land class images.

The first image is the supervised classification information of agriculture and forestry land object hyperspectral images. The hyperspectral data include 16 classes except for the background. From Fig. 5.3, it can be seen that, in the distribution of 16 classes of agriculture and forestry land objects, the vast majority of them are straight line types, including horizontal, vertical, and oblique line directions. Also, there are some right-angle fold line shapes and another tiny minority of irregular shapes.

The second image is the pseudocolor image of Band 28, Band 19, and Band 10 of the San Diego military image. (b) is the local enlarged graph of some area in the

Fig. 5.3 Distribution diagram of Indian agriculture and forestry land class

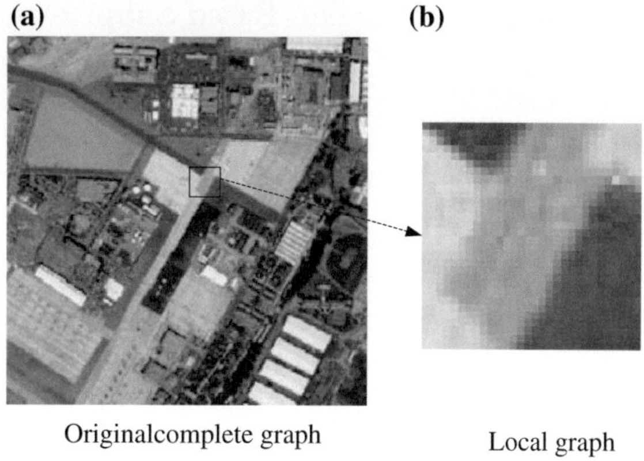

Fig. 5.4 Land class distribution of San Diego military image. **a** Original complete graph. **b** Local graph

middle. From Fig. 5.4, it can be observed that most large-scale land objects involving buildings and roads in the image show obvious linear and right-angle distributions.

In allusion to these geometric characteristics, we put forward a geometric method of synthesizing the training samples below, without any prior information regarding the overall process. Figure 5.5a is the sample synthesis mode of the straight line shape. In the local 3 × 3 window, a straight line (the expression is $y = \tan\theta \cdot x + b$) gets through the central pixel. Given Class 1 is covered under the straight line, whereas Class 0 is covered above it, the central pixel is divided into S^2 subpixels. $S_1, S_2, \ldots, S_9$, respectively, show the ratios of Class 1 in nine low-resolution pixels, similar to the component value acquired after the spectral unmixing. It is obtained by calculating the polygon area of the shaded part in each pixel. These nine component values are used for constituting the input vectors of the training samples upon normalization. The class values (i.e., for each subpixel, mark as Class 1 when the ratio of 1 is more than 50 %) of the quantized S^2 subpixels in the center pixel constitute the output of the training samples. When the angle $\theta(\theta \in [0, 360°))$ and the intercept b are changed, it can obtain more training samples. It is noted that, for each θ, b should be set in a corresponding interval, to ensure that the straight line goes by the central pixel. The synthetic method of the right-angle shaped sample is shown in Fig. 5.5b. The principle is the same as the straight line method. We do not discuss it here.

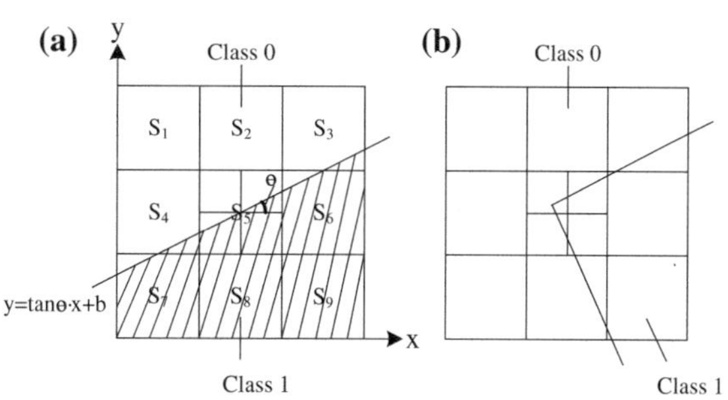

Fig. 5.5 Artificially synthesized method of training sample. **a** Straight line shape. **b** Right-angle shape

5.2 Spatial Attraction-Based Subpixel Mapping (SPSAM)

The SPSAM (Mertens et al. 2006) assumed that similar land objects attract each other, with specific physical significance, and realized the spatial correlation theory simply and effectively. The SPSAM directly calculated the spatial attraction between each subpixel and the neighborhood low-resolution pixel, and distributed the land object class for the subpixel as per the attraction, which was a simple subpixel mapping method easy to implement without a priori information. Above all, the model provided a powerful spatial attraction tool for the expression of spatial correlation.

Nevertheless, the SPSAM has two deficiencies. First, this method regards each neighborhood low-resolution pixel as an integral whole, without considering the specific spatial distribution of the subpixel in the neighborhood pixel. This attraction description is inaccurate. Second, this model only considers the correlation between the pixels, but neglects the correlation between the subpixels inside the pixel. These two points result in more noisy pixels and lower overall accuracy in the result of the subpixel mapping.

For this purpose, this section deems spatial attraction as the theoretical tool, first puts forward the subpixel mapping method based on a modified SPSAM (MSPSAM), and considers the spatial correlation between each subpixel in the neighborhood low-resolution pixel and the subpixel in the central pixel. Then it fuses the interior pixel attraction model, puts forward the MSAM in consideration of the interpixel and intrapixel correlation at the same time, fully describes the spatial correlation, and enhances the subpixel mapping accuracy.

5.2.1 *Subpixel Mapping Based on the Modified Subpixel/Pixel Spatial Attraction Model (MSPSAM)*

1. **Subpixel/Pixel Spatial Attraction Model (SPSAM)**

The mathematical model extracting the spatial correlation theory (Verhoeye and De Wulf 2002) formalizes the subpixel mapping problem into a linear optimization problem. It is given that the number of pixels in the low-resolution image is L, and the amplification ratio is S (i.e., each pixel is divided into S^2 subpixels). Then given the number of classes is Nc, and the number of subpixels assigned in Class c is C_c, define x_{cf} as

$$x_{cf} = \begin{cases} 1, & \text{if } p_f \text{ belongs to class } c \\ 0, & \text{otherwise} \end{cases} \tag{5.5}$$

The mathematical model of the problem is:

$$\max \quad J = \sum_{Nc=1}^{Nc} \sum_{f=1}^{LS^2} x_{cf} SD_{cf}$$

$$s.t. \quad \sum_{Nc=1}^{Nc} x_{cf} = 1, \quad f = 1, 2, \ldots, LS^2 \tag{5.6}$$

$$\sum_{f=1}^{LS^2} x_{cf} = C_c, \quad c = 1, 2, \ldots, C$$

Here, SD_{cf} is the correlation measure between the Class c land object in x_{cf} and the subpixel p_f, and it can be expressed as the weighted linear synthesis of the unmixing component value relevant to the Class c land object of the neighborhood pixel of the subpixel:

$$SD_{cf} = \sum_{k=1}^{N_A} w_k F_c(P_k) \tag{5.7}$$

where N_A is the number of neighborhood pixels, w_k is the spatial correlation weight of each neighborhood pixel, and $F_c(P_k)$ is the component value of Class c in the k neighborhood pixel P_k. The meaning of the model representation is that the solved subpixel mapping result should maximize the correlation of similar components between each subpixel in the mixed pixel and the neighborhood pixel on the whole.

Mertens et al. (2006) utilize the SPSAM to give a simple and effective solution for the above model. In the SPSAM, it determines the spatial distribution of the subpixel class by calculating the spatial attraction between each subpixel in the pixel and the corresponding neighborhood pixel. Note $p_{ij}\,(i,j = 1, 2, \ldots, S)$ as a subpixel in the pixel $P_{ab}(a = 1, 2, \ldots, L_a;\ b = 1, 2, \ldots, L_b;\ L_a$ and L_b are, respectively, the numbers of low-resolution image raster rows and columns). The sum of attraction of the subpixel p_{ij} by Class c components in the neighborhood pixel is:

$$D_{c,ij} = \sum_{k=1}^{N_A} w_k F_c(P_k) = \sum_{k=1}^{N_A} \frac{F_c(P_k)}{d_k h} \tag{5.8}$$

where h is a parameter, and d_k is the Euclidean distance between the geometric center of the subpixel p_{ij} and the geometric center of the pixel P_k. As shown in Fig. 5.6, it is the schematic diagram of distance calculation of the SPSAM method when the amplification ratio $S = 4$.

Finally, determine the subpixel assigned to Class c in P_{ab} in accordance with the size of $D_{c,ij}$. If $F_c(P_{ab})$ is the component value of Class c in P_{ab}, the $F_c(P_{ab}) \cdot S^2$ subpixels with the maximum corresponding attraction value $D_{c,ij}$ are assigned to Class c. In this way, it can guarantee that the sum of the spatial attraction of the similar components between each subpixel in the mixed pixel and the neighborhood, that is, J in formula (5.6), is the maximum.

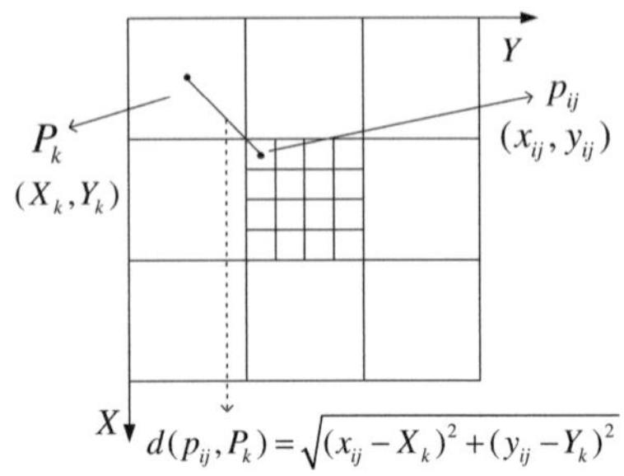

Fig. 5.6 Schematic diagram of distance calculation of the SPSAM method

2. **Modified Subpixel/Pixel Spatial Attraction Model (MSPSAM)**

From the above descriptions, it can be seen that the SPSAM method fails to consider the specific spatial distribution of the subpixel in the neighborhood pixel during the attraction solution, and this attraction description is inaccurate. As shown in Fig. 5.6, SPSAM regards all subpixels assigned to Class c in each neighborhood pixel as the centroid only with the mass (the corresponding class component value) without volume locating in the center of the pixel. While solving the attraction of some subpixel in the central pixel, it will directly calculate the centroid and the attraction of the subpixel. However, the volume of the centroid (the area of the corresponding pixel is S^2) is significantly larger than the volume of the other object (the area of the corresponding pixel is 1), which cannot be omitted. On the other hand, the distribution of the large object may not be uniform. For example, all subpixels in the interior Class c are distributed near some side or the right-angle corner of the square pixel, so it is improper to handle it as the centroid. While solving such problems of the nonignorable volume in physics, the calculus method is applied: that is, first dividing the object into many tiny elements (referred to as infinitesimal), solving the attraction of each infinitesimal on the other object, and then accumulating and summing. The reasonableness relevant to such a calculus method in physics has no lack of correlation theoretical exposition and experimental verification. The MSPSAM is proposed in accordance with this principle. Due to the digital image in the form of the raster, during the subpixel mapping, each subpixel after the super-resolution partition will store a gray value, corresponding to the class mark to be assigned. But, if considering an infinitesimal smaller than the subpixel, the calculation of the attraction will involve massive calculus changes of the mathematics, and the calculated amount is increased, which cannot add accuracy for the descriptions of the attraction. Thus, we can simply regard each subpixel as an infinitesimal for the purpose of calculating the spatial attraction.

Note p_m is the subpixel in the neighborhood pixel P_k of P_{ab}, and d_m is the Euclidean distance between the geometric center of the subpixel p_{ij} and the geometric center of p_m. In the MSPSAM, SD_{cf} in formula (5.8) turns into:

$$\begin{aligned} SD_{cf} &= \sum_{k=1}^{N_A} w_k F_c(P_k) = \sum_{k=1}^{N_A} w_k \frac{F_c(P_k)S^2}{S^2} \\ &= \frac{1}{S^2}\sum_{m=1}^{N_A S^2} w_k x_{cm} = \frac{1}{S^2}\sum_{m=1}^{N_A S^2} \frac{x_{cm}}{d_m} \end{aligned} \tag{5.9}$$

Furthermore, formula (5.6) can be equivalently transformed into the following problem solution for each pixel in the low-resolution image,

$$\max \quad J_{\text{between}} = \sum_{i=1}^{S}\sum_{i=1}^{S} A_{ij} \tag{5.10}$$

where

$$\begin{aligned} A_{ij} &= \sum_{m=1}^{N_A S^2} \frac{Z(p_m, p_{ij})}{d_m} \\ Z(p_m, p_{ij}) &= \begin{cases} 1, & \text{if } p_m \text{ and } p_{ij} \text{ belong to the same class;} \\ 0, & \text{otherwise} \end{cases} \end{aligned} \tag{5.11}$$

The constraint condition is the same as formula (5.6).

The solution steps of the MSPSAM-based spatial attraction are as follows:

1. Initialization: The SPSAM method can gain the spatial distribution of various subpixels without iteration and is a fast subpixel mapping method, therefore it can be used for the initialization step of the MSPSAM method.
2. Process of spatial attraction solution. Select some mixed pixel P_{ab} in order. Calculate the sum of the attraction of the subpixel p_{ij} by the subpixel assigned in Class c in the neighborhood pixel:

$$A_{c,ij} = \frac{1}{S^2}\sum_{m=1}^{N_A S^2} \frac{x_{cm}}{d_m} \tag{5.12}$$

   The obtained $A_{c,ij}$ is used for determining the class of each subpixel in the mixed pixel P_{ab}, and the specific step is the same as SPSAM.
3. Carry out Step 2 for each mixed pixel in the low-resolution image.
4. Repeat Steps 2 and 3, and iterate over and over again, until two subpixel mapping results have very small gaps or until the definite iterations.

Figure 5.7 shows the schematic diagram for the description of the spatial attraction of the subpixel p_{ij} by some neighborhood pixel P_k while utilizing the MSPSAM method, and the dark spot is the center of each subpixel. According to the specific distribution of various subpixels in the neighborhood pixel, MSPSAM solves the spatial attraction, equivalent to making the "differential" for the original low-resolution pixel to a certain degree for the purpose of solving the spatial attraction.

5.2.2 *Subpixel Mapping Based on the Mixed Spatial Attraction Model (MSAM)*

The MSPSAM has improved the description method of the SPSAM spatial attraction to a certain degree, but it still considers the spatial correlation between the pixels, without considering the intersubpixel spatial correlation in the pixel. It is mainly embodied in formula (5.6), without including the description for the correlation between the subpixels in the pixel. This section first extracts a spatial attraction model describing the correlation between the subpixels in the pixel based on deeply analyzing the pixel-swapping algorithm (PSA). Then it puts forward a MSAM of considering the spatial correlation between the pixels and between the subpixels in the pixel at the same time.

1. **Pixel-Swapping Algorithm**

The PSA maximizes the spatial correlation between the subpixel and the neighborhood subpixel through changing the spatial distribution of various subpixels in the pixel (Atkinson 2005). As a classic subpixel mapping technique, the PSA has been widely applied to various aspects such as landscape index calculation and agricultural land drawing. The original PSA randomly distributes the subpixels assigned to various classes in accordance with the component value of various classes in the mixed pixel. Shen et al. (2009) put forward a modified PSA (MPS) (i.e., regarding the subpixel mapping SPSAM result as the initial state of the PSA) and proved that the MPS gave a significant improvement in mapping accuracy and efficiency compared with the original PSA in a large number of experiments.

After initializing and gaining the initial spatial distribution of various subpixels, if considering the $c(c = 1, 2, \ldots, Nc)$ class, for each subpixel p_{ij}, the sum of all p_k attractions by the subpixel in the neighborhood is (for the sake of narration, here we assume that different classes have the attraction; i.e., whether p_{ij} is assigned to Class c or not, it will receive such attraction):

$$B_{c,ij} = \sum_{k=1}^{N_B} \lambda_k x_{ck} \tag{5.13}$$

where N_B is the number of neighborhood subpixels, and λ_k is the spatial correlation weight between the subpixels, with the expression of:

$$\lambda_k = \exp(\frac{-d_k}{a_0}) \tag{5.14}$$

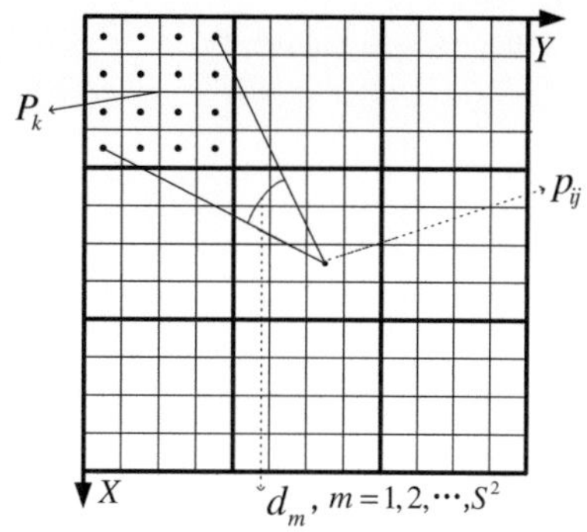

Fig. 5.7 Principle of spatial attraction solution by the MSPSAM

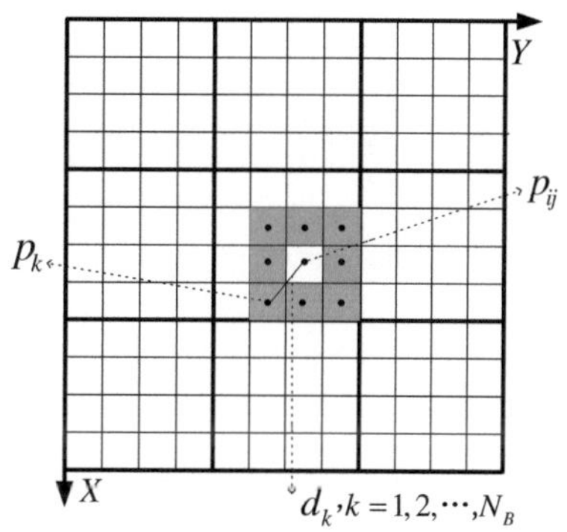

Fig. 5.8 Description of PSA attraction

In the formula, a_0 is a parameter, and d_k is the Euclidean distance between the geometric center of the subpixel p_{ij} and the geometric center of p_k. Figure 5.8 is the schematic diagram for the description of the attraction of the subpixel p_{ij} by the neighborhood pixel P_k (gray part) when the PSA solves the spatial attraction.

After obtaining $B_{c,ij}$ of various subpixels p_{ij}, operate with the pixel as the unit. The specific steps are:

1. Select the pixel P_{ab}, and find all spatial positions with the minimum attraction value in the subpixel assigned to Class c and all spatial positions with the maximum attraction value in the subpixel assigned to non-Class c. If the minimum attraction value is less than the maximum attraction value, the two subpixel classes are swapped; otherwise, they are not swapped.
2. Repeat Step 1 for each pixel.
3. Repeat the above process until the definite iterations, or until two iterative results have very small differences.

In fact, the above Step 1 can be equivalent to arranging each p_{ij} in P_{ab} in the order of $B_{c,ij}$ from the largest to the smallest, and gain the sequence of $sequence_0$. It has S^2 elements, and each element is the class attribute of each p_{ij} and is composed of the elements of "1" and "0", corresponding to Class c and non-Class c,

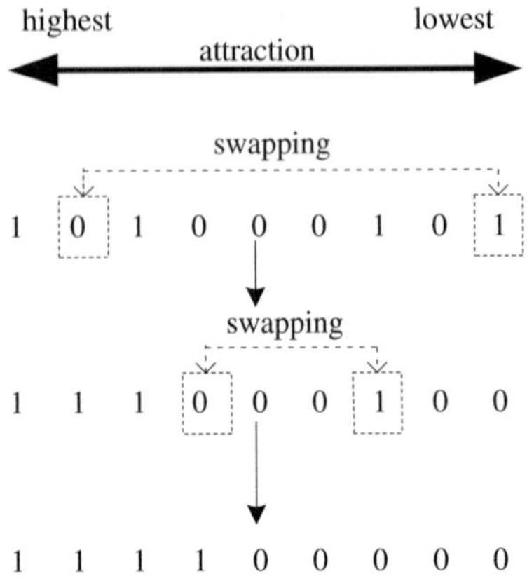

Fig. 5.9 PSA iteration-swapping process

respectively. In $sequence_0$, find the first "0" from left to right, and find the first "1" from right to left. If "0" is in front of "1", both are swapped mutually; otherwise, they are not swapped. We explain the PSA iteration swapping process by a simple example. As shown in Fig. 5.9, it is the iteration-swapping process of the sequence when the amplification ratio $S = 3$. It is easy to see from this figure that the ultimate aim of the PSA is that "1" is in the front, and "0" is in the rear in the final $sequence_0$. At the moment, in P_{ab}, the sum of the similar attraction $B_{c,ij}$ in the neighborhood by p_{ij} assigned to Class c (i.e., "1"), that is, $\sum_{i=1}^{S}\sum_{i=1}^{S} x_{c,ij} \cdot B_{c,ij}$, will be the maximum, thus is non-Class c (i.e., "0").

Upon considering all Nc classes comprehensively, it can gain the spatial attraction model of describing the intrapixel spatial correlation, and maximize the sum of the spatial attraction between the subpixels in the pixel:

$$\max \quad J_{\text{within}} = \sum_{i=1}^{S}\sum_{i=1}^{S} B_{ij} \tag{5.15}$$

where

$$B_{ij} = \sum_{k=1}^{N_B} Z(p_k, p_{ij}) \exp\left(\frac{-d_k}{a_0}\right) \tag{5.16}$$

2. **Mixed Spatial Attraction Model (MSAM)**

The intrapixel and interpixel spatial correlation of the earth surface class distribution assumed in accordance with the spatial correlation theory, and two spatial attraction models described, respectively, in the previous two sections can gain the MSAM of fusing the interpixel and intrapixel spatial attractions. The MSAM will solve the following problems for each pixel in the low-resolution image:

Fig. 5.10 Description of MSAM attraction

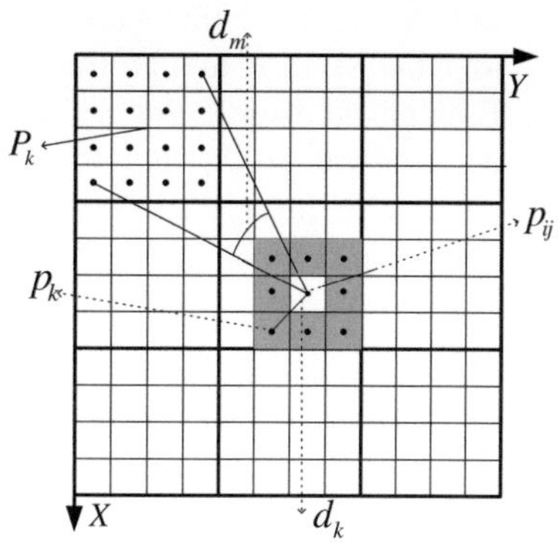

$$\max \quad J_{\text{intergration}} = \alpha_1 J_{\text{between}} + \alpha_2 J_{\text{within}} = \sum_{i=1}^{S}\sum_{i=1}^{S}(\alpha_1 A_{ij} + \alpha_2 B_{ij}) \tag{5.17}$$

where α_1 and α_2 are two weight coefficients. Figure 5.10 is the schematic diagram for the description of the spatial attraction of the subpixel p_{ij} by the neighborhood pixel P_k and the neighborhood subpixel p_k in the new model. Formula (5.17) can gain the subpixel distribution of maximizing $J_{\text{intergration}}$ by the "enumeration method" when S is smaller, whereas it needs to search the optimal solution of the formula with the help of the intelligent optimization algorithm when S is larger.

The genetic algorithm is an optimization calculation method based on the biological evolution theory of genetic selection and natural elimination, and searching the quasi-optimal solution of the problem by simulating the natural evolution process. Because such a classic optimization calculation is proposed, the theoretical and applied research on it have never stopped. At present the algorithm has been extensively applied to pattern recognition, image processing, and machine learning. The basic concept is that various possible solutions are randomly generated for a problem, and each solution is regarded as a chromosome, with the gene on the chromosome, and these chromosomes constitute a population. According to the expression of the optimization problem, calculate the fitness of each chromosome, adopt the selection operator to retain the large fitness and eliminate the small fitness, and update the population by the crossover and variation genetic operator, to make the population evolve towards the direction of the large fitness. After multigeneration evolution, the optimal individual in the last generation population will be deemed the optimal solution of the problem. In order that the optimal individuals in the successive generations are retained, it adopts an "elitism strategy," to ensure that the algorithm gradually converges towards the direction of the large fitness.

First, we illustrate the corresponding relation between the chromosome and the solution of the problem. It is given that the chromosome is expressed as $\mathbf{g} = [g_1, g_2, \ldots, g_{S^2}]$, and S^2 genetic values are the class values of S^2 subpixels in the pixel. Now, the solution steps of the MSAM are described as follows.

1. Initialization: It is the same as the MSPSAM and MPS methods. Here it also uses the SPSAM to gain the initial subpixel class distribution state.
2. Each mixed pixel P_m $(m = 1, 2, \ldots, M)$ in the low-resolution image generates an elite individual $\boldsymbol{G}_m$. It is the chromosome expressed by each subpixel class distribution in the P_m corresponding to the SPSAM result.
3. Select the mixed pixel P_m from the low-resolution image, and do the following steps.

   3.1. Randomly and initially generate the population with R chromosomes. The number of genes assigned to various classes in each chromosome will be distributed as per various component values in P_m.
   3.2. Calculate the fitness of each chromosome in the population as per formula (5.17), and find the optimal individual. If the fitness is greater than $\boldsymbol{G}_m$, it will become the elite individual on behalf of $\boldsymbol{G}_m$; otherwise, $\boldsymbol{G}_m$ will not be changed. $\boldsymbol{G}_m$ will cover the σR individuals with the smallest fitness in ratio of σ.
   3.3. Do the crossover and variation for each chromosome in the population. The number of genes assigned to various classes in the chromosome is strictly limited as per various component values, so the traditional mutation operator is undesirable. Here, it adopts the crossover operator, and randomly swaps the gene on the chromosome. The determination mode for the number of the gene swapping on each chromosome is as follows.
   3.4. Calculate the fitness of each chromosome in the population after crossover. For each individual, if the fitness after crossover is greater than the fitness before crossover, the crossover is allowed; otherwise, the original individual is not changed.
   3.5. Select the optimal individual from the population after crossover. If the fitness is greater than the fitness of $\boldsymbol{G}_m$, it will become the elite individual on behalf of $\boldsymbol{G}_m$; otherwise, $\boldsymbol{G}_m$ will not be changed.
   3.6. Carry out the evolution T times as per Steps 3.3–3.5.
   3.7. Upon evolution for T times, update the subpixel class distribution in the mixed pixel P_m by $\boldsymbol{G}_m$.

4. Do Step 3 for each mixed pixel in the low-resolution image.
5 After all mixed pixels are updated as per Step 3, the subpixel in the neighborhood pixel will be distributed again; that is, $J_{\text{intergration}}$ will be changed. Therefore, it is necessary to repeat Steps 3–4, and iterate H times. Finally it gains the subpixel mapping result in the new model. Figure 5.11 provides the overall flowchart of the algorithm.

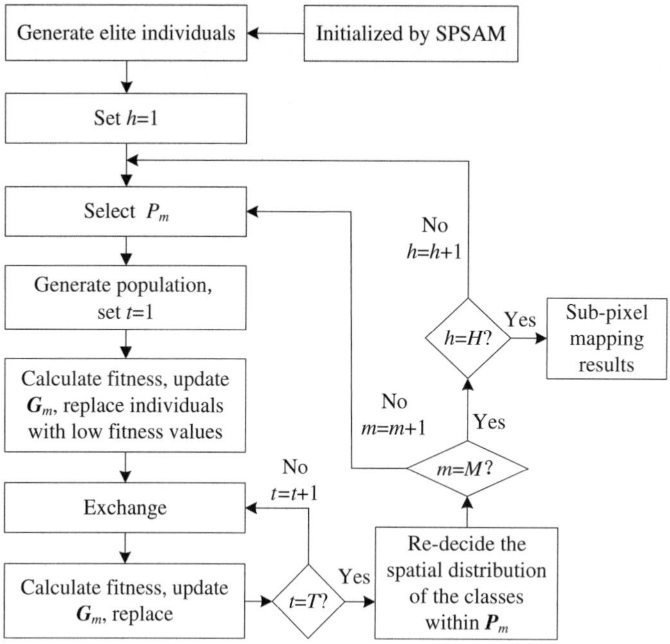

Fig. 5.11 Flowchart of the MSAM solution

5.3 Subpixel Mapping Using Markov Random Field with Subpixel Shifted Remote Sensing Images

5.3.1 Markov Random Field-Based Subpixel Mapping

Most of the existing subpixel mapping methods are similar to the postprocessing of the spectral unmixing technique; that is, these methods are strictly limited to belong to various types of subpixels based on various types of component values gained by spectral unmixing, which makes very high requirements on the accuracy of spectral unmixing. However, in fact, in the current development of the spectral unmixing technique, the accuracy of various techniques has not yet reached a fully dependable degree. As a result, spectral unmixing errors will inevitably be passed into the subsequent subpixel mapping process. Take a simple example to illustrate this problem. Suppose that Fig. 5.12a is the real component of a certain land class in each pixel in a 3 × 3 image, Fig. 5.12b is the true distribution (shaded area) of the corresponding certain land class after 4 × 4 division. If the size of the land class component obtained by the spectral unmixing technique in 9 pixels is shown as in Fig. 5.12c, the corresponding subpixel mapping results are displayed in Fig. 5.12d. It can be seen that gray subpixels in 3 pixels in the first line of (d) are errors introduced by the spectral unmixing, but these errors cannot be eliminated by the subpixel mapping technique fully dependent on those spectral unmixing results.

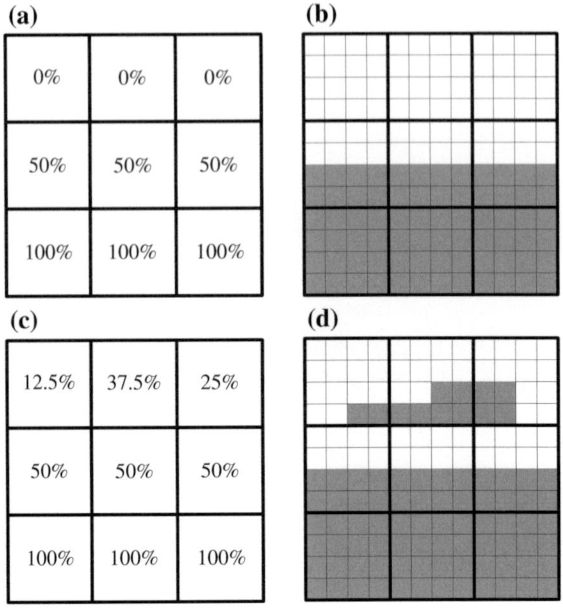

Fig. 5.12 Spectral unmixing error is passed into the process of subpixel mapping: **a** real component; **b** true distribution; **c** spectral unmixing results; **d** and **c** corresponding subpixel mapping results

Teerasit et al. (2005) proposed a MRF-based subpixel mapping method, which has the following characteristics and advantages.

1. It is used directly for multiband remote-sensing data and does not rely on the results of spectral unmixing (or spectral unmixing is just to get the initial random distribution of subpixels, but the result will be corrected in the MRF processing).
2. Take into account the spatial information and undetermined multiband remote-sensing data spectral information, so as to achieve both constraints.
3. The spectrum constraint term takes into account the differences within the class spectrum with the help of the covariance matrix, which is more sufficient compared to the general spectral unmixing technique on the excavation of spectral information.

Now the principle of MRF-based subpixel mapping is described below (Teerasit et al. 2005; Tolpekin and Hamm 2008; Tolpekin and Stein 2009).

Shilling:

Y	Original low-resolution observation image; the size is $M \times N$
X	High-resolution image; the size is SM × SN
$P\ (X)$	Priori probability
$P\ (Y \mid X)$	Conditional probability; the probability of Y when X is given
$P\ (X \mid Y)$	Posterior probability; the probability of X when Y is given

and:

$$P(X) = \frac{1}{Z_p}\exp(-U(X)/M) \tag{5.18}$$

$$P(Y|X) = \frac{1}{Z_l}\exp(-U(Y|X)/M) \tag{5.19}$$

$$P(X|Y) = \frac{1}{Z}\exp(-U(X|Y)/M) \tag{5.20}$$

of which, Z_p, Z_l, and Z are normalization constants, M is a parameter, $U(X)$, $U(Y|X)$, and $U(X|Y)$ are, respectively, corresponding energy functions of $P(X)$, $P(Y|X)$, and $P(X|Y)$. According to Bayesian theory, $U(X|Y) = U(X) + U(Y|X)$.

When Teerasit et al. (2005) use the MRF for subpixel mapping, $U(X)$ will be expressed with the Gibbs energy function, but this approach needs to extract the Gibbs energy function parameter from a priori information (such as the land class classification map under high resolution). But the information is generally difficult to obtain in the actual situation.

Tolpekin and Hamm (2008), Tolpekin and Stein (2009) use a space isotropic operator to describe $U(X)$, and the method does not require a priori information. Suppose that Q_j ($j = 1, 2, \ldots, B$, and B is the total number of pixels in the Y image) is a low resolution pixel in the observation image Y and is divided into S^2 subpixels q_{ij} ($i = 1, 2, \ldots, S^2$); $U(X)$ can be expressed as

$$U(X) = \sum_{j=1}^{B}\sum_{i=1}^{S^2}\frac{\alpha}{A_0}\sum_{k=1}^{n}\frac{\delta\left[c(q_{ij}), c(q_k)\right]}{d_k} \tag{5.21}$$

$$\delta[c(q_{ij}), c(q_k)] = \begin{cases} 0, & \text{if class attribute value } c(q_{ij}) = c(q_k) \\ 1, & \text{others} \end{cases} \tag{5.22}$$

of which $\alpha \in (0, +\infty)$ is a weight coefficient, n is the number of neighborhood subpixels, d_k is the Euclidean distance between the geometric center of q_{ij} and its neighborhood subpixel q_k, and A_0 is the normalization constant:

$$A_0 = \sum_{k=1}^{n}\frac{1}{d_k} \tag{5.23}$$

Each pixel Q_j corresponds to a spectral vector $\boldsymbol{y}_j$, which consists of gray values of Q_j under multiband remote-sensing data. It is usually supposed that $\boldsymbol{y}_j$ obeys the mean vector $\boldsymbol{\mu}_j$; the covariance is the normal distribution of $\boldsymbol{C}_j$. And the expression of $\boldsymbol{\mu}_j$ and $\boldsymbol{C}_j$ is:

$$\boldsymbol{\mu}_j = \sum_{l=1}^{L} e_{lj}\boldsymbol{\mu}_l \tag{5.24}$$

$$\boldsymbol{C}_j = \frac{1}{S^2}\sum_{l=1}^{L} e_{lj}\boldsymbol{C}_l \tag{5.25}$$

of which e_{lj} is the component value in the class i, and $\boldsymbol{\mu}_l$ and $\boldsymbol{C}_l$ are, respectively, the mean vector and covariance matrix of class i. The change of $c(q_{ij})$ will result in a change of e_{lj}, thus getting different $\boldsymbol{\mu}_l$ and $\boldsymbol{C}_l$. The expression of $U(Y|X)$ is:

$$U(Y|X) = \sum_{j=1}^{B}\left[\frac{1}{2}(\boldsymbol{y}_j - \boldsymbol{\mu}_j)'\boldsymbol{C}_j^{-1}(\boldsymbol{y}_j - \boldsymbol{\mu}_j) + \frac{1}{2}\ln|\boldsymbol{C}_j|\right] \tag{5.26}$$

It can be seen that the spectral differences in the class can be described with $\boldsymbol{C}_j$, which has important significance on the excavation of multiband remote-sensing data spectral information. $U(X|Y)$ can be obtained by formula (5.20):

$$\begin{aligned} U(X|Y) &= U(X) + U(Y|X) \\ &= \sum_{j=1}^{B}\sum_{i=1}^{S^2}\frac{\alpha}{A_0}\sum_{k=1}^{n}\frac{\delta\left[c(q_{ij}), c(q_k)\right]}{d_k} + U(Y|X) \\ &= (1+\alpha)\left[\frac{\alpha}{1+\alpha}\sum_{j=1}^{B}\sum_{i=1}^{S^2}\frac{1}{A_0}\sum_{k=1}^{n}\frac{\delta\left[c(q_{ij}), c(q_k)\right]}{d_k} + \frac{1}{1+\alpha}U(Y|X)\right] \end{aligned} \tag{5.27}$$

According to maximum posteriori probability theory, the optimal solution needs to make the maximum posteriori probability $P(X|Y)$, that is, the minimum energy function $U(X|Y)$:

$$\begin{aligned} X^{\text{opt}} &= \arg\left\{\max_X[P(X|Y)]\right\} = \arg\left\{\min_X[U(X|Y)]\right\} \\ &= \arg\left\{\min_X\left[\beta\sum_{j=1}^{B}\sum_{i=1}^{S^2}\frac{1}{A_0}\sum_{k=1}^{n}\frac{\delta\left[c(q_{ij}), c(q_k)\right]}{d_k} + (1-\beta)U(Y|X)\right]\right\} \\ &= \arg\left\{\min_X\left[\beta U_{\text{spatial}} + (1-\beta)U_{\text{spectral}}\right]\right\} \end{aligned} \tag{5.28}$$

of which $\beta = \frac{\alpha}{1+\alpha} \in (0,1)$ is the weight coefficient, which controls the balance between spatial and spectral constraint items. It is thus clear that the MRF can be used directly to handle multiband remote-sensing data.

5.3.2 Markov Random Field-Based Subpixel Mapping with Subpixel Shifted Remote-Sensing Images

It can be seen from Sect. 5.3.1 that in the original MRF-based subpixel mapping method, only the undetermined low-resolution image spectral information is used. In fact, the subpixel mapping is an under-constrained problem, with only spectral information of a single image for too few constraint conditions, which will generate multiple solutions to meet the existing constraint conditions. The additional information can be used to increase constraint conditions, which is an effective way to solve such problems.

Currently, there are some techniques to make use of additional information to increase constraint conditions and improve subpixel mapping accuracy. We roughly divide these techniques into two categories. One is to integrate additional multiband remote-sensing images, and the other is to make use of other forms of data and embed component values, that is, spectral unmixing results obtained from blending the image into the HNN proportion constraint item to provide multiple constraint conditions. However, the effect of HNN depends heavily on the results of spectral unmixing, and the component value errors will inevitably be passed into the subpixel mapping.

This section presents a method to make use of SSRSI spectral information to increase the MRF spectrum constraint conditions. The SSRSI is obtained by the observation satellite shooting the same scene repeatedly at different times. However, due to the track offset, the subpixel level displacement will inevitably exist between SSRSIs. Thus, SSRSI spectral information can be embedded into MRF-based spectral constraint items to provide multiple spectral constraints. Different from Ling et al. (2010) regarding the method of utilizing component value information after the SSRSI spectral unmixing to provide multiple proportion constraints of the HNN model, the method in this section does not need to carry out SSRSI spectral unmixing and can make use of multiband SSRSI spectral information directly.

This method is now described below. With SSRSI spectral information, (5.28) becomes:

$$X^{\mathrm{opt}} = \arg\left\{\min_X\left[\beta U_{\mathrm{spatial}} + \frac{1}{T}\sum_{t=1}^{T}(1-\beta)U^t_{\mathrm{spectral}}\right]\right\} \tag{5.29}$$

$$U^t_{\mathrm{spectral}} = \sum_{j=1}^{B}\left[\frac{1}{2}(\mathbf{y}^t_j - \boldsymbol{\mu}^t_j)'(\mathbf{C}^t_j)^{-1}(\mathbf{y}^t_j - \boldsymbol{\mu}^t_j) + \frac{1}{2}\ln|\mathbf{C}^t_j|\right] \tag{5.30}$$

of which T is the number of widths of the SSRSI image (including the undetermined observation image Y); Q^t_j is the pixel in t width of SSRSI; $\boldsymbol{y}^t_j$ is the corresponding spectral vector of Q^t_j and obeys the mean vector $\boldsymbol{\mu}^t_j$; and the covariance matrix is the Gaussian distribution of $\boldsymbol{C}^t_j$.

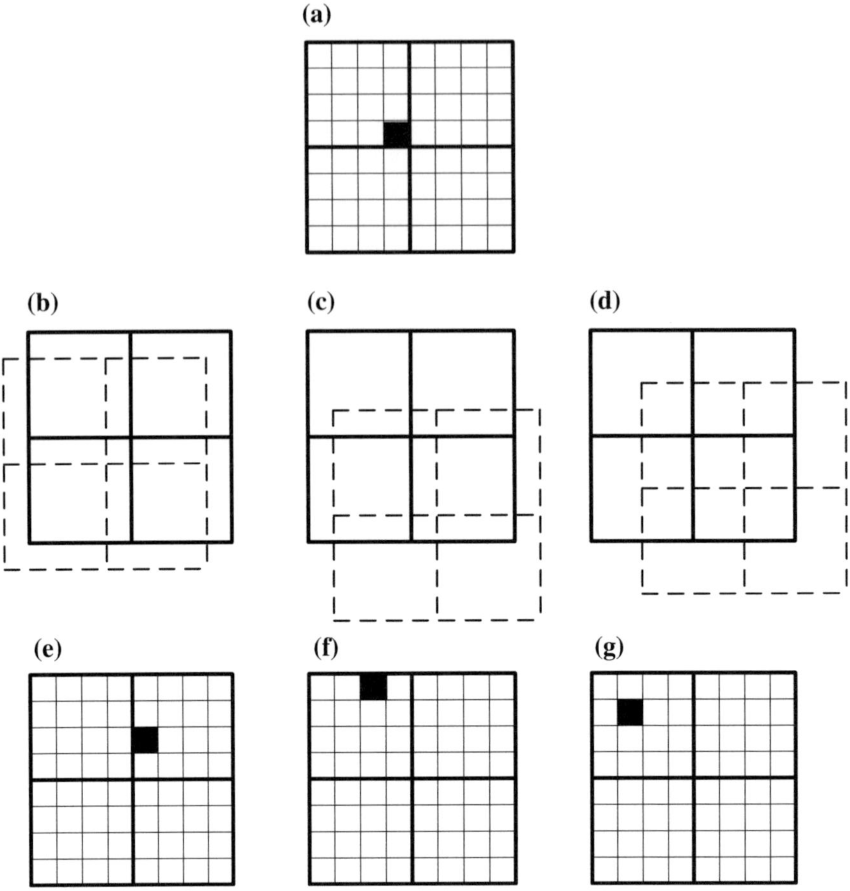

Fig. 5.13 Four 2 × 2 size of SSRSI (single band display): **a** the observation image; **b**–**d** respectively, with (–1,1), (1,3), and (2,2) subpixel displacement SSRSI; **b**–**d** the position of the *black subpixel* in three pieces of SSRSI, that is, (**b**)–(**d**)

Next take a simple example to illustrate how SSRSI provides multiple spectrum constraint conditions. It is assumed that there are four 2 × 2 size low-resolution SSRSIs. Figure 5.13 shows the four images in two-dimensional form (namely, single-band). (a) shows that each low-resolution pixel in the observation image is divided into 4 × 4 subpixels. The subpixel displacement exists between Figs. 5.13b–d and (a), with the displacement size, respectively, as (–1, 1), (1, 3), and (2, 2). Here (f_a, f_b) means to move f_a subpixels to the right and f_b subpixels down. In (a), the black subpixel is located within the pixel $Q^1_{(1,1)}$, and in (b)–(d) three SSRSIs, the black subpixel is located, respectively, within the pixel $Q^2_{(1,2)}$, $Q^3_{(1,1)}$, and $Q^4_{(1,1)}$, shown as (e)–(g). Thus, when using SSRSI, the black subpixel needs to meet not only the spectral constraint of $\boldsymbol{y}^1_{(1,1)}$, but also the spectral constraint of $\boldsymbol{y}^2_{(1,2)}$, $\boldsymbol{y}^3_{(1,1)}$, and $\boldsymbol{y}^4_{(1,1)}$.

Suppose that (f_c, f_d) is the coordinate of certain subpixels in X; it is known that the subpixel displacement of the t SSRSI is (f_{at}, f_{bt}). Then the corresponding low-resolution pixel coordinate (F_e, F_f) of the subpixel (f_c, f_d) in the t SSRSI can be calculated by the formula:

$$\begin{aligned} F_e &= \text{floor}((f_c - f_{at} - 1)/S) + 1 \\ F_f &= \text{floor}((f_d - f_{bt} - 1)/S) + 1 \end{aligned} \tag{5.31}$$

of which, floor (•) is a rounding function, and its value is the closest to the "•" but not more than its integer. That is, the subpixel at (f_c, f_d) needs to meet the spectral constraint from the pixel $Q^t_{(F_e,F_f)}$ in the t SSRSI.

Now the implementation steps combining the SSRSI and MRF subpixel mappings are described below:

1. Carry out spectral unmixing of the observation image so as to obtain component values of various types of land class.
2. Initialization: different from random initialization of Teerasit et al. (2005) and Tolpekin and Hamm (2008) and Tolpekin and Stein (2009); here make use of SPSAM to get the initial subpixel distribution status. In this way, uncertainty introduced by the random initialization can be eliminated, and the reasonable initial state can accelerate the convergence of the MRF model (Teerasit et al. 2005).
3. Select the pixel Q_j from the observation image, and for the subpixel q_{ij}, the probability belonging to category c_r ($r = 1, 2, \ldots, R$, and R is the total number of land class category) is:

$$P\left[c(q_{ij}) = c_r\right] = \frac{1}{Z}\exp\left(-\frac{1}{M}U_r\right) \tag{5.32}$$

$$U_r = \frac{\beta}{A_0}\sum_{k=1}^{n}\frac{\delta[c_r, c(q_k)]}{d_k} + \frac{1-\beta}{T}\sum_{t=1}^{T}U^t_{\text{spectral}} \tag{5.33}$$

   Different c_r corresponds to different $\boldsymbol{\mu}^t_j$ and $\boldsymbol{C}^t_j$ in U^t_{spectral}. The standard class of subpixel q_{ij} is carried out in accordance with probabilities.
4. Carry out Step 3 for each pixel in the observation image.
5 Repeat Steps 3 and 4. If after the first iteration, the percentage value Ra of the updated subpixel number accounting for the total number of pixels in the image is less than a set value D, then we can determine that the algorithm has converged. Meanwhile, in order to prevent the slow convergence process from consuming time, a certain number H of iterations can be set. Figure 5.14 shows the flowchart of the algorithm.

Note that the spectral unmixing in Step 1 is not necessary, and its use in combination with SPSAM is just to get the subpixel distribution status in the initial

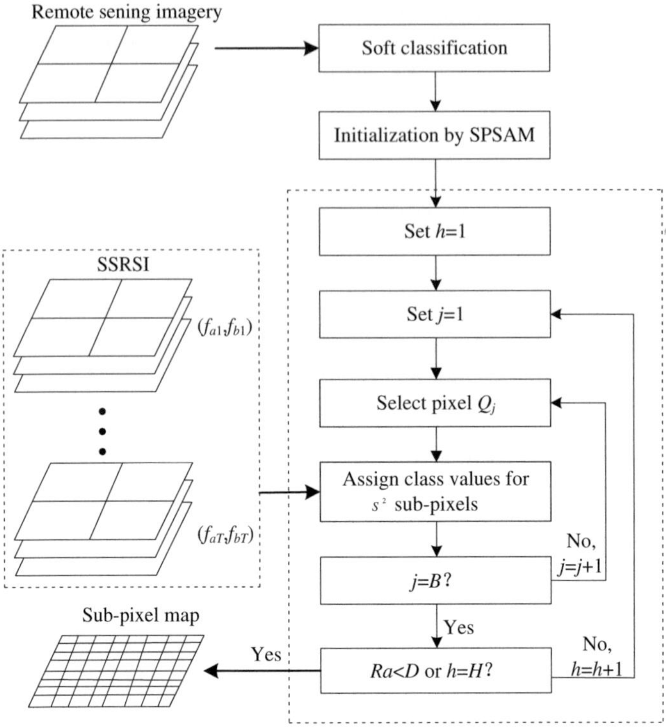

Fig. 5.14 Subpixel mapping flowchart combining the SSRSI and MRF

state. For the MRF, the excavation of spectral information is completed through (5.33). In other words, the corresponding components of various types of land class in the ultimate results of subpixel mapping will likely differ from the component values obtained by the spectral unmixing in the initial state.

5.4 Accuracy Assessment

5.4.1 Subpixel Mapping for Land Class with Linear Features Using the Least Squares Support Vector Machine (LSSVM)

In order to eliminate errors introduced in the rectification and spectral unmixing process, the proposed subpixel mapping method for a land class with linear features based on LSSVM has a more objective verification and evaluation. The experiment is carried out for a synthetic component diagram, which stems from resampling fuzzy processing on high-resolution diagrams and can be achieved through the mean filter. The sampling proportion is taken as 2; namely, each pixel in the

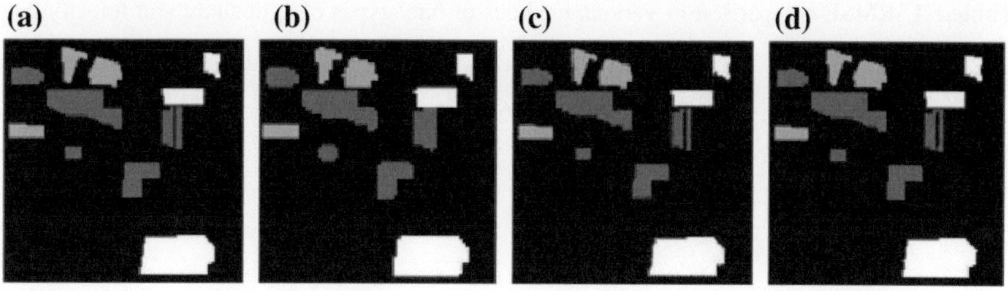

Fig. 5.15 Experimental results of various methods of three kinds of agriculture and forestry land class: **a** true high resolution diagram; **b** HC results; **c** BPNN results; **d** LSSVM results

low-resolution component diagram contains 2 × 2 pixels in high resolution, and similarly, the zoom ratio of subpixel mapping is taken as $S = 2$. In this way, refer to the known true figure to compare and evaluate the results of various methods. In this section, two sets of experiments were conducted. In each set of experiments, compare the HC method, BPNN, and LSSVM methods. The HC method determines each pixel as the corresponding category of the maximum mixing ratio value. The BPNN training sample method is the same as the LSSVM method, also stemming from synthesis. The LSSVM adopts the Gaussian kernel function. The input of the test sample is extracted from the low-resolution diagram (component diagram) with the same mode of construction as formula (5.4).

1. **Two Types of Experiments**

The set of experiments selects three categories from Indian agriculture and forestry land class figures: Class 2, Class 12, and Class 14. Each experiment contains two categories: target and background. Now the agriculture and forestry land class figures of three categories are put together in one figure, and the original true distribution of three categories is shown in Fig. 5.15a. Among them, black is the background, dark gray is Class 2, light gray is Class 12, and white is Class 14. Figure 5.15b is the result of the HC method, and it is clear to see that the edge is very rough with very low resolution compared to the original real figure. Figure 5.15c is the subpixel mapping result of the BPNN method, and many edge portions in the figure incur a "sawtooth" shape. Figure 5.15d shows the proposed result of subpixel mapping based on the LSSVM compared with the BPNN and HC; the effective and ideal recovery is carried out for the vast majority of straight lines and right-angle edge details, with the results closer to the original true figure and the effect better.

One indicator to be used more commonly for quantitative evaluation of two types of land class subpixel mapping is the root mean square error (RMSE), which is expressed:

$$\mathrm{RMSE} = \sqrt{\frac{\sum_{q=1}^{Np} (y_q - x_q)^2}{Np}} \tag{5.34}$$

Table 5.1 RMSE comparison of various methods of three types of agriculture and forestry land class

RMSE	HC	BP	SVM
Class 2	0.108	0.064	**0.034**
Class 12	0.062	0.034	**0.018**
Class 14	0.087	0.044	**0.009**

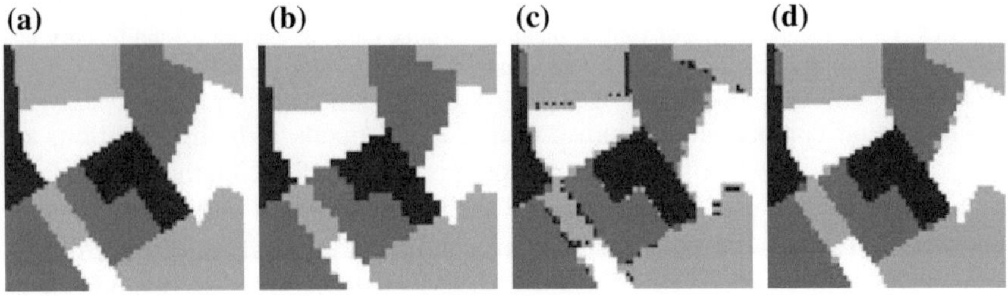

Fig. 5.16 Experimental results of various methods of multiclass land class diagram: **a** true high-resolution diagram; **b** HC results; **c** BPNN results; **d** LSSVM results

of which *Np* is the total number of the high-resolution image, y_q is the grey value of the pixel in the real figure, and x_q is the grey value of the pixel in subpixel mapping results. RMSE is a measure of the differences between the two figures: the smaller, the closer the two figures, indicating the better the subpixel mapping effect. Table 5.1 lists the RMSE comparison of various methods. The data in the table show that the RMSE of the BPNN and LSSVM methods is significantly lower than that of the HC method, indicating the reliability for learning of the synthetic training samples used. Meanwhile, in several ways, the LSSVM has the lowest RMSE (bold black data); compared to the BPNN; three categories of other RMSEs reduce 0.030, 0.016, and 0.035, respectively, proving that the effectiveness of the LSSVM is the best from the perspective of quantitative analysis.

2. **Multiclass Experiments**

In the second set of experiments, the land class distribution diagram of Tolpekin and Stein (2009) is carried out in a simulation experiment. The land class distribution diagram is shown in Fig. 5.16a. Figure 5.16b–d show, respectively, subpixel mapping results of the HC, BPNN, and LSSVM methods. Similarly, it can be observed that the HC method loses almost all edge details with the worst effect. In the BPNN method results, part of the junction of each category appears to be a misjudgment phenomenon, thus showing obvious noise. The LSSVM method has fewer misjudgments, but relatively speaking, to a large extent, improves the noise phenomenon in the BPNN method, with the results closer to the original true figure.

Accuracy evaluation of multiclass land class subpixel mapping generally uses the Kappa and PCC (percentage of correctly classified pixels) coefficient, and the greater the coefficient value, the better the effect. However, the existence of the

Table 5.2 Comparison of four coefficients of various methods of multiclass land class

Coefficient	HC	BPNN	LSSVM
Kappa	0.596	0.583	**0.626**
PCC	0.705	0.694	**0.726**
Kappa	0.310	0.220	**0.524**
PCC	0.491	0.415	**0.647**

large number of pure pixels (namely, unmixed pixels) in the original low-resolution diagram will make these two coefficient values too large, which cannot provide any useful information for the accuracy evaluation of subpixel mapping. In order to exclude the impact of these pure pixels, the Kappa and PCC coefficients are also used for evaluation, and these two coefficients only give statistics of the subpixel mapping accuracy of mixed pixels in the low-resolution diagram.

Table 5.2 lists the coefficient comparison of the Kappa, PCC, and Kappa and PCC of various methods. Comparative analysis shows that the effect of the BPNN and LSSVM methods is better than the HC method and four coefficients of LSSVM are the highest. Especially for the Kappa and PCC coefficients, the gap between several methods is reflected more clearly.

Also, in two sets of subpixel mapping experiments, in addition to the error and accuracy evaluation of the results, the time consumed is also an important factor of the two learning methods in LSSVM and BPNN evaluation. The test of learning algorithms (the decision process of the subpixel category) takes a very short time, therefore here only training times can be compared. The BPNN takes about 2 min to obtain network weights, but the LSSVM only takes not less than 10 s to calculate a and b in formula (5.1). Therefore, the LSSVM-based subpixel mapping method has certain realtime features.

5.4.2 *MSPSAM and MSAM*

Similarly, in order to eliminate the errors introduced in the rectification and spectral unmixing process, the experiments in this section are also carried out for the synthesized component diagram. This section provides a total of three experiments. In each set of experiments, compare SPSAM, MSPSAM, MPS, and MSAM methods.

1. **Experiment 1**

The first set of experiments is carried out on three analogue pictures: ring, cross line, and letter picture. Three pictures are shown in Fig. 5.17a. Each picture contains 112 × 112 pixels, including white target and black background, and a quadruple resampling process is conducted, namely, every 4 × 4 high-resolution pixel can be blurred as a low-resolution pixel. The fuzzy picture in Fig. 5.17b contains 28 × 28 pixels.

To facilitate visual comparison, subpixel mapping results are displayed in the error map. The gray represents the correct positioning pixel, the white means that

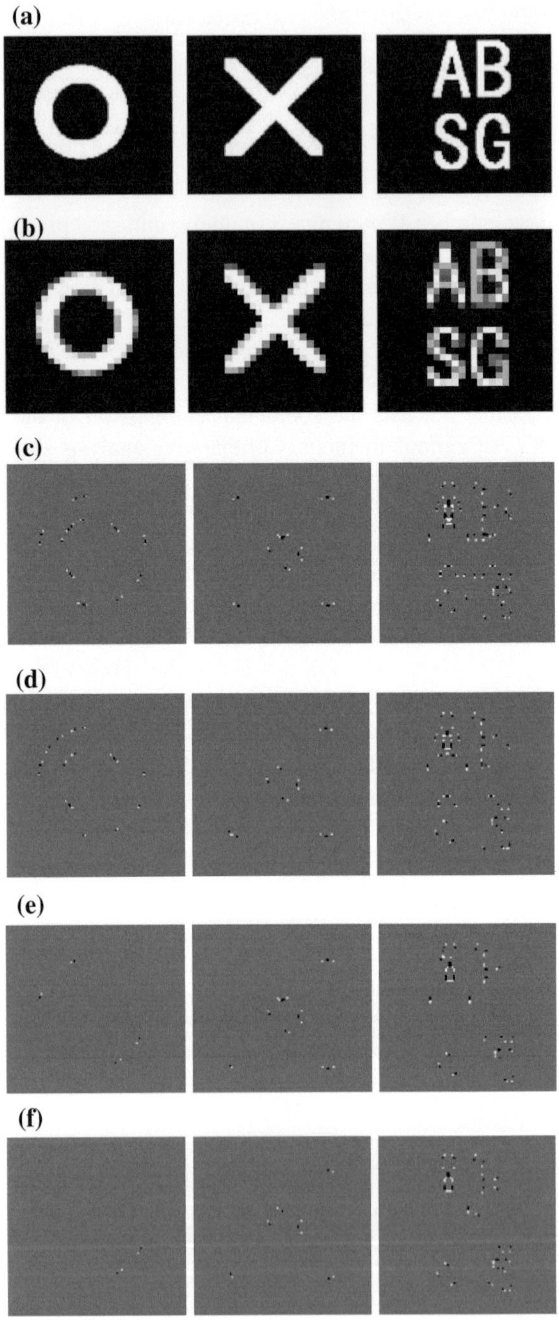

Fig. 5.17 Subpixel mapping errors of four methods in three analogue pictures: **a** real high-resolution map; **b** blurring pictures; **c** SPSAM error; **d** MSPSAM error; **e** MPS error; **f** MSAM error. The first column: *concentric circle*; the second column: *cross line*; the third column: *letter picture*

Table 5.3 EMP of three pictures in four methods

EMP	SPSAM	MSPSAM	MPS	MSAM
Ring	36	32	8	**4**
Cross line	26	22	24	**14**
Letter picture	146	118	80	**72**

the target is wrongly divided as background, and the black indicates that the background is wrongly divided as the target. Error maps of the SPSAM, MSPSAM, MPS, and MSAM methods are shown in Fig. 5.17c–f. When using SPSAM for positioning, more errors are generated at the junction of the target and the background, particularly in the letter picture. MSPSAM and MPS errors are smaller than those of SPSAM, but still larger than MSAM's.

We can also compare various methods by quantitative analysis. Here we directly get statistics of error-mapping pixels (EMP) between subpixel mapping results and the real figure. EMP of three pictures in the four methods as shown in Table 5.3. It can be seen that the EMP gap between three pictures is large, which is decided by the spatial structure of the target. In three shapes, the ring structure is most consistent with the spatial correlation. Thus, for the ring, the EMP of various methods is smaller than those of the other two shapes. In addition, when three simulation pictures use MSPSAM for subpixel mapping, compared to SPSAM, EMP reduces 4, 4, and 28, respectively, showing that the proposed MSPSAM is feasible. However, in four methods, the MSAM EMP is the smallest (black coarse data). The data evaluation shows that the proposed MSAM can produce more precise subpixel mapping results compared to the other three methods.

2. **Experiment 2**

The second set of experiments selects the land class distribution diagram from the article of Tolpekin and Stein (2009). This diagram is 60 × 60 pixels, showing the distribution of four categories of land class in a certain agricultural area of Flevoland in the Netherlands. For simplicity, two types of land class (Class 1 and Class 2) in the original figure are merged here. After such treatment, the new land class figure covers three categories of land class, shown as Fig. 5.18a, labeled C0, C1, and C2, and corresponding to black, gray, and white portions, respectively. The quadruple and sixfold resampling is carried out to obtain the low-resolution image, shown in Fig. 5.18b, c, respectively. Figure 5.18d–g show subpixel mapping results of SPSAM, MSPSAM, MPS, and MSAM, respectively, under quadruple magnification. Figure 5.18h–k show subpixel mapping results of SPSAM, MSPSAM, MPS, and MSAM, respectively, under sixfold magnification.

Refer to the true figure in 5.18a; the SPSAM method results have many misclassified pixels, but the situation is slightly weakened in the results of the MSPSAM and MPS methods. Compared to the real figure, it is not difficult to find that these two methods still have many errors, especially in the sixfold amplification process. However, in the MSAM method results, the vast majority of land class edges in two scales obtain a satisfactory recovery, and the resulting land class distribution diagram is closest to the real figure among the four methods.

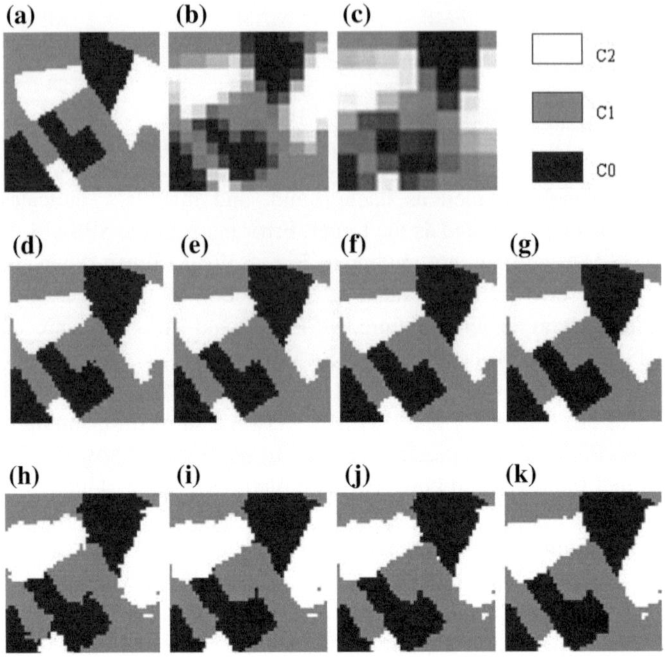

Fig. 5.18 Subpixel mapping results of Experiment 2: **a** original reference diagram; **b** low-resolution figure after quadruple resampling; **c** low-resolution figure after sixfold resampling; **d**–**g** results of the SPSAM, MSPSAM, MPS, and MSAM methods when $S = 4$; **h**–**k** results of the SPSAM, MSPSAM, MPS, and MSAM methods when $S = 6$

In addition to the visual comparison, Fig. 5.19 shows Kappa, PCC, and Kappa and PCC of various subpixel mapping methods under two scales. When comparing Kappa and PCC coefficients, the gap between several methods is reflected more clearly. When the sixfold proportion of the subpixel mapping is carried out, the accuracy of the four methods is reduced, which is due to a more complex subpixel mapping process along with a larger scale; in each low-resolution pixel the large number ($S^2 = 36$) of subpixels needs to be classified, with increasing uncertainty. Comparing the data in the two figures, it is easy to see that compared to the SPSAM, MSPSAM improves positioning accuracy to a certain extent, proving once again that the proposed MSPSAM is reasonable and feasible. Furthermore, the proposed MSAM has the highest accuracy in the four subpixel mapping methods.

3. **Experiment 3**

In order to validate further the advantages of the proposed method, the third set of experiments selects a real remote-sensing image for analysis. The land class picture shows part of the Nanjing urban area, with its color original picture shown on the website http://www.ceode.cas.cn/txzs/dxyy/. Figure 5.20a is the distribution diagram of the land class in the region. The size of this figure is 80 × 80 pixels,

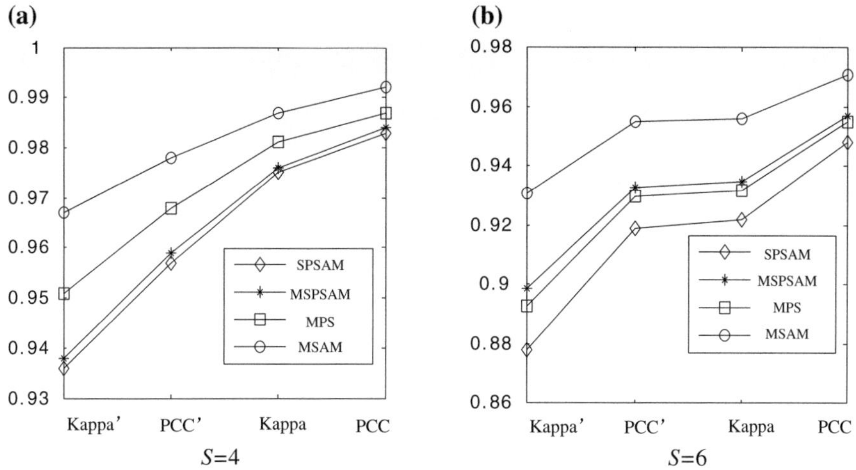

Fig. 5.19 Quantitative evaluation of various subpixel mapping methods in Experiment 2

covering three categories of land class: vegetation, lake water, and urban area. The same as in Experiment 2, the quadruple and sixfold resampling are carried out to obtain the low-resolution images shown in Fig. 5.20b, c, respectively. Under quadruple magnification, the subpixel mapping results of SPSAM, MSPSAM, MPS, and MSAM are shown in Fig. 5.20d–g, respectively; under sixfold magnification, the subpixel mapping SPSAM, MSPSAM, MPS, and MSAM results are shown in Fig. 5.20h–k, respectively.

A similar conclusion to the front can be obtained from the comparison between Fig. 5.20d, h, namely, the SPSAM effect is worse and many pixels are misclassified. MSPSAM improves this phenomenon to some extent, but the effect is limited. In real remote-sensing images, the distribution of land class is more complex, and as the distance increases, the correlation between pixels decreases rapidly. That is, for real remote-sensing images, the spatial correlation of land class among low-resolution pixels may not be as good as that in the previous two sets of experiments. Under such circumstances, MPS has a greater advantage than MSPSAM, because MPS considers the correlation with its nearest neighbor subpixels, which is also one reason for the effects of Fig. 5.20f, j being better than Fig. 5.20e, i. However, the spatial correlation between low-resolution pixels still exists and cannot be ignored, even if this correlation is weak. Based on the observation and comparison of results of MPS and MSAM, it can be found that MSAM can produce better subpixel mapping results, especially in the observation process of distribution of the vegetation type of land class; differences between the two methods are more obvious.

Figure 5.21 shows Kappa, PCC, and Kappa and PCC of various subpixel mapping methods under two scales. By quantitative evaluation, we can see that the accuracy of the MSPSAM is better than that of the SPSAM and the MSAM accuracy is the highest.

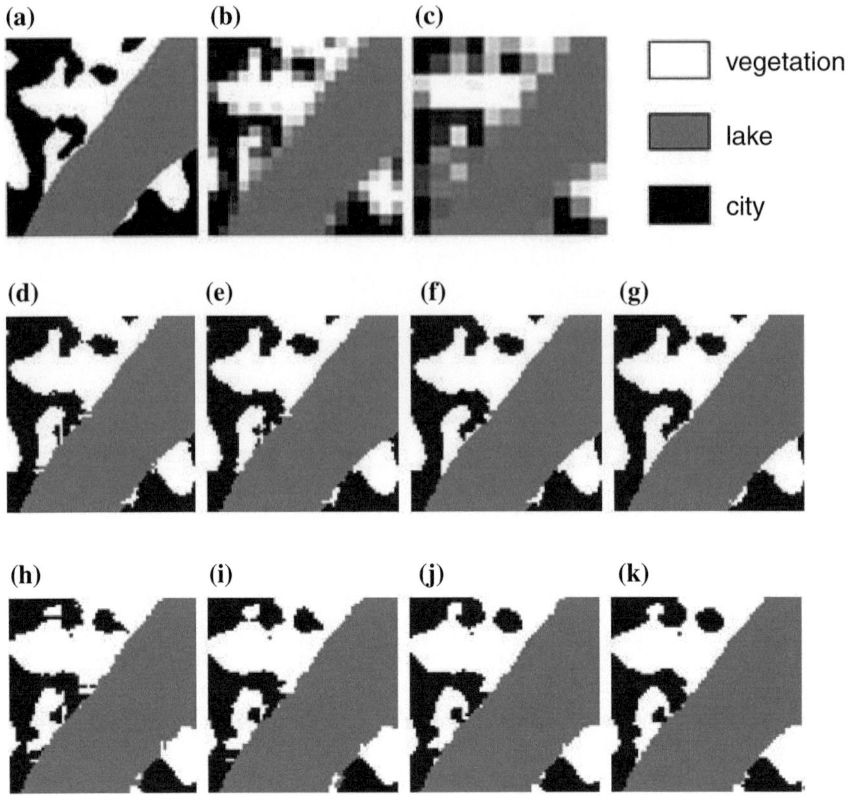

Fig. 5.20 Subpixel mapping results of Experiment 3: **a** original reference diagram; **b** low-resolution figure after quadruple resampling; **c** low-resolution figure after sixfold resampling; **d–g** results of the SPSAM, MSPSAM, MPS, and MSAM methods when $S = 4$; **h–k** results of the SPSAM, MSPSAM, MPS, and MSAM methods when $S = 6$

5.4.3 MRF-Based Subpixel Mapping with Subpixel Shifted Remote-Sensing Images

Experiments are carried out on simulated hyperspectral data, which is convenient for the proposed algorithm to conduct an objective evaluation and analysis. The distribution diagram of land class still uses the land class map in Sect. 5.4.2, Experiment 2, as shown in Fig. 5.22. Three categories of land class C0, C1, and C2 cover 1011, 1710, and 879 pixels, respectively. The spectral data are selected from part of the hyperspectral remote-sensing test area of the Indiana Forestry in the northwestern part of Indiana shot in June 1992. It can be inferred from formula (5.30) that the computational complexity will give rise to quadratic growth along with the number of bands. Therefore, it is necessary to conduct appropriate band selection to reduce computational complexity. Here we use the band selection

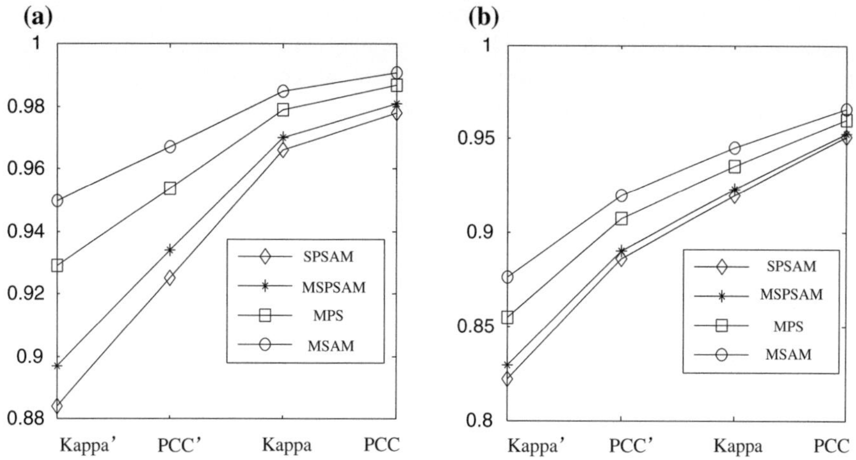

Fig. 5.21 Quantitative evaluation of various subpixel mapping methods in Experiment 3: **a** $S = 4$; **b** $S = 6$

Fig. 5.22 Distribution diagram of three categories of land class

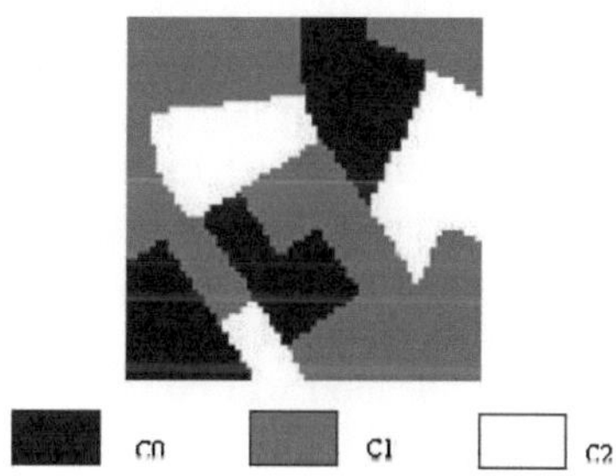

method of Wang et al. (2007) to select five bands: 17, 29, 41, 97, and 200. Then, we randomly select 1011, 1710 and 879 spectral samples from three categories of land class (newborn corn, baby corn, and newborn corn) of hyperspectral images, respectively, corresponding to C0, C1, and C2 pixels. In this way, simulated hyperspectral data can be synthesized.

We use the mean filter to carry out the resampling band by band on the original high-resolution analogue hyperspectral images, so as to obtain hyperspectral images under low resolution. Here we discuss four scales: $S = 4$, 5, 6, and 10. The original high-resolution land class distribution diagram can be used to evaluate the performance of the subpixel mapping method. The SSRSI is obtained by manually setting subpixel displacement on the original high-resolution hyperspectral images.

Five subpixel mapping methods are compared in the experiment, namely, SPSAM, HNN, HNN combining SSRSI, traditional MRF, and MRF combining SSRSI. T is set to 4, that is, choosing 4 pieces of SSRSI, and the corresponding subpixel displacement is set at (0, 0), [floor (S/2), 0], [0, floor (S/2)], and [floor (S/2), floor (S/2)], respectively. First, conduct spectral unmixing to get all kinds of land class components. The linear spectral mixture analysis (LSMA) method is widely used because of its clear physical meaning and convenient operation. We use LSMA for spectral unmixing, and the resulting component diagram of three categories of land class under four scales is shown in Fig. 5.23. Pure white corresponds to 100 % of components and pure black corresponds to 0 % of components. It can be observed from these component diagrams that the proportion information under low resolution cannot describe the spatial distribution of land class, which also shows the necessity of the subpixel mapping technique. After soft classification, the land class distribution diagram obtained by using a variety of subpixel mappings is shown in Fig. 5.24.

Comparing the land class distribution reference diagrams in Fig. 5.22, it can be found that many wrong subpixels exist in the SPSAM result; these pixels appear as noise and are pit-like, and the boundary line between various types of land class is not easy to identify. Especially when S increases, the boundary information is almost lost (e.g., $S = 10$). This is because the SPSAM strictly depends on the results of soft classification, making errors generated by soft classification be transmitted directly to the subpixel mapping and not be eliminated. Slightly different from the SPSAM, in the result of the HNN method, the noise pixels decrease. The HNN is an optimization tool in essence. In the subpixel mapping model based on HNN, various types of component values are used to limit the number of subpixels belonging to various types of land class as the constraint entry. In the iteration process, the attribute value of each class (between 0 and 1) changes constantly, and the HNN energy function gradually decreases in stability. Ultimately, quantify all types of attribute values to obtain the land class HC map under high resolution, namely, the subpixel mapping result. In this result, the ratio constraint conditions provided by soft classification will be satisfied to a greater degree but not strictly. Thus, some significant noise and pit-like shapes will disappear. However, by comparison with the reference diagram, many wrong divisions still exist in the HNN method results. After all, the HNN still takes the proportion information of soft classification as the constraint condition. In addition, it is clear that the HNN combines SSRSI results and the SSRSI can produce more satisfactory results, which also stresses the importance of SSRSI on the subpixel mapping technique. The subpixel mapping method based on the MRF model eliminates most isolated pixels. Compared to the SPSAM on the resulting land class diagram, HNN and HNN combining the SSRSI method are more reasonable and accurate. However, in the conventional MRF model results, there is still part of the pit shape and isolated pixels. Especially when $S = 10$, the region covered by the C0 and C2 land class at the lower left corner is now classified into C1. This is because that as S becomes large, the subpixel mapping process becomes more complicated, and more subpixels in each low-resolution pixel need to be classified to make uncertainty increase. Obviously in this case, the effect of the traditional MRF model is limited to the single spectral constraint. It is observed that the effect of the positioning result generated by the MRF combining SSRSI method is very significant. Compared with the traditional MRF, more land class boundary lines have been effectively restored. In $S = 10$, the effect of this improvement is the most prominent. The result of the proposed method

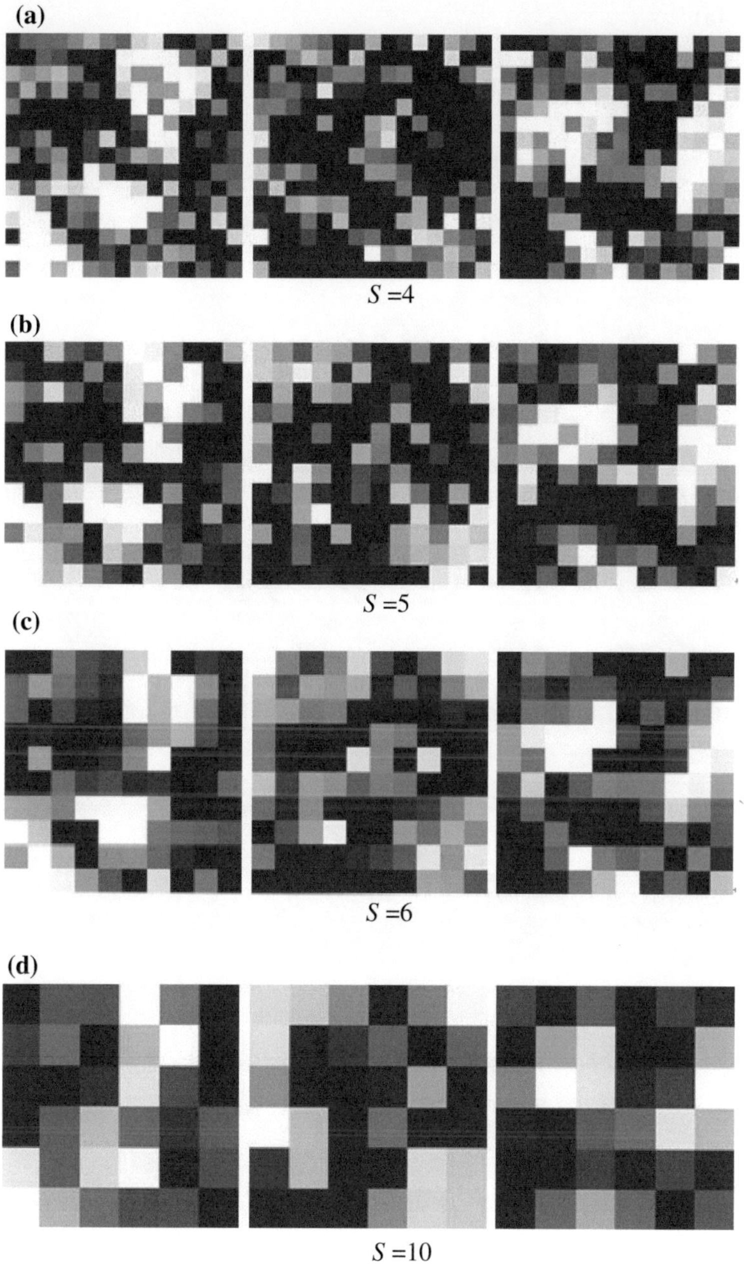

Fig. 5.23 LSMA spectral unmixing results. From left to right: C0, C1, and C2

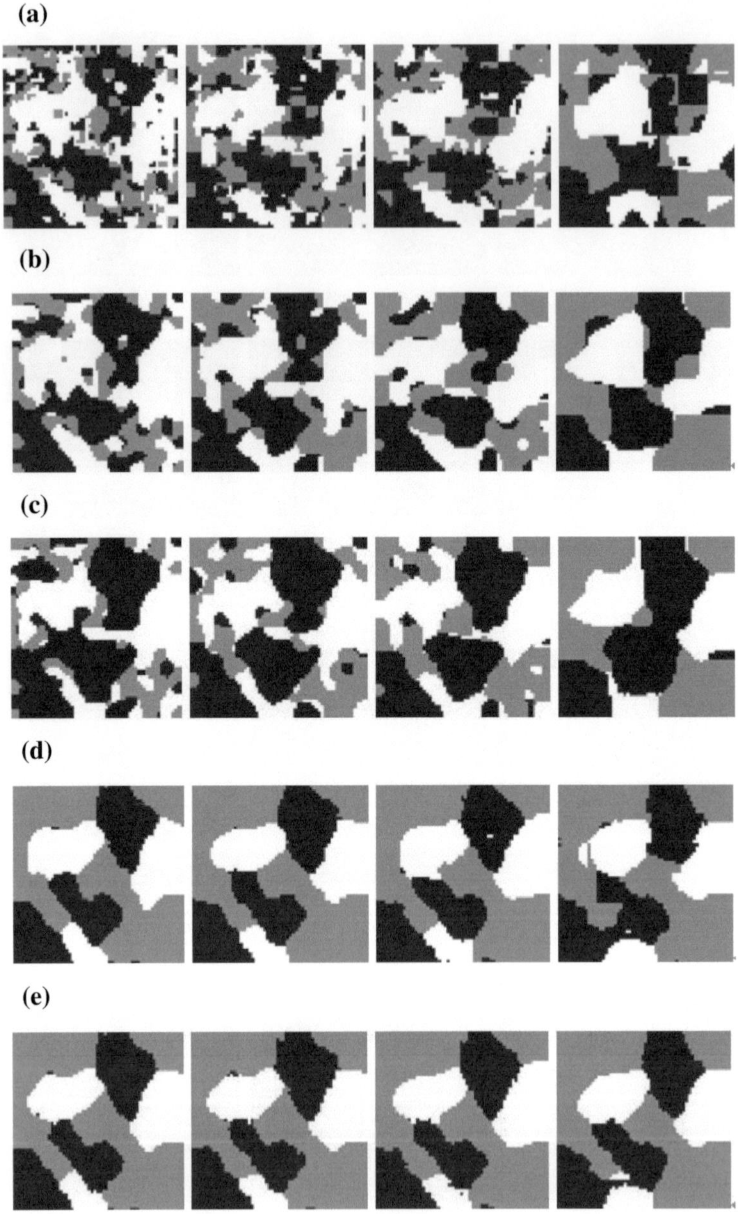

Fig. 5.24 Results of five kinds of subpixel mapping methods in four kinds of fuzzy proportion. From left to right: S = 4, 5, 6, 10; **a** SPSAM; **b** HNN; **c** HNN combining SSRSI; **d** traditional MRF; **e** MRF combining SSRSI

Table 5.4 Kappa coefficient evaluation of results of five kinds of subpixel mapping methods

Proportion	SPSAM	HNN	HNN + SSRSI	MRF	MRF + SSRSI
$S = 4$	0.434	0.494	0.490	0.887	**0.889**
$S = 5$	0.491	0.568	0.571	0.865	**0.875**
$S = 6$	0.524	0.589	0.609	0.854	**0.873**
$S = 10$	0.537	0.635	0.690	0.772	**0.823**

Table 5.5 PCC (%) evaluation of results of five kinds of subpixel mapping methods

Proportion	SPSAM	HNN	HNN + SSRSI	MRF	MRF + SSRSI
$S = 4$	60.9	65.0	64.5	92.8	**92.9**
$S = 5$	65.4	70.7	70.9	91.4	**92.1**
$S = 6$	68.0	72.4	73.7	90.7	**91.9**
$S = 10$	69.4	76.1	79.6	85.6	**88.8**

among five kinds of methods is closest to the true distribution diagram of land class with the best effect.

Here we compare five kinds of methods by quantitative evaluation. Tables 5.4 and 5.5, respectively, list the Kappa coefficient and PCC of five kinds of methods under four kinds of proportion. The Kappa of the SPSAM, HNN, and HNN combining SSRSI methods is lower than 0.700 under four kinds of proportions and the PCC is lower than 80.0 %, significantly lower than both MRF methods. Specifically, the accuracy of the SPSAM method is the lowest among the five kinds of methods. By using SSRSI, the accuracy of the HNN model has been increased, but is still limited. Compared to traditional MRF methods, the MRF combining SSRSI has a higher Kappa and PCC. For example, by adding SSRSI, at $S = 5$, Kappa increases from 0.865 of traditional MRF methods to 0.875 and PCC increases from 91.4 to 92.1 %; at $S = 10$, Kappa improves 0.051 and PCC improves 3.2 %.

5.5 Chapter Conclusions

Aiming at the deficiencies of the BPNN learning algorithm, this chapter proposes a LSSVM-based subpixel mapping method. Meanwhile, according to the linear distribution features of buildings, agriculture and forestry land class, and roads and others, an artificially synthetic sample training method is also proposed without prior information.

SPSAM has a clear physical meaning and has the advantage without iterations, but this model cannot describe the spatial attraction in a precise way. This chapter adopts calculus thought used to solve the attraction between objects in physics to propose the MSPSAM. However, although the description of attraction of the

MSPSAM is more accurate than SPSAM, the attraction among subpixels in the pixels is ignored. And then the MSAM is proposed on the basis of MSPSAM.

MRF does not depend on the results of spectral unmixing, taking into account the spatial information and undetermined multiband spectral information of remote-sensing data, which can be used directly for multiband remote-sensing data. However, the subpixel mapping is an under-constrained problem and tends to produce multiple solutions to meet the existing constraint conditions. In order to solve this problem, one method of using SSRSI spectral information to increase spectrum constraint entry of the MRF model is proposed.

Several proposed methods are carried out with sufficient validation in experiments, showing the rationality of the proposed method and the advantage in subpixel mapping accuracy.

References

Atkinson PM (1997) Mapping subpixel boundaries from remotely sensed images. Innovations GIS 4:166–180

Atkinson PM (2005) Subpixel target mapping from soft-classified, remotely sensed imagery. Photogram Eng Remote Sens 71(7):839–846

Ge Y, Li S, Lakhan VC (2009) Development and testing of a subpixel mapping algorithm. IEEE Trans Geosci Remote Sens 47(7):2155–2164

Ling F, Du Y, Xiao F, Xue H, Wu S (2010) Super-resolution land-cover mapping using multiple subpixel shifted remotely sensed images. Int J Remote Sens 31(19):5023–5040

Mertens KC, Verbeke LPC, De Wulf RR (2003a) Sub-pixel mapping with neural networks: Real-world spatial configurations learned from artificial shapes. In: Proceedings of 4th international symposium on remote sensing of Urban Areas, Regensburg, Germany pp 117–121

Mertens KC, Verbeke LPC, Ducheyne EI, De Wulf RR (2003b) Using genetic algorithms in sub-pixel mapping. Int J Remote Sens 24(21):4241–4247

Mertens KC, Basets BD, Verbeke LPC, De Wulf RR (2006) A subpixel mapping algorithm based on subpixel/pixel spatial attraction models. Int J Remote Sens 27(15):3293–3310

Nguyen MQ, Atkinson PM, Lewis HG (2005) Superresolution mapping using a Hopfield neural network with LIDAR data. IEEE Geosci Remote Sens Lett 2(3):366–370

Shen Z, Qi J, Wang K (2009) Modification of pixel-swapping algorithm with initialization from a subpixel/pixel spatial attraction model. Photogram Eng Remote Sens 75:557–567

Tatem AJ, Lewis HG, Atkinson PM, Nixon MS (2001) Super-resolution target identification from remotely sensed images using a hopfield neural network. IEEE Trans Geosci Remote Sens 39 (4):781–796

Tatem AJ, Lewis HG, Atkinson PM, Nixon MS (2002) Super-resolution land cover pattern prediction using a Hopfield neural network. Remote Sens Environ 79(1):1–14

Tatem AJ, Lewis HG, Atkinson PM, Nixon MS (2003) Increasing the spatial resolution of agricultural land cover maps using a hopfield neural network. Int J Geogr Inf Sci 17(7):647–672

Teerasit K, Arora MK, Varshney PK (2005) Super-resolution land-cover mapping using a Markov random field-based approach. Remote Sens Environ 96(3–4):302–314

Tolpekin VA, Hamm NAS (2008) Fuzzy super resolution mapping based on Markov random fields. In: Proceedings of international geoscience and remote sensing symposium, pp 875–878

Tolpekin VA, Stein A (2009) Quantification of the effects of land-cover-class spectral separability on the accuracy of markov-random-field-based superresolution mapping. IEEE Trans Geosci Remote Sens 47(9):3283–3297

Verhoeye J, De Wulf RR (2002) Land-cover mapping at subpixel scales using linear optimization techniques. Remote Sens Environ 79(1):96–104

Wang L, Zhang Y, Li J (2006) BP neural network-based sub-pixel mapping method. In: International conference on intelligent computing, pp 755–760

Wang L, Jia X, Zhang Y (2007) A novel geometry-based feature-selection technique for hyperspectral imagery. IEEE Geosci Remote Sens Lett 4(4):171–175

Zhang L, Wu K, Zhong Y, Li P (2008) A new subpixel mapping algorithm based on a BP neural network with an observation model. Neurocomputing 71(10–12):2046–2054

Chapter 6
Super-Resolution Technique of HSI

Hyperspectral imagery has been more widely used, but lower spatial resolution seriously affects its application effect. How to improve the spatial resolution has been highly valued by the academic world, but has not been solved yet. To this end, this chapter is based on the classic theory to establish new model and new method to improve the spatial resolution, thus further improving the application effect of hyperspectral imagery.

6.1 POCS Algorithm-Based Super-Resolution Recovery

6.1.1 Basic Theory of POCS

POCS algorithm expresses the super-resolution problem with a group of airspace algebraic equations. The definition reflects the prior information of unknown image or closed convex sets of constraints, to reconstruct high-resolution images through successive projection iterations of projection operators. Relationship model between the high-resolution image and the low-resolution image is as follows:

$$g_i(m_1, m_2) = \sum_{n_1, n_2} f(n_1, n_2)h(m_1, m_2; n_1, n_2) + v(m_1, m_2) \tag{6.1}$$

Of which, $g_i(m_1, m_2)$ is the i frame of low-resolution image, $f(n_1, n_2)$ is the original high-resolution image, $v(m_1, m_2)$ is the additive noise, and $h(m_1, m_2; n_1, n_2)$ is the i space-variant point spread function. Under-sampling of the low-resolution sensor and motion blur between imaging system and scenery result in the degradation of actually observed images.

POCS algorithm regards the reconstruction image $f(n_1, n_2)$ as a element of Hilbert space, and the relevant prior information or constraints of $f(n_1, n_2)$ limits the solution of a closed convex set in the Hilbert space. M messages will correspond to m closed convex sets C_i, and closed convex sets C_i and their respective projection operator P_i produces the following projection sequence:

L. Wang and C. Zhao, *Hyperspectral Image Processing*,
DOI 10.1007/978-3-662-47456-3_6

$$\hat{f}_{k+1} = P_m P_{m-1} \ldots P_1 \hat{f}_k, k = 1, 2, \ldots, m \tag{6.2}$$

Assuming that the statistical property of the degradation function and the noise process is known, each pixel of the degraded image can define restrictions of the following closed convex sets:

$$\begin{gathered} C_{n_1,n_2,i,k} = \left\{ f_i(m_1, m_2) : |r_k^{(f_i)}(n_1, n_2) \le \delta_0 \right\} \\ 0 \le n_1 \le N - 1, 0 \le n_2 \le N - 1, k = 1, 2, \ldots, L \end{gathered} \tag{6.3}$$

Of which,

$$r_k^{(f_i)}(n_1, n_2) = g_k(n_1, n_2) - \sum_{m_1=0}^{M-1} \sum_{m_2=0}^{M-1} f_i(m_1, m_2) h_{ik}(m_1, m_2; n_1, n_2) \tag{6.4}$$

Parameter δ_0 is a priori boundary to reflect statistical confidence coefficient. Of which, σ_v is the standard deviation of the noise, and $c \ge 0$ determines an appropriate range of statistical confidence coefficient. These settings are used to define high-resolution images. Limit the estimated value at each iteration, making that the projection error of each pixel from the low-resolution to high-resolution image is less than a predetermined boundary. The definition of the projection $P_{n_1,n_2;i,k}[f_i(m_1, m_2)]$ on $C_{n_1,n_2,i,k}$ of $f_i(m_1, m_2)$ is as follows:

$$P_{n_1,n_2;i,k}[f_i(m_1, m_2)] = \begin{cases} f_i(m_1, m_2) + \frac{r_k^{(f_i)}(n_1,n_2) - \delta_0}{\sum_o \sum_p h_{ik}^2(o,p,n_1,n_2)} h_{ik}(m_1, m_2; n_1, n_2), r_k^{(f_i)}(n_1, n_2) > \delta_0 \\ f_i(m_1, m_2), \delta_0 < r_k^{(f_i)}(n_1, n_2) < -\delta_0 \\ f_i(m_1, m_2) + \frac{r_k^{(f_i)}(n_1,n_2) + \delta_0}{\sum_o \sum_p h_{ik}^2(o,p,n_1,n_2)} h_{ik}(m_1, m_2; n_1, n_2), r_k^{(f_i)}(n_1, n_2) < -\delta_0 \end{cases} \tag{6.5}$$

Amplitude constraints can be defined as:

$$C_A = \{f_i(m_1, m_2) : \alpha \le f_i(m_1, m_2) \le \beta,\ 0 \le m_1, m_2 \le M - 1\} \tag{6.6}$$

The projection P_A on amplitude constraints C_A is as follows:

$$P_A[f_i(m_1, m_2)] = \begin{cases} \alpha & f_i(m_1, m_2) < \alpha \\ f_i(m_1, m_2) & \alpha \le f_i(m_1, m_2) \le \beta \\ \beta & f_i(m_1, m_2) > \beta \end{cases} \tag{6.7}$$

The essence of POCS algorithm is to explain and achieve algebraic iteration of super-resolution processing problems from the perspective of the space geometry. Here make further analysis for the algorithm. Supposed that the estimated and unknown high-resolution image $\boldsymbol{f}(x,y)$ and low-resolution image $\boldsymbol{g}(x,y)$ compose the following algebraic equation set:

$$\begin{cases} h_{11}f_1 + h_{12}f_2 + \cdots h_{1N}f_N = g_1 \\ h_{21}f_1 + h_{22}f_2 + \cdots h_{2N}f_N = g_2 \\ \vdots \\ h_{M1}f_1 + h_{M2}f_2 + \cdots h_{MN}f_N = g_M \end{cases} \tag{6.8}$$

The sampled data of image $\boldsymbol{f}(x,y)$ and $\boldsymbol{g}(x,y)$ is respectively as N and M, and $h_{i,j}$ is a constant. $\boldsymbol{f} = [f_1, f_2, \ldots, f_N]$ can be seen as a vector of N dimensional space, and each equation in equation set (6.8) represents a hyperplane. Select the initial value of the iteration as $\boldsymbol{f}^{(0)}$, and the next estimated value $\boldsymbol{f}^{(1)}$ is taken as the projection of $\boldsymbol{f}^{(0)}$ on the first hyperplane $h_{11}f_1 + h_{12}f_2 + \cdots h_{1N}f_N = \boldsymbol{g}_1$, namely:

$$\boldsymbol{f}^{(1)} = \boldsymbol{f}^{(0)} + \frac{g_1 - \sum \boldsymbol{H}_1 \boldsymbol{f}^{(0)}}{\sum \boldsymbol{H}_1^2} \sum \boldsymbol{H}_1 \tag{6.9}$$

Of which, $\boldsymbol{H}_1 = [h_{11}, h_{12}, \ldots, h_{1N}]^T$ Then, take the projection of $\boldsymbol{f}^{(1)}$ on the second hyperplane as the estimated value $\boldsymbol{f}^{(2)}$, and so on, until $\boldsymbol{f}^{(M)}$ meets the last hyperplane equation in (6.8), completing the first loop of iteration. So, loop iteration will get a series of vectors $\boldsymbol{f}^{(0)}$, $\boldsymbol{f}^{(M)}$, $\boldsymbol{f}^{(2M)}$, ... For any given N, M, and $h_{i,j}$, the vector $\boldsymbol{f}^{(kM)}$ will converge to f. If the equation set (6.8) has a unique solution, f is the solution; if the equation set (6.8) has infinitely many solutions, f is the solution to make the following formula to obtain the minimum:

$$\|\boldsymbol{f} - \boldsymbol{f}^{(0)}\|^2 = \sum_{i=1}^{N} (f - \boldsymbol{f}_i^{(0)})^2 \tag{6.10}$$

POCS algorithm represents super-resolution problem as a set of airspace algebraic equation set, using the projection iterative method to seek feasible solutions consistent with the low-resolution images, and easily introducing the priori information and additional constraint conditions.

6.1.2 POCS Algorithm-Based Super-Resolution Recovery

Supposed that $\boldsymbol{f}[n_1, n_2, \lambda_0]$ is the high-resolution target image of the wave length as λ_0 and the high-resolution image spatial sampling meets the Nyquist sampling law, thereby the space continuous reconstruction image $\boldsymbol{f}(x_1, x_2, \lambda_0)$ can be recovered. In the recovery of the space continuous reconstruction image, the sampling impulse array $\boldsymbol{f}_s(x_1, x_2, \lambda_0)$ is:

$$\boldsymbol{f}_s(x_1, x_2, \lambda_0) = \sum_{n_1=0}^{N_1-1}\sum_{n_2=0}^{N_2-1} \boldsymbol{f}_j[n_1, n_2, \lambda_0] \times \delta(x_1 - \frac{n_1}{L_1}, x_2 - \frac{n_2}{L_2}) \tag{6.11}$$

Here, $L1$ and $L2$ represent the sampling density relationship between low-resolution images and high-resolution images. If the number of sampling per unit area of the low-resolution image is 1, then the number of sampling per unit area of the high-resolution image is $L_1 \times L_2$. Thus, the recovered space continuous reconstruction image $\boldsymbol{f}(x_1, x_2, \lambda_0)$ can be expressed as:

$$\begin{aligned}
\boldsymbol{f}(x_1, x_2, \lambda_0) &= \iint \boldsymbol{f}_s(x_1 - u_1, x_2 - u_2, \lambda_0)\boldsymbol{h}_r(u_1, u_2)\mathrm{d}u_1\mathrm{d}u_2 \\
&= \iint \sum_{n_1=0}^{N_1-1}\sum_{n_2=0}^{N_2-1} \boldsymbol{f}[n_1, n_2, \lambda_0] \times \delta(x_1 - \frac{n_1}{L_1}, x_2 - \frac{n_2}{L_2})\boldsymbol{h}_r(u_1, u_2)\mathrm{d}u_1\mathrm{d}u_2 \\
&= \sum_{n_1=0}^{N_1-1}\sum_{n_2=0}^{N_2-1} \boldsymbol{f}[n_1, n_2, \lambda_0] \times \iint \delta(x_1 - \frac{n_1}{L_1}, x_2 - \frac{n_2}{L_2})\boldsymbol{h}_r(u_1, u_2)\mathrm{d}u_1\mathrm{d}u_2 \\
&= \sum_{n_1=0}^{N_1-1}\sum_{n_2=0}^{N_2-1} \boldsymbol{f}[n_1, n_2, \lambda_0] \times \boldsymbol{h}_r(x_1 - \frac{n_1}{L_1}, x_2 - \frac{n_2}{L_2})
\end{aligned} \tag{6.12}$$

Furthermore, the reconstruction image $\boldsymbol{f}(x_1, x_2, \lambda_0)$ will become the image $\boldsymbol{f}_c(x_1, x_2, \lambda_0)$ by spatial filtering:

$$\begin{aligned}
\boldsymbol{f}_c(x_1, x_2, \lambda_0) &= \iint \boldsymbol{f}(v_1, v_2, \lambda_0) \times \boldsymbol{h}(x_1 - v_1, x_2 - v_2)\mathrm{d}v_1 v_2 \\
&= \sum_{n_1=0}^{N_1-1}\sum_{n_2=0}^{N_2-1} \boldsymbol{f}[n_1, n_2, \lambda_0] \times \iint \boldsymbol{h}_r(x_1 - \frac{n_1}{L_1}, x_2 - \frac{n_2}{L_2})\boldsymbol{h}(x_1 - v_1, x_2 - v_2)\mathrm{d}v_1 v_2 \\
&= \sum_{n_1=0}^{N_1-1}\sum_{n_2=0}^{N_2-1} \boldsymbol{f}[n_1, n_2, \lambda_0] \times \boldsymbol{h}_b(x_1, x_2, n_1, n_2)
\end{aligned} \tag{6.13}$$

In previous description, we have been fixing the wave length of image as λ_0. If it is regarded as the continuous variable λ, for the fixed (x_1, x_2) or (n_1, n_2) at any space position, the result $\boldsymbol{f}_c(x_1, x_2, \lambda_0)$ of spatial filtering after spectral domain filtering will become:

$$\begin{aligned} g_c(x_1, x_2, \lambda_i) &= \int_0^{\infty} f_c(x_1, x_2, \lambda) r_i(\lambda) d\lambda \\ &= h_b(x_1, x_2, n_1, n_2) \times \sum_{n_1=0}^{N_1-1} \sum_{n_2=0}^{N_2-1} \int_0^{\infty} h[n_1, n_2, \lambda] r_i(\lambda) d\lambda \\ &= \sum_{n_1=0}^{N_1-1} \sum_{n_2=0}^{N_2-1} \Psi_{i,n_1,n_2}\{f[n_1, n_2, \lambda]\} \times h_b(x_1, x_2, n_1, n_2) \end{aligned} \tag{6.14}$$

In the formula, $r_i(\lambda)$ is the spectral domain filtering spectral response function to obtain the resource image at the wave length λ. The image wave length and spatial position are sometimes regarded as the fixed value and sometimes regarded as a variable, just to facilitate the analysis and understanding, but in fact, they can be always regarded as the reference variables, at this time, the single point $g_c(x_1, x_2, \lambda_i)$ can be regarded as the two-dimensional continuous image. With the above results, the relationship between the low-resolution discrete observation image $g(m_1, m_2, \lambda_i)$ and spectral domain filtering result image $g_c(x_1, x_2, \lambda_i)$ can be established as follows:

$$\begin{aligned} g(m_1, m_2, \lambda_i) &= g_c(x_1, x_2, \lambda_i)|_{x_1=m_1, x_2=m_2} \\ &= \sum_{n_1=0}^{N_1-1} \sum_{n_2=0}^{N_2-1} \Psi_{i,n_1,n_2}\{f[n_1, n_2, \lambda]\} \times h_b(m_1, m_2, n_1, n_2) \end{aligned} \tag{6.15}$$

For the operator Ψ_{i,n_1,n_2}, in actual operation, the discrete λ can be converted to a point multiplication matrix with $f[n_1, n_2, \lambda]$.

Equation (6.15) is the imaging model of hyperspectral images. In the algorithm process, if making use of spectral endmembers to map the original high-dimensional data into low-dimensional transform space and carrying out the above super-resolution process, the complexity of the algorithm and the protection interest category will be greatly reduced. Therefore, in the above imaging model, we can additionally implement the mapping operator from the original high-dimensional data to low-dimensional data and the mapping operator from the low-dimensional data to the original high-dimensional data so as to achieve this goal. Then, this model is simply marked again as follows.

The spatial filtering operator, spectral domain filtering operator, mapping operator from the original high-dimensional data to low-dimensional data and mapping operator from low-dimensional data to the original high-dimensional data are respectively marked as H_{spa}, H_{spe}, Φ_{inv}, and Φ (Φ is formed with spectral endmembers for column vectors, and for $\Phi_{\text{inv}} = (\Phi^T\Phi)^{-1}\Phi^T$, the role of operators Φ_{inv} and Φ is equivalent to the corresponding left multiplication operation of the matrix), and the relationship between the high-resolution image f and low-resolution image g can be simply marked as:

$$\boldsymbol{g} = \Phi_{\text{inv}} H_{\text{spe}} H_{\text{spa}} \Phi \boldsymbol{f} \tag{6.16}$$

Furthermore, $\Phi_{\text{inv}} H_{\text{spe}} H_{\text{spa}} \Phi$ is marked as the integrated operator H, namely:

$$\text{g} = \text{Hf} \tag{6.17}$$

The above model is the necessary calculation model in the increase process of the resolution. It should be noted that some work orders of the four operators can be exchanged and some cannot. For example, spatial filtering operator H_{spa} and spectral domain filtering operator H_{spe} can be exchanged; Φ and H_{spa} can be exchanged, but Φ and H_{spe} cannot be exchanged, because H_{spe} cannot be directly applied to low-dimensional transform space without continuity of the spectral domain. The above exchange only belongs to the feasibility in the principle but is not strictly equivalent, and the specific effect should be determined according to the theory and experiment. In general, in the POCS algorithm optimization process, the recovery error of the object image and the observation image in the transform domain is not always synchronized convergence with the recovery error in the original domain, but they may be identical to tend to converge in general. Furthermore, the cost function based on the recovery error in the transform domain and the cost function based on the recovery error in the original domain will be different to some extent. The previous criterion can make the algorithm complexity reduce, but the latter criterion is directly based on the actual problem to be solved. In specific application, they can be compromised, namely, to establish the correspondence between the transform domain and the original domain images.

If the super-resolution algorithm is implemented in the sliding local window, because of the time-invariant characteristics, the spatial filtering operator and spectral domain filtering operator do not vary with sliding local window.

The integrated operator H can be deduced to determine by all the levels of operators H_{spa},H_{spe},Φ_{inv}, and Φ, but the process is relatively complicated. As the data input operations of all levels of operators are linear, the resulting data can be expressed as the weighted sum of all the discrete data. Thus, each element of the input data in the order is 1 and all other elements is 0, and the results obtained after being taken into the super-resolution equation should be the corresponding weight of the position element. After obtaining all the weights, the operator H_0 can be easily formed with the same size as input data, which acts on the input data equivalent to the corresponding multiplication of point-by-point elements, and then composite summation operation can get the integrated operator H. As mentioned earlier, the integrated operator H does change with the spatial position, so it can be applied to the whole algorithm iterative process, which will greatly improve the algorithm execution speed.

On the basis of the above imaging model, we can use POCS algorithm to conduct super-resolution processing of hyperspectral image. This process can be summarized as follows:

1. Build the model. First, recover the low-resolution discrete resource image (namely observation image) as spatial continuous image according to the sampling theorem by the pulse function. By simulating the actual imaging process and the principle, establish the relationship model between successive images and low-resolution discrete resource images, including spatial filtering, spectral sampling, spatial sampling, and additional noise and other processes. In addition to setting reasonable levels of filtering operators, operators should strive to integrate and simplify to obtain concise super-resolution model. Unlike conventional processing scheme, during the establishment process of the relationship model between continuous image and low-resolution discrete resource image, we do not directly apply the resulting successive images, but make use of the resulting interested spectrum end element to recover as the continuous spectrum base function by the sampling theorem, representing as space—spectrum consecutive images in the transform domain. Furthermore, establish the reasonable integral filter for spectral information fusion sampling, and the sampling process will be transformed into a reasonable weighting of spectrum base function. Such a conversion not only can reduce the amount of computation of model inversion, the more important but also can achieve the purpose to protect the interested categories.
2. Invert the model. Apply efficient super-resolution recovery algorithm to obtain optimal estimation of high-resolution target images. In the establishment of model, the recovery process from the low-resolution discrete resource image to the spatial continuous image needs the impact response of the imaging system, namely, the point spread function; the establishment process of the relationship model between the successive image and low-resolution discrete resource image (namely, observation image) involves spatial constant fuzzy operators in the characterization degradation process, and in the inversion process, the cascade operators need to be dealt with, and the changes of the associated spectral endmembers also need to be considered. The calculation of integrated operators is extremely complex; taking into account the space invariance of integrated operators in the hyperspectral imaging process, the complexity of the algorithm can be reduced by local analysis and the discrete calculation method as mentioned earlier.

6.2 MAP Algorithm-Based Super-Resolution Recovery

6.2.1 Basic Theory of MAP

In the spatial algorithm of the super-resolution processing, MAP algorithm is one of the most popular hotspot algorithms. It has a strong ability to contain spatial priori constraints, achieving good application results in practical processing. The algorithm regards the additive noise N, the measured image Y and the required ideal image Z as smooth random field, according to the Bayes rule, to obtain the MAP estimate $\hat{Z}$ of the unknown image Z by maximizing the conditional probability function $P(Z/Y)$.

MAP algorithm takes into account the image shift, blurring, under-sampling, and noise pollution, and other image degradation problems. High-resolution original images get a few frames of low-resolution degraded images after translation, blurring, and under-sampling and noise pollution:

$$\boldsymbol{Y}_k = \boldsymbol{D}_k\boldsymbol{C}_k\boldsymbol{F}_k\boldsymbol{Z} + \boldsymbol{\eta}_k, k = 1, 2, \ldots, p \tag{6.18}$$

In the formula, p is the frames of degraded images of under-sampling low resolution, $\boldsymbol{Y}_k$ is the k frame of low-resolution image in the arrangement of $N \times 1$, N is the total number of pixels, L is the multiples of resolution increase after treatment in each direction, Z is a frame of high-resolution image in the arrangement of $L^2N \times 1$, $\boldsymbol{F}_k$ is the translation matrix with the size of $L^2N \times L^2N$, which is used to describe the relative movement between the k frame and the reference frame, $\boldsymbol{C}_k$ is the fuzzy matrix with the size of $L^2N \times L^2N$, $\boldsymbol{D}_k$ is the uniform under-sampling matrix of $N \times L^2N$, $\boldsymbol{\eta}_k$ is the vector of additive noise of $N \times 1$.

Each frame of low-resolution degraded image provides different information about the original high-resolution images. Formula (6.18) constitutes an equation set consisting of p algebraic equations, to combine p equations to get:

$$\begin{bmatrix} \boldsymbol{Y}_1 \\ \boldsymbol{Y}_2 \\ \vdots \\ \boldsymbol{Y}_p \end{bmatrix} = \begin{bmatrix} \boldsymbol{D}_1\boldsymbol{C}_1\boldsymbol{F}_1 \\ \boldsymbol{D}_2\boldsymbol{C}_2\boldsymbol{F}_2 \\ \vdots \\ \boldsymbol{D}_p\boldsymbol{C}_p\boldsymbol{F}_p \end{bmatrix} \boldsymbol{Z} + \begin{bmatrix} \boldsymbol{N}_1 \\ \boldsymbol{N}_2 \\ \vdots \\ \boldsymbol{N}_p \end{bmatrix} = \begin{bmatrix} \boldsymbol{W}_1 \\ \boldsymbol{W}_2 \\ \vdots \\ \boldsymbol{W}_p \end{bmatrix} \boldsymbol{Z} + \eta \tag{6.19}$$

$$\boldsymbol{Y} = \boldsymbol{W}\boldsymbol{Z} + \boldsymbol{\eta} \tag{6.20}$$

P frames of low-resolution images $\boldsymbol{y} = [y_1^T, y_2^T, \ldots, y_p^T]^T = [y_1, y_2, \ldots, y_{pM}]^T$, of which, $M = N_1N_2$. Assuming, the high-resolution image to be reconstructed as $z = [z_1, z_2, \ldots, z_N]^T$, of which, $N = L_1N_2L_1N_2$. The relationship between the k frame of low-resolution image and high-resolution image can be expressed with the following mathematical model:

$$y_{k,m} = \sum_{r=1}^{N} w_{k,m,r}(s_k) z_r + \eta_{k,m} \tag{6.21}$$

Of which, $m = 1, 2, \ldots M$; $k = 1, 2, \ldots, p$. $w_{k,m,r}(s_k)$ represents the relationship between the r pixel of high-resolution image z and the m pixel of the k frame of low-resolution image y_k. The vector $\boldsymbol{s}_k = [s_{k,1}, s_{k,2}, \ldots, s_{k,K}]^T$ contains the displacement parameters of the k frame of low-resolution image. $\eta_{k,m}$ is the additive

Gaussian noise of the variance as σ_η^2, and the multivariate probability density function is expressed as follows:

$$\begin{aligned} Pr(n) &= \frac{1}{(2\pi)^{pM/2}\sigma_\eta^{pM}} \exp\left\{ -\frac{1}{2\sigma_\eta^2} n^T n \right\} \\ &= \frac{1}{(2\pi)^{pM/2}\sigma_\eta^{pM}} \exp\left\{ -\frac{1}{2\sigma_\eta^2} \sum_{m=1}^{pM} \eta_m^2 \right\} \end{aligned} \tag{6.22}$$

The direct estimate of the high-resolution image z by the low-resolution degraded image y is a usually morbid inverse problem, leading to excessive noise amplification. The probability density function $Pr(z)$ appropriately selected can play the role of regularization. Reflect the statistical characteristics of the random field with a Gaussian model, and the probability density function is as follows:

$$Pr(z) = \frac{1}{(2\pi)^{N/2}|\boldsymbol{C}_z|^{1/2}} \exp\{ -\frac{1}{2} z^T \boldsymbol{C}_z^{-1} z\} \tag{6.23}$$

Of which, $\boldsymbol{C}_z$ is the $N \times N$ covariance matrix of z.

Under the given low-resolution image y, the high-resolution image z, and the displacement s are simultaneously estimated and can be expressed by the following formula:

$$\hat{z}\cdot\hat{s} = \arg\max_{z,s} Pr(z, s|\mathbf{y}) \tag{6.24}$$

By Bayesian criterion:

$$\hat{z}\cdot\hat{s} = \arg\max_{z,s} \frac{Pr(\mathbf{y}|z, s)Pr(z, s)}{Pr(\mathbf{y})} \tag{6.25}$$

The denominator $Pr(y)$ is not the function of z or the displacement s and z and s are mutually independent, so the MAP estimate of the high-resolution image z can be rewritten as:

$$\hat{z}\cdot\hat{s} = \arg\max_{z,s} Pr(\mathbf{y}|z, s)Pr(z)Pr(s) \tag{6.26}$$

The above formula is equivalent to the following formula:

$$\begin{aligned} \hat{z}\cdot\hat{s} &= \arg\max_{z,s} \{L(z, s)\} \\ &= \arg\max_{z,s} \{-\log[Pr(y|z, s)] - \log[Pr(z)] - \log[Pr(s)]\} \end{aligned} \tag{6.27}$$

According to the mathematical model and the noise probability density function, the conditional probability density function can be written as follows:

$$Pr(\mathbf{y}|z,s) = \frac{1}{(2\pi)^{pM/2}\sigma_\eta^{pM}} \exp\left\{ -\frac{1}{2\sigma_\eta^2}\sum_{m=1}^{pM}\left(\mathbf{y}_m - \sum_{r=1}^{N} w_{m,r}(s)z_r\right)^2 \right\} \tag{6.28}$$

Consider the priori density model and ignore unrelated items with z or s, and the above estimate can be rewritten as:

$$\begin{aligned}\hat{z}\cdot\hat{s} &= \underset{z,s}{\arg\max}\ \{L(z,s)\} \\ &= \underset{z,s}{\arg\max}\left\{ -\frac{1}{2\sigma_\eta^2}\sum_{m=1}^{pM}\left(y_m - \sum_{r=1}^{N} w_{m,r}(s)z_r\right)^2 + \frac{1}{2\lambda}\sum_{i=1}^{N}\left(\sum_{j=1}^{N} d_{i,j}z_j\right)^2 \right\}\end{aligned} \tag{6.29}$$

In the n iteration, estimate the current value of the motion parameters, and then determine the image estimation descent gradient based on the following equation:

$$\begin{aligned}g_k(z,s) &= \frac{\partial L(z,s)}{\partial z_k} \\ &= -\frac{1}{2\sigma_\eta^2}\sum_{m=1}^{pM} w_{m,k}(s)\left[\sum_{r=1}^{N} w_{m,r}(s_k)z_r - y_m\right] + \frac{1}{\lambda}\sum_{i=1}^{N} d_{i,k}\left(\sum_{j=1}^{N} d_{i,j}z_j\right)\end{aligned} \tag{6.30}$$

The $n+1$ high-resolution image is estimated as:

$$\hat{z}^{n+1} = \hat{z}^n - \mu^n \nabla_z L(z,s)\big|_{z=\hat{z}^n} \tag{6.31}$$

Of which, the step size parameter formula is as follows:

$$\mu^n = \frac{\frac{1}{\sigma_\eta^2}\sum_{m=1}^{pM}\sum_{r=1}^{N} w_{m,r}(\hat{s}^n)g_r(\hat{z}^n,\hat{s}^n)\left(\sum_{r=1}^{N} w_{m,r}(\hat{s}^n)\hat{z}_r^n - y_m\right) + \frac{1}{\lambda}\sum_{i=1}^{N}\sum_{j=1}^{N} d_{i,j}g_j(\hat{z}^n,\hat{s}^n)\left(\sum_{j=1}^{N} d_{i,j}\hat{z}_j^n\right)}{\frac{1}{\sigma_\eta^2}\sum_{m=1}^{pM}\left(\sum_{r=1}^{N} w_{m,r}(\hat{s}^n)g_r(\hat{z}^n,\hat{s}^n)\right)^2 + \frac{1}{\lambda}\sum_{i=1}^{N}\left(\sum_{j=1}^{N} d_{i,j}g_j(\hat{z}^n,\hat{s}^n)\right)^2} \tag{6.32}$$

$\hat{z}^n$, $\hat{s}^n$, and μ^n is respectively, the n image estimation, motion parameter estimation, and iterative step size. So far, all super-resolution iterative formulas of MAP algorithm are obtained.

From the foregoing, MAP estimate of the high-resolution image is translated into the extremum problem to solve the cost function. $L(z,s)$ is mutually depended by the projection error and image priori error between low-resolution image and high-resolution image:

$$L(z, s) = \frac{1}{2\sigma_\eta^2}(y - W_s z)^T (y - W_s z) + \frac{1}{2} z^T C_z^{-1} z \tag{6.33}$$

If the motion parameter can be given, $L(z, s)$ constitutes a quadratic function on z, and can relatively easily minimize z. In the iterative solution process, the calculation of the cost function $L(z, s)$ on the gradient $\nabla_z L(z, s)$ of z is the key content, which is calculated as follows:

$$\nabla_z L(z, s) = \left[\frac{\partial L(z, s)}{\partial z_1}, \frac{\partial L(z, s)}{\partial z_2}, \ldots, \frac{\partial L(z, s)}{\partial z_N}\right]^T \tag{6.34}$$

Set $\nabla_z L(z, s) = 0$, the optimal estimate $\hat{z}$ of z is as follows:

$$\hat{z} = \left[\boldsymbol{W}^T \boldsymbol{W} + \sigma_\eta^2 \boldsymbol{C}_z^{-1}\right]^{-1} \boldsymbol{W}^T \boldsymbol{y} \tag{6.35}$$

The iterations should be repeated until the norm of estimated difference between the adjacent two is less than the specified threshold or the iterations achieve the specified number of times. The description of the basic theory of the above two parts mainly refers to the postdoctoral report of Meng (2004).

6.2.2 MAP Algorithm-Based Super-Resolution Recovery

As mentioned earlier, make use of spectral endmembers for low-dimensional projection of the original hyperspectral data, and then carry out super-resolution processing on the data on the transform domain on the space, which will reduce the computational complexity and protect interest category. based on this premise and the imaging model described earlier, make use of the traditional MAP algorithm to construct the super-resolution algorithm mainly presented in this chapter, specifically described below.

Apply again the mapping operator Φ_{inv} from the original high-dimensional data to low-dimensional data and the mapping operator Φ from the low-dimensional data to the original high-dimensional data, and they can connect the original high-dimensional spatial hyperspetral data g^G with low-dimensional spatial hyperspetral data g^D within the transform domain by the following formula:

$$g^G = \Phi g^D, \Phi_{\text{inv}} g^G = g^D \tag{6.36}$$

The input data is a, the output is estimated as $\hat{\boldsymbol{a}}$, the corresponding estimation error is v, and empty spectral synthesis filter operator is H, then the relationship formula is:

$$\hat{\boldsymbol{a}} = H\Phi\boldsymbol{a} + \boldsymbol{v} \tag{6.37}$$

At this time, priori probability $p(v)$ of the error v is:

$$p(\boldsymbol{v}) = \frac{1}{\boldsymbol{Z}}\exp\left(-\boldsymbol{v}^T\boldsymbol{K}^{-1}\boldsymbol{v}\right) \tag{6.38}$$

And the conditional probability $p(\boldsymbol{v}|\boldsymbol{a})$ of v is:

$$p(\boldsymbol{v}|\boldsymbol{a}) = \frac{1}{\boldsymbol{Z}}\exp\left(-(\hat{\boldsymbol{a}} - H\Phi\boldsymbol{a})^T K^{-1}(\hat{\boldsymbol{a}} - H\Phi\boldsymbol{a})\right) \tag{6.39}$$

The pixel at any location in the image only has a larger relationship with adjacent pixels, and as the distance increases, this correlation rapidly reduces even completely disappears, so in the actual super-resolution processing, often use local analysis instead of global analysis, with its benefits is that it can avoid large-scale matrix operations to reduce the complexity of the algorithm. Set the local pixel of all low-resolution observation images related with the pixel $\boldsymbol{a}_{n_1,n_2}$ in the high-resolution images as $\boldsymbol{a}^{(i)}$, $i = 1, 2, \ldots, M$. The spatial comprehensive filtering operator is described as the corresponding weight α_i of the input vector, $i = 1, 2, \ldots, M$, then the relationship between $\boldsymbol{a}_{n_1,n_2}$ and $\boldsymbol{a}^{(i)}$ can be derived as follows:

$$\boldsymbol{a}_{n_1,n_2} = \sum_{i=1}^{M} \alpha_i H_{\text{spe}}\Phi\boldsymbol{a}^{(i)} \tag{6.40}$$

The above relational expression is derived based on the pixels (and its related inputs) in the high-resolution target images as the basic unit of analysis; equally the similar relational expression can also be derived based on the pixel components in the high-resolution target images as the basic unit of analysis, but at this time, the relevant inputs have no change. It is easy to analyze that the former method will have the smaller computational complexity, so here, we use the pixel basic analysis unit. On the other hand, the analysis is about the problem at the output returning the low-dimensional transformation spatial data or the original high-dimensional data. It can be seen at the input that low-dimensional transform spatial input data greatly reduce the computational complexity of spectral domain filtering, as shown in Eq. (6.40); while the output returning to the high-dimensional transformation data space adds unnecessary calculations. So, we adopt the super-resolution method returning to the original high-dimensional data.

According to the previous expression of the probability density function and input–output relational expression, it can be derived that the optimal estimation $\hat{\boldsymbol{a}}$ based on MAP super-resolution restoration algorithm meets the following optimization expression:

$$\hat{\boldsymbol{a}} = arg \min_{s}\left(-\left(\hat{\boldsymbol{a}} - \sum_{i=1}^{M}\alpha_i H_{\text{spe}}\Phi\boldsymbol{a}^{(i)}\right)^T \boldsymbol{K}^{-1}\left(\hat{\boldsymbol{a}} - \sum_{i=1}^{M}\alpha_i H_{\text{spe}}\Phi\boldsymbol{a}^{(i)}\right) + \sum_{i=1}^{M}\alpha_i(\boldsymbol{a}^{(i)})^T\boldsymbol{\Lambda}^{-1}(\boldsymbol{a}^{(i)})\right) \tag{6.41}$$

It is assumed that the spectral domain filtering operator H_{spe} does not change with the spatial position. This optimization process will be accomplished by an iterative manner. To this end, set the cost function $E(a)$ of the optimal estimation process as follows:

$$E(\boldsymbol{a}) = \frac{1-\lambda}{2}\left(\hat{\boldsymbol{a}} - \sum_{i=1}^{M}\alpha_i H_{\text{spe}}\Phi\boldsymbol{a}^{(i)}\right)^T \boldsymbol{K}^{-1}\left(\hat{\boldsymbol{a}} - \sum_{i=1}^{M}\alpha_i H_{\text{spe}}\Phi\boldsymbol{a}^{(i)}\right) + \frac{\lambda}{2}\sum_{i=1}^{M}\alpha_i(\boldsymbol{a}^{(i)})^T\boldsymbol{\Lambda}^{-1}(\boldsymbol{a}^{(i)}) \tag{6.42}$$

The iterative formula from the $(n-1)$ optimal estimation $\boldsymbol{a}_{n-1}$ to the n optimal estimation $\boldsymbol{a}_n$ is:

$$\boldsymbol{a}_n = \boldsymbol{a}_{n-1} - \alpha\nabla E(\boldsymbol{a}_{n-1}) \tag{6.43}$$

Of which, the calculation formula of gradient estimate $\nabla E(\boldsymbol{a}_{n-1})$ and step length parameter α of the cost function $E(a)$ can be inferred as follows:

$$\nabla E(\boldsymbol{a}^{(i)}) = (\lambda - 1)\left(\sum_{i=1}^{M}\alpha_i H_{\text{spe}}\Phi\boldsymbol{a}^{(i)}\right)^T \boldsymbol{K}^{-1}\left(\hat{\boldsymbol{a}} - \sum_{i=1}^{M}\alpha_i H_{\text{spe}}\Phi\boldsymbol{a}^{(i)}\right) + \lambda\alpha_i\boldsymbol{\Lambda}^{-1}\boldsymbol{a}^{(i)} \tag{6.44}$$

$$\alpha_i = \frac{(\nabla E(\boldsymbol{a}_{n-1}^{(\mathrm{i})}))^T(\nabla E(\boldsymbol{a}_{n-1}^{(\mathrm{i})}))}{(\nabla E(\boldsymbol{a}_{n-1}^{(\mathrm{i})}))^T[(1-\lambda)(H_{\text{spe}}\Phi)^T\boldsymbol{K}^{-1}H_{\text{spe}}\Phi + \lambda\alpha_i\boldsymbol{\Lambda}^{-1}\boldsymbol{a}^{(\mathrm{i})}](\nabla E(\boldsymbol{a}_{n-1}^{(\mathrm{i})}))} \tag{6.45}$$

6.3 Resolution Enhancement Method for Single Band

As an important method to improve the spatial resolution, interpolation has been concerned by many scholars. Typically, traditional interpolation methods such as bilinear interpolation and cubic interpolation method focus on the image smoothing, to obtain a better visual effect. But, these methods have obvious drawbacks: while maintaining the image smoothing, often leading to blurred image edges. The edge information of the image is not only an important factor affecting the visual effect but also the critical factor in the target recognition and tracking, image matching, image registration and other image processing issues. Therefore, interpolation technique based on edge has become a hot research in recent years. This chapter focuses on the geometric dual model and single spectral band image interpolation method with edge characteristics.

6.3.1 Construction of Geometric Dual Model and Interpolation Method

Rodrigues et al. (2002) proposed a locally adaptive nonlinear interpolation algorithm (Locally adaptive nonlinear interpolation, LAI). For a point to be interpolated, by calculating the local standard deviation and comparing the result with the predetermined threshold value, to determine which method should be adopted to complete the interpolation calculation at this point. The algorithm is completed through four steps: image expansion (namely the up-sample), edge preservation, image smoothing, and image filling. The method keeps both image smoothing and edge preservation. Li et al. (2001a, b) proposed an edge interpolation method based on local covariance (Edge-Directed Interpolation technique based on Local Covariance, EDILC). The method carries out local covariance estimate on original low-resolution image according to the geometric dual feature between low-resolution covariance and high-resolution covariance, and makes use of the obtained low-resolution local covariance estimation to carry out high-resolution adaptive interpolation.

One contribution of Li et al. (2001a, b) articles is to creatively put forward the low-resolution statistics instead of high-resolution statistics, so as to solve a difficult problem in the interpolation problem. Now EDILC algorithm is summarized below.

Set the two-dimensional optical image X, and mark all pixel points as $X(i)$, $i = 1, 2, \ldots, M$. We use the gray value of N neighboring points $\{X(n-k)\}$, $k = 1, 2, \ldots, N$ with $X(n)$ to predict the gray value of the point $X(n)$, and the linear prediction equation can be written as:

$$\hat{X}(n) = \sum_{k=1}^{N} \alpha(k) X(n-k) \tag{6.46}$$

Use the least square of adaptive prediction method to define a $M \times 1$ training window:

$$\boldsymbol{y} = [X(n-1), \ldots, X(n-M)]^T \tag{6.47}$$

The forecast neighborhood of y is a $M \times N$ matrix:

$$C = \begin{bmatrix} X(n-1-1) & \cdots & X(n-1-N) \\ \vdots & & \vdots \\ X(n-M-1) & \cdots & X(n-M-N) \end{bmatrix} \tag{6.48}$$

According to the traditional linear prediction theory, the estimate of the minimum mean square error (MMSE) in the steady-state random process can be decided by the second-order statistics, namely, the weight coefficient vector $\boldsymbol{\alpha} = [\alpha(1)\ldots\alpha(N)]^T$ can be expressed as:

$$\boldsymbol{\alpha} = \boldsymbol{R}_{XX}^{-1}\boldsymbol{r}_X \tag{6.49}$$

Of which,

$$\begin{cases} \boldsymbol{r}_X = [r_1 \ldots r_k \ldots r_N], r_k = Cov\{X(n)X(n-k)\}, 1 \le k \le N \\ \boldsymbol{R}_{XX} = [R_{kl}], R_{kl} = Cov\{X(n-k)X(n-l)\}, 1 \le k, l \le N \end{cases} \tag{6.50}$$

If using the predictive neighborhood matrix C mentioned earlier to express $\boldsymbol{r}_X$ and $\boldsymbol{R}_{XX}$, then $\boldsymbol{R}_{XX} = \boldsymbol{C}^T\boldsymbol{C}/M^2, \boldsymbol{r}_X = \boldsymbol{C}^T\boldsymbol{y}/M^2$, so that we can get another form of expression of the formula (6.49):

$$\boldsymbol{\alpha} = (\boldsymbol{C}^T\boldsymbol{C})^{-1}(\boldsymbol{C}^T\boldsymbol{y}) \tag{6.51}$$

This method is actually a linear interpolation, namely, solving the synthetic weights of interpolation pixels in each direction by statistical theory. In specific interpolation process, if the two-dimensional space coordinates are marked as pixel points, then the algorithm makes use of the four nearest points in a diagonal direction for interpolation on the even coordinate points, and then makes use of the four nearest points in horizontal and vertical directions for interpolation on the odd–even coordinate points. EDILC interpolation method achieves good results on the edge preservation. The algorithm adopts the interpolation method mixed with linear interpolation method, but the calculated amount is still large.

Inspired by EDILC interpolation thought, this chapter introduces a new adaptive edge interpolation method based on linear prediction model and adopts an adaptive image interpolation algorithm based on the edge detection. Its main idea comes from the digitized image consisting of each pixel gradation and color value ordered arrangement, and the described object and the lighting effect of the image is expressed mainly by the changes of gray values. Most of the region pixel gray values in the image have the continuous transition feature between the adjacent pixels, reflecting that the outside surface geometric shape and illumination conditions of most objective objects in the image have continuous attributes. In the image interpolation, the bilinear or higher order interpolation can be used to make the pixel gray value generated by the interpolation to continue the continuation of the original image gray changes, so that the variable density of the enlarged image is natural and smooth. However, the gray values between some pixels and adjacent pixels exists the mutation, namely, the gradation discontinuity. These pixels with gray value mutation are edge pixels of the contour or texture image of the description object in the image, and in the image interpolation, if these discontinuous grey pixels are generated new added pixels with the conventional interpolation algorithm, which is bound to make the contour and texture of enlarged image blurred and reduce the image quality.

The correspondence location between the low-resolution image and its corresponding high-resolution image has similar edge feature, so there exists similar linear prediction model between them, and we call this feature as the geometric dual feature of the low-resolution linear prediction model and high-resolution linear prediction model. Now the algorithm is described below.

One low-resolution image $X_{i,j}$ with the size of $H \times V$ is planned as the high-resolution image $Y_{i,j}$ with the interpolation as $2H \times 2V$, to meet $Y_{2i-1,2j-1} = X_{i,j}, 1 \leq i \leq H,\ 1 \leq j \leq V$ then the interpolation task is to complete the assignment of other points in $Y_{i,j}$.

First, consider to make use of four nearest neighboring points in the diagonal direction to interpolate the unknown interlaced dot matrix $Y_{2i,2j}$ with the known interlaced dot matrix $Y_{2i+1,2j+1}$. Figure 6.1a shows the geometric dual in this case. At the moment, the high-resolution linear prediction model is:

$$Y_{2i,2j} = \sum_{k=0}^{1}\sum_{l=0}^{1} \alpha_h^{2k+l+1} Y_{2(i+k)-1,2(j+l)-1} \tag{6.52}$$

Once the high-resolution interpolation coefficient vector $\boldsymbol{\alpha}_h = [\alpha_h^1, \alpha_h^2, \alpha_h^3, \alpha_h^4]^T$ is calculated, the value of $Y_{2i,2j}$ can be confirmed.

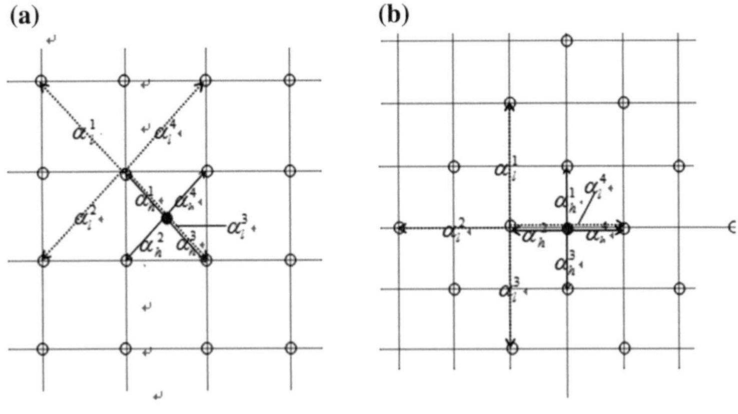

Fig. 6.1 Geometric dual. **a** interpolating the interlacing dot matrix $Y_{2i,2j}$ with the interlacing dot matrix $Y_{2i+1,2j+1}$, **b** interpolating the interlacing dot matrix $Y_{i,j}$ $(i+j = \text{odd})$ with the interlacing dot matrix $Y_{i,j}$ $(i+j = \text{even})$

Obviously, we cannot solve high-resolution interpolation coefficient vector $\boldsymbol{\alpha}_h$ through the high-resolution linear prediction model (6.52). According to the geometric dual feature mentioned earlier, our idea is to transform this issue to solve the low-resolution interpolation coefficient vector $\boldsymbol{\alpha}_l = [\alpha_l^1, \alpha_l^2, \alpha_l^3, \alpha_l^4]^T$ through the low resolution linear prediction model and then replace $\boldsymbol{\alpha}_h$ with $\boldsymbol{\alpha}_l$. As can be seen from Fig. 6.1a, there is no the low-resolution model completely coinciding with the high resolution model center, so we choose the low-resolution model with a difference of only half a pixel from the center for approximate replacement. Only the low-resolution interpolation coefficient vector is unknown in the low-resolution model, but it is just an underdetermined equation and can not solve only low-resolution interpolation coefficient vector. The solving of underdetermined problem requires at least four simultaneous low-resolution models, for this end, to select four nearest low-resolution models from the high-resolution model (with the center respectively as four points used for interpolation) to form the following simultaneous solving formula:

$$\boldsymbol{R}_l \boldsymbol{\alpha}_l = \boldsymbol{r}_l \tag{6.53}$$

Here, $\boldsymbol{r}_l = [r_1, r_2, r_3, r_4]^T$ contains four nearest neighboring points along the diagonal direction used for the interpolation $Y_{2i,2j}$, and we might assume that they are arranged in a clockwise direction. $\boldsymbol{R}_l = [\boldsymbol{R}_1, \boldsymbol{R}_2, \boldsymbol{R}_3, \boldsymbol{R}_4]^T$ is a matrix of 4×4, and its column vector $\boldsymbol{R}_k, k = 1, 2, 3, 4$ is formed by four nearest neighboring points along the diagonal direction with r_k as the center. After the low-resolution interpolation coefficient $\boldsymbol{\alpha}_l$ is calculated, replacing the high-resolution interpolation coefficient $\boldsymbol{\alpha}_h$, then the evaluation of the interpolated point $Y_{2i,2j}$ is completed.

The previously described condition is with the interlacing dot matrix $Y_{2i+1,2j+1}$ to interpolate the interlacing dot matrix $Y_{2i,2j}$, and when we use the interlacing dot matrix $Y_{i,j}$ $(i+j = \text{even})$ to interpolate the interlacing dot matrix $Y_{i,j}$ $(i+j = \text{odd})$, we can use the technology similar to the previous process, except that the operation becomes the operations in the horizontal and vertical direction from diagonal operation (45° of rotation transformation). Figure 6.1b describes the geometric dual in this case.

6.3.2 *Mixed Interpolation Method*

1. **Use of the mixed interpolation method**

In the smoother area in the image, four low-resolution linear interpolation models are prone to repeating the same model, and the formula (6.53) will also become the underdetermined equation set. Therefore, the effect of the above method is only obvious in the region with obvious edge feature, and for the smooth area, the effect is not ideal. Mixed interpolation method can effectively solve this problem. Specifically, adopt the method presented in this chapter in the region with obvious edge feature, and still adopt the bilinear interpolation method in the smooth region.

This operation is performed point-by-point. Local standard deviation can reflect the extent of the local gray scale change, and can therefore better reflect the degree of smoothness. For each pixel point $Y(i,j)$ to be interpolated, determine whether it is an edge point with the method as follows: calculate the standard deviation of the four nearest neighboring points of the pixel $Y(i,j)$ and also calculate the rank of $\boldsymbol{R}_l$, and only when the obtained standard deviation exceeds the preset threshold and $\boldsymbol{R}_l$ is at full rank, then $Y(i,j)$ is considered to be an edge pixel point. Increase the "full rank" restrictions to better ensure the existence of solutions of interpolation model. Mixed interpolation method not only overcomes the underdetermined problem of the equation set but also reduces the complexity of the calculations.

2. **The effectiveness of the edge preservation**

It is easy to know that our proposed method is similar to EDIULC method in principle, so they have the similar function and effect. Now, take a simple example to illustrate their same effectiveness for the edge preservation.

As shown in Fig. 6.2, the pixel $X(n)$ to be predicted is on a vertical edge $((|p-q| \gg 0))$. For simplicity, we only consider the second-order prediction model in two directions shown by the arrow:

$$\begin{aligned}\hat{X}(n) &= \alpha(1)X(n-1) + \alpha(2)X(n-2) \\ &= \alpha(1)p + \alpha(2)q,\end{aligned} \tag{6.54}$$

Apply EDIULC method to obtain the covariance $\hat{R}$ and $\overset{\frown}{\vec{r}}$:

$$\begin{cases} \hat{\boldsymbol{R}} = \begin{bmatrix} 8p^2+4q^2 & 6p^2+2pq+4q^2 \\ 6p^2+2pq+4q^2 & 6p^2+6q^2 \end{bmatrix} \\ \hat{\boldsymbol{r}} = \begin{bmatrix} 6p^2+2pq+4q^2 \\ 6p^2+6q^2 \end{bmatrix} \end{cases} \tag{6.55}$$

The optimal prediction coefficient vector can be obtained by calculation:

$$\boldsymbol{\alpha} = \begin{bmatrix} \alpha_1 \\ \alpha_2 \end{bmatrix} = \begin{bmatrix} 0 \\ 1 \end{bmatrix}$$

For the same problem, apply the proposed method by us:

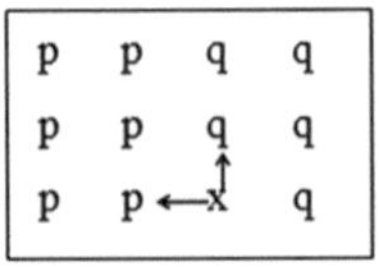

Fig. 6.2 Vertical edge prediction

$$\boldsymbol{R}_l = \begin{bmatrix} q & p \\ p & p \end{bmatrix}, \boldsymbol{r}_l = \begin{bmatrix} q \\ p \end{bmatrix} \tag{6.56}$$

We only need a smaller amount of calculations to obtain the same solution vector as the former (for the same point to be interpolated, the latter is reduced about 80 times of multiplications and several additions compared to the former). Obviously, the obtained result can reflect the edge feature of the image, which is not possible for traditional interpolation method.

3. **The limit of interpolation coefficient and the process of interpolation algorithm**

The value of the pixel point is within a range [a, b], so the value of $Y_{2i,2j}$ obtained by the above method should be between [a, b]. The matrix consisting of the image pixel points exists some singular matrices, and the solutions of these matrices obtained by numerical solution are usually divergent, thus the calculated pixel value exceeds [a, b] range. In order to solve this problem, the interpolation coefficient should be limited between [0, 1], namely, $0 < \alpha_i < 1, i = 1, 2, 3, 4$; and $\sum_{i=1}^{4} \alpha_i = 1$, to determine the interpolation coefficients by solving linear equations with constraint conditions:

$$\boldsymbol{R}_l \boldsymbol{\alpha}_l = \boldsymbol{r}_l \text{ and } 0 < \alpha_i < 1, i = 1, 2, 3, 4, \sum_{i=1}^{4} \alpha_i = 1 \tag{6.57}$$

The least squares can be used to solve the above linear equations with constraint conditions. The solved coefficient replaces the high-resolution interpolation coefficient $\boldsymbol{\alpha}_h$ into equation:

$$Y_{2i,2j} = \sum_{k=0}^{1} \sum_{l=0}^{1} \alpha_h^{2k+l+1} Y_{2(i+k)-1,2(j+l)-1} \tag{6.58}$$

The pixel values obtained are within the range [0, 255] and the singular points due to linear equation numerical solution divergence will not appear.

The main process of the image interpolation algorithm can be summarized as follows:

1. The original image pixel can be divided into two types: continuity or discontinuity of gray changes with neighboring pixels, namely, the extraction of the image edge by certain algorithm.
2. The interpolation of points not at the edge of the image is adopted linear interpolation, not only reducing computational complexity but also making the images become smoother.

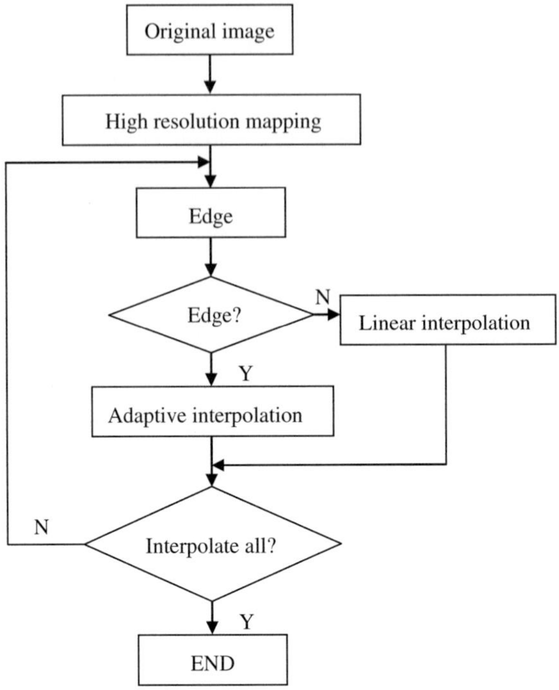

Fig. 6.3 Implementation diagram of adaptive image interpolation algorithm

3. For the point to be interpolated at the edge of the image, make use of similarity of surrounding pixels and adopt an interpolation algorithm of adaptive coefficients. The implementation process of interpolation algorithm is shown in Fig. 6.3.

6.4 Performance Assessment

6.4.1 POCS and MAP-Based Super-Resolution Methods

Firstly, the experiment adopts the Indian agriculture and forestry hyperspectral remote sensing image. We respectively compare the treatment effect of the linear interpolation, POCS-based super-resolution reconstruction method and MAP-based super-resolution reconstruction method; for the latter two methods, we will separately take the raw data and transform data as input data. In order to obtain supervision and evaluation, regard the original image down-sampling as low-resolution observation image and the original image as the reference image of high-resolution target image. Make use of the following formula for down-sampling:

$$x_l[n_1, n_2] = \frac{1}{N_1 N_2} \sum_{m_1=N_1 n_1}^{N_1(n_1+1)-1} \sum_{m_2=N_2 n_2}^{N_2(n_2+1)-1} x_h[m_1, m_2] \tag{6.59}$$

The obtained low-resolution image is shown in Fig. 6.4a, b gives the resolution improvement effect obtained by the linear interpolation, which is used for the comparison with the algorithm presented in this chapter. Note that all methods in this chapter are full spectrum processing simultaneously, and in order to facilitate the display and comparison, the following refers to the comparison of processing results of the 10th spectral band, with no special instructions later. Statistical results indicate that the relative error between linear interpolation effect and the reference image is 2.29 %. Relative error is calculated as the specific value of the entire pixel absolute error sum and the original image entire pixel sum.

In the first set of experiments, we implement POCS-based super-resolution recovery method, and the input and output are applied raw data. The obtained high-resolution target image is shown in Fig. 6.5a, and the relative error between the reference image and the resulting image is 2.25 %.

The second set of experiments adopts the input–output format from the transform domain to the transform domain and still uses the POCS algorithm. The obtained high-resolution target image is shown in Fig. 6.5b, and the relative error between the reference image and the resulting image is 2.13 %. Compared with the first set of experiments, the computing speed improves dozens of times but the error decreases. This result suggests that the data input based on the transform domain has the ability to lower the computation volume of the algorithm and a stronger image recovery capability. Two sets of experiments carry out the category statistics for error images, and the results are shown in Table 6.1. As can be seen by comparison, the data input based on the transform domain also plays a role in protecting the treatment effect of interest categories.

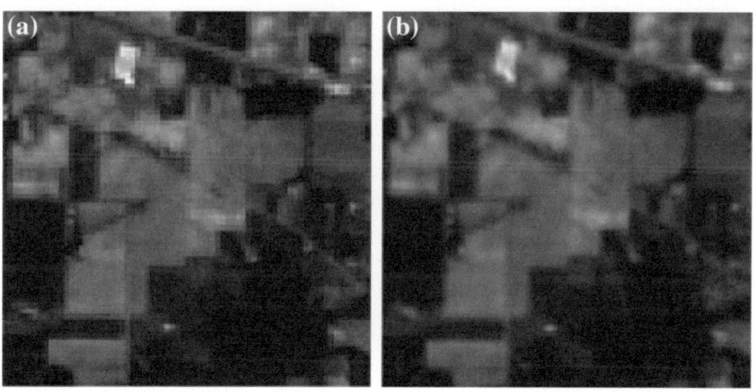

Fig. 6.4 The original low-resolution image and its linear interpolation result. **a** Low-resolution image. **b** Linear interpolation image

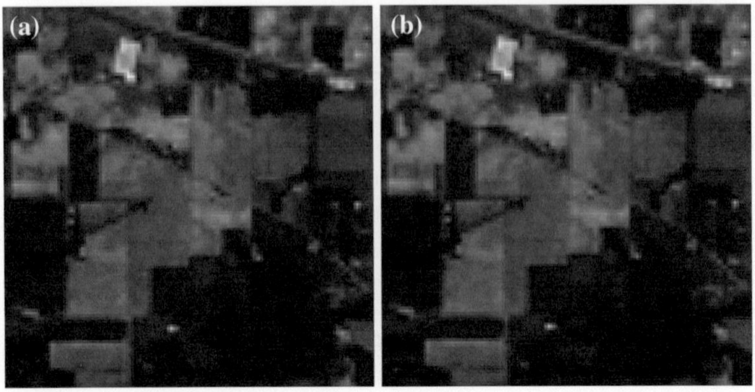

Fig. 6.5 Image recovery results based on POCS. **a** From the original domain to the original domain. **b** From the transform domain to the transform domain

Table 6.1 Different categories of recovery errors

	Category 2	Category 3	Category 10	Category 11	Category 14	Background	Other categories
Original domain input (%)	1.73	1.97	1.57	1.66	2.20	2.66	2.01
Transform domain input (%)	1.47	1.72	1.28	1.37	1.81	2.60	2.11

In the third set of experiments, we implements MAP-based super-resolution recovery method, and the input and output are applied raw data. The obtained high-resolution target image is shown in Fig. 6.6a, and the relative error between the reference image and the resulting image is 2.50 %.

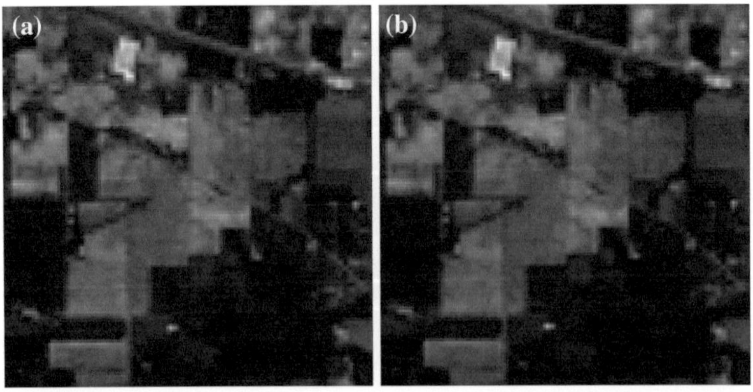

Fig. 6.6 Comparison of resolution improvement results in two ways. **a** From the original domain to the original domain. **b** From the transform domain to the transform domain

Table 6.2 Different categories of recovery errors

	Category 2	Category 3	Category 10	Category 11	Category 14	Background	Other categories
Original domain input (%)	1.97	2.18	1.82	1.91	2.47	2.91	2.27
Transform domain input (%)	1.67	2.10	1.47	1.55	2.10	3.00	2.39

The fourth set of experiments adopts the input–output format from the transform domain to the original domain and still uses the MAP algorithm. The obtained high-resolution target image is shown in Fig. 6.6b, and the relative error between the reference image and the resulting image is 2.22 %. The condition is similar in contrast with the previous two sets of experiments, and the computing speed in the fourth set of experiments improves dozens of times but the error decreases compared to the third set of experiments. This result suggests again that the data based on the transform domain has the ability to lower the computation volume of the algorithm and a stronger image recovery capability. Two sets of experiments carry out the category statistics for error images, and the results are shown in Table 6.2. As can be seen by comparison, the data input based on the transform domain indeed plays a role in protecting the treatment effect of interest categories.

In addition, this section also carries out experimental demonstration on the Santiago military hyperspectral image. Figure 6.7 shows a comparison of the 10th spectral band corresponding original image, down-sampling image, linear interpolation image and MAP transform domain image recovery image, and the visual effect (especially the treatment effect of three small airplanes in the upper left corner) further indicates the effectiveness of the proposed algorithm.

These results have small differences in the subjective evaluation, but the local enlarged image and objective evaluation can reflect the different effects of various methods. In general, the treatment effect of the original data as the input is significantly less than the treatment effect of the transform data as the input; the effect difference of POCS method and MAP method is not large, and both are superior to the linear interpolation effect; when the transform data is as the input, the algorithm can effectively protect the super-resolution effect of interest categories; when the transform data is as the input, the output with raw data or transform data have a small effect on the calculation processing effect, but often has a great impact on the amount of calculation, and therefore, the reasonable choice is necessary.

6.4.2 Dual Interpolation Method

On the one hand, the proposed algorithm is used in image compression system as the core technology to evaluate the effect; on the other hand, it is also directly

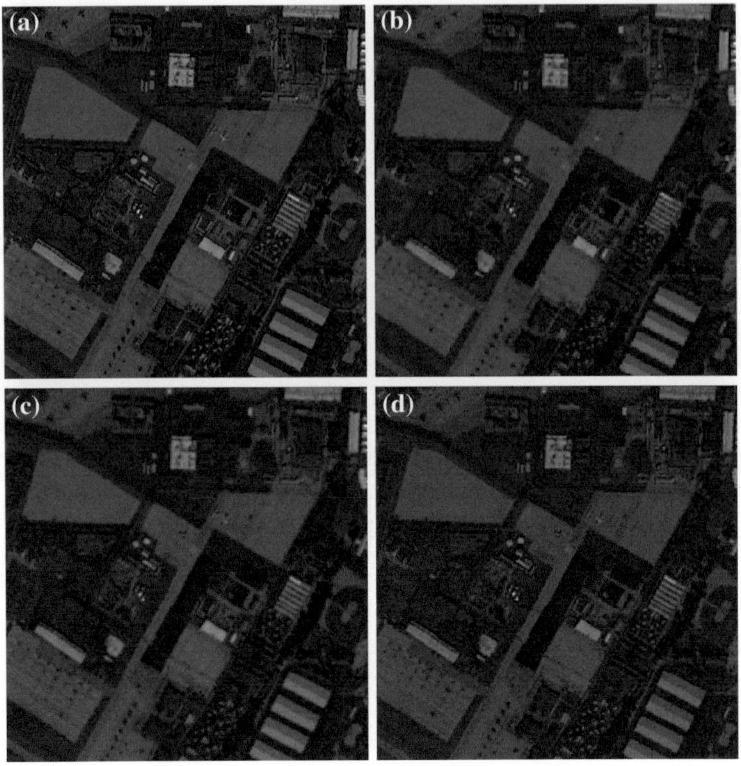

Fig. 6.7 Comparison of the resolution improvement effect of military images. **a** Original image. **b** Down-sampling image. **c** Linear interpolation image. **d** MAP transform domain recovery image

applied to remote sensing image processing to carry out subjective and objective evaluation.

1. **Application in the image compression system**

The experimental content in this section is assisted in the completion by the collaborator Chen Gang who is a graduate of Harbin Institute of Technology Imagery Institute. For the transmission of high quality images on the limited channels, this chapter uses a down-sampling compression method to improve the compression ratio. This method requires carrying out down-sampling of the source image at the compression end and reverts to the original size of image through the image super-resolution at the decompression end. This method can improve the effect of the compression ratio increase on the image quality, and the key to implement this method is the image super-resolution algorithm.

(a) Achievement of sampling in the video image
In order to increase the compression ratio and reduce the processing amount of data by the processor, the image down-sampling is carried out at the

compression end after the video A/D conversion. This chapter discusses the television guiding system with 400 kbps of channel bandwidth, which requires the source image compressed for more than 53 times. According to articles of Alfred et al. (2003), here select the down-sampling of the original image to the quarter, and then reuse H.26L algorithm to compress 14 times. Use the formula (6.60) for the image down-sampling:

$$x_l[n_1, n_2] = \frac{1}{N_1 N_2} \sum_{m_1=N_1 n_1}^{N_1(n_1+1)-1} \sum_{m_2=N_2 n_2}^{N_2(n_2+1)-1} x_h[m_1, m_2] \tag{6.60}$$

Here, $N1$ and $N2$ are taken as 2, namely, the down-sampling of the original image in the horizontal and vertical direction is to half of the original sample. Thus, the amount of data that needs to be compressed reduces to the original quarter, reducing the demands on the processor and ensuring the system real-time compression source video image.

(b) Specific applications of the interpolation algorithm
The compression end makes the original source image down-sampling to a quarter before the compression transmission and the decompression end decompresses a quarter of the original image. In order to display on the monitor, the decompression image needs the super-resolution operation to restore to its original size of image. According to the analysis of the foregoing, the conventional image super-resolution algorithm does not keep the edge very well with the blurred interpolated image; some mathematical tools are too complex, making the real-time achievement of the algorithm is difficult. Here, adopt the image super-resolution algorithm introduced in this chapter.
In order to reduce the amount of calculation of DSP processor, the algorithm is implemented in the FPGA. DSP puts the decompression data into the dual-port RAM, and FPGA reads the image data from the dual-port RAM and carries out super-resolution processing and then sends to the video D/A for display. The super-resolution process of one image is shown in Fig. 6.8. First, determine whether it is the edge of the image content, and adopt the linear

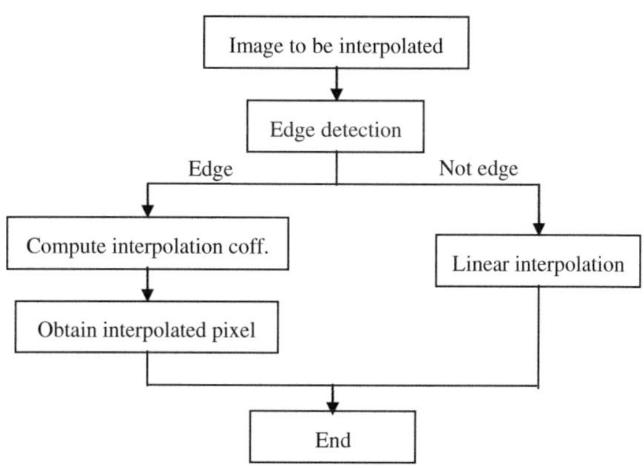

Fig. 6.8 Image super-resolution algorithm flowchart

interpolation with small amount of calculation for non-edge points; adopt the adaptive image super-resolution algorithm based on the spatial dual introduced in this chapter for the edge points. In the implementation process of the algorithm, first calculate the interpolation coefficient in the low-resolution space, and then map this set of coefficients to the high-resolution space, to calculate the pixel value of the point to be interpolated.

(c) Improvement on the image display quality of the interpolation

The image super-resolution algorithm in this chapter is applied to the image down-sampling compression process, not only reducing the amount of calculation of the compression algorithm but also improving significantly the recovered display image quality. This chapter selects three representative scenes to compare the image display quality in direct compression and down-sampling compression. Table 6.3 lists all evaluation indicators of the decompression image, when three pieces of remote sensing images are compressed more than 50 times directly and all evaluation indicators of the decompression image when the image down-sampling is to a quarter and then is decompressed 14 times. Comparison of data in the table shows that the image quality transmitted by the down-sampling compression method is better than the image quality transmitted by the direct compression method under the same compression ratio.

From Figs. 6.9, 6.10 and 6.11 below, we can visually see the improvement of down-sampling compression on the image quality with the super-resolution algorithm of this chapter. Of which, figure a is the recovered image after 52 times of direct compression, and we can see the block effect produced by too large compression ratio; figure b is the recovered image after 52 times of down-sampling compression, and compared to a image, b image is more delicate with clearer outline of the object.

2. **Other experiment evaluations of the interpolation effect**

In the first set of experiments, make use of the proposed algorithm to carry out the interpolation processing for a spectral band of military hyperspectral image of the San Diego airport, so as to obtain the subjective evaluation of the algorithm. The experimental results and the local area are shown in Fig. 6.12, which reflects good interpolation effect of the proposed method.

Table 6.3 Image quality comparison of direct compression and down-sampling compression

Image	Compression method	Compression ratio	SNR (dB)	PSNR (dB)	MSE
Flight	Direct compression	53.05	28.76	32.46	6.08
	Down-sampling compression	52.10	31.75	35.45	4.30
Ship	Direct compression	54.86	19.56	33.47	5.41
	Down-sampling compression	55.61	20.44	34.34	4.89
Tank	Direct compression	55.04	23.47	29.17	8.87
	Down-sampling compression	54.21	32.50	38.20	3.14

Fig. 6.9 Comparison of different compression methods of aircraft remote sensing images. **a** Direct compression. **b** down-sampling compression

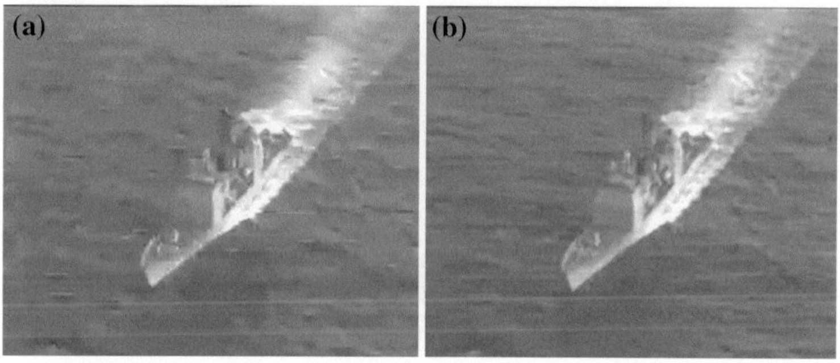

Fig. 6.10 Comparison of different compression methods of vessel remote sensing images. **a** Direct compression. **b** down-sampling compression

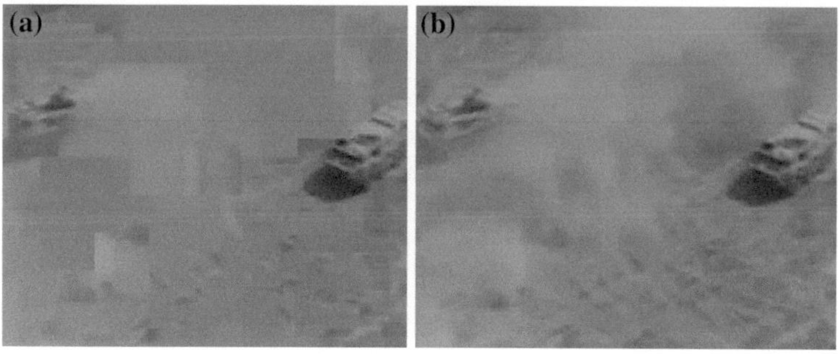

Fig. 6.11 Comparison of different compression methods of tank remote sensing images. **a** Direct compression. **b** Down-sampling compression

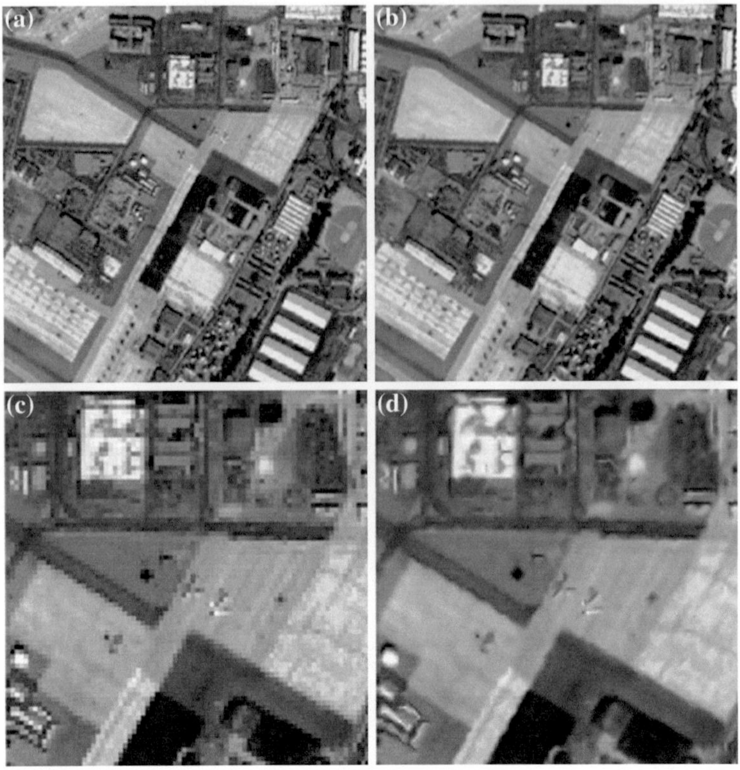

Fig. 6.12 Edge-based adaptive super-resolution result map. **a** Original image. **b** Interpolated image. **c** Original image (part). **d** Interpolated image (part)

Hyperspectral single spectral image is similar to the two-dimensional image of ordinary optical image, so the interpolation method in this chapter can be applied to the processing of ordinary optical image. In the second set of simulation experiments, we also apply this method to deal with the Lena optical image, which is widely used in the literature so as to compare and demonstrate with scholars in the same field. Select four algorithms and three kinds of evaluation criteria for comparison. These four algorithms are linear interpolation, EDIULC method, LAI method, and new method proposed by us. Three kinds of evaluation criteria are the peak signal noise ratio (PSNR), cross-correlation (CC) coefficient and combination factor (the combination of three different factors: loss of correlation, mean distortion, and variance distortion, CCMV) (Wang et al. 2002). First, introduce three kinds of evaluation criteria.

Set two equal-sized images to be compared respectively as X and Y, x_i and y_i is respectively the corresponding pixel point of X and Y, with the total number as N. Then, the three kinds of evaluation criteria are calculated as follows:

Table 6.4 Evaluation and comparison of several interpolation methods

Lena	Linear	LAI	EDIULC	New
PSNR	29.8931	29.4604	28.0456	29.9098
CC	0.9884	0.9872	0.9839	0.9884
CCMV	0.9882	0.9870	0.9838	0.9882

$$\mathrm{PSNR} = -10\log_{10}\left[\left(\sum_{i=1}^{N}\left(\frac{X(i)-Y(i)}{255}\right)^2\right)/N\right] \tag{6.61}$$

$$\mathrm{CC} = \left|\frac{\sum_{i=1}^{N} x_i y_i - N\bar{x}\bar{y}}{\sqrt{\left(\sum_{i=1}^{N} x_i^2 - N\bar{x}^2\right)\left(\sum_{i=1}^{N} y_i^2 - N\bar{y}^2\right)}}\right| \tag{6.62}$$

$$\mathrm{CCMV} = \frac{4\sigma_{xy}\bar{x}\bar{y}}{(\sigma_x^2+\sigma_y^2)[(\bar{x})^2+(\bar{y})^2]} \tag{6.63}$$

Of which, $\bar{x} = \frac{1}{N}\sum_{i=1}^{N} x_i$, $\bar{y} = \frac{1}{N}\sum_{i=1}^{N} y_i$, $\sigma_{xy} = \frac{1}{N-1}\sum_{i=1}^{N}(x_i-\bar{x})(y_i-\bar{y})$, $\sigma_x^2 = \frac{1}{N-1}\sum_{i=1}^{N}(x_i-\bar{x})^2$, $\sigma_y^2 = \frac{1}{N-1}\sum_{i-1}^{N}(y_i-\bar{y})^2$.

Select the standard optical image Lena and intercept part of sub region of the original image, and use the formula (6.60) for the down-sampling, and use a variety of methods to carry out interpolation for the down-sampling image, and the obtained image and a variety of indicators are shown in Table 6.4. The experimental results show that the method proposed in this chapter not only has better visual effects but also three kinds of measurement indicators have achieved the best results.

6.5 Chapter Conclusions

According to the imaging particularity of hyperspectral image and spectral information compensatory, this chapter researches and builds the relationship model between the low-resolution resource image and the high-resolution target image; according to the limited land class categories and the particularity of hyperspectral imaging applications, the pixels in the super-resolution model are characterized as weighted linear combination of interest spectral endmembers, thereby reducing the complexity of the algorithm, while focusing on the protection of the super-resolution effect of interest categories; in order to ensure the effective implementation of the super-resolution algorithm, this chapter researches the high efficiency and low complexity integrated operator offline calculation method and model inversion algorithm. The proposed super-resolution method makes use of hyperspectral image spectral information complementarity rather than relying on auxiliary information, through the associated interest spectrum endmember to achieve the super-resolution method with the interest category preservation property. The study is worth learning on dealing with mixed pixel interpretation, target detection and recognition as well as image matching and other problems, and is also easy to combine with hyperspectral image compression/decompression process to form an integrated algorithm.

The edge interpolation method of the linear prediction model application proposed in this chapter is based on the geometric duality of the model with linear prediction model of the low-resolution image to approximately correspond to the high-resolution linear prediction model, with self-adaptation, edge preservation feature, and simplicity, and efficiency and other advantages. The use of mixed interpolation method is not only necessary but also can reduce the amount of calculation and improve the interpolation effect, and can be easily applied to multispectral and hyperspectral image super-resolution processing.

References

Alfred M, Bruckstein ME, Ron K (2003) Down-scaling for better transform compression. IEEE Trans Image Process 12(9):66–81

Li X, Orchard MT (2001a) Edge-directed prediction for lossless compression of natural images. IEEE Trans Image Process 10(6):813–817

Li X, Orchard MT (2001b) New edge-directed interpolation. IEEE Trans Image Process 10 (10):1521–1527

Meng QW (2004) Satellite image restoration and treatment of MAP estimation algorithm based on super resolution[D]. Postdoctoral research report on the work of the Harbin Institute of Technology

Rodrigues L, Borges DL, Goncalves LM (2002) A locally adaptive edge-preserving algorithm for image interpolation. Proc Braz Symp Comput Graph Image Process XV: 7–10

Wang Z, Bovik AC, Lu L (2002) Why is image quality assessment so difficult. IEEE Int Conf 4:13–17

Chapter 7
Anomaly Detection Technique of HSI

Due to the significant value in both military area and civil area, hyperspectral remote sensing imagery detection has been widely developed in recent 20 years. Anomaly detection algorithms could detect pixels which is distinct from the surrounding background without any prior knowledge. In real applications, anomaly detection has received considerable interest in hyperspectral imaging since it's very difficult to obtain target information, and on the other hand, it is influenced by atmospheric transport, background and imaging equipment error.

7.1 Kernel Detection Algorithm Based on the Theory of the Morphology

In order to analyze hyperspectral remote sensing image accurately and stably, it is very necessary to comprehensively consider the spectral and spatial information of the hyperspectral data. In this section, the Morphology-Based Kernel RX detection algorithm is described. The algorithm uses the extended morphological closed operation for hyperspectral image feature extraction, in order to reach the purpose of reducing dimensions. By closed transformation of band selection in removing redundant at the same time, it also can be able to smooth spectrum data, avoid the discontinuity of the band information, effectively combine the spatial information, and spectral information. Then, anomalies are detected in hyperspectral imagery after the dimension reduction. KRX operator is used to detect anomalies in imagery to get the gray image of test results, and Area Close–Open of Gray morphology is used to filter the results above to get the final test results. KRX operator can make good use of spectral characteristic information of the image data to mine nonlinear statistical characteristics between bands. Area Close–Open of Gray morphology can have both closed to the spatial characteristics of image data, and effectively restrain background noise and interference, fill holes of images, in order to achieve lower false alarm probability and improve the detection probability. KRX operator and Area Close–Open of Gray morphology are combined in order to achieve good results (You 2011).

L. Wang and C. Zhao, *Hyperspectral Image Processing*,
DOI 10.1007/978-3-662-47456-3_7

7.1.1 Band Selection Based on Morphology

Mathematical morphology is a special kind of image processing technology. Its description language is set theory. It designs a set of concepts and methods that based on set operations and provides a unified and powerful tool to deal with the problems in the image. It describes the image features of various research objects and the relationship between objects by studying the geometrical characteristics of the objects in the image. Therefore, several basic concepts and operations of mathematical morphology are employed to achieve the purpose of processing and analysis through a flexible combination and decomposition of structural elements and an application of morphological transform sequences. The basic idea of mathematical morphology for image processing is that the structural elements are employed for displacement, intersection, and union operation to the original image, and then output the processed image. The thought of mathematical morphology algorithm is simple and intuitive, and the characteristic of geometric description is very suitable for processing and analyzing information related with visual information. Two basic operations of mathematical morphology are eroding and dilating. The two operations are defined initially in binary image, but have now been extended to the gray image. In the grayscale morphology, image is processed as a continuous value set. It is assumed that $f(x,y)$ is the input image, $b(x,y)$ is a structure element, and also is a sub-image function. The structural element b is employed for gray images dilating and eroding as shown in formulas (7.1) and (7.2).

$$d(x,y) = f \oplus b = \mathrm{Max}\{f(x-s,y-t) + b(s,t) | x-s, y-t \in D_f; s,t \in D_b\} \quad (7.1)$$

$$e(x,y) = f \otimes b = \mathrm{Min}\{f(x+s,y+t) - b(s,t) | x+s, y+t \in D_f; s,t \in D_b\} \quad (7.2)$$

where D_f and D_b are respectively the definition domain of f and b, and $x-s$, $x+s$, $y-t$, and $y+t$ must be within the domain of f, s, and t must be within the domain of b.

The main task of gray morphological operation is how to calculate the maximum or minimum value within the neighborhood of each pixel in the image, and that is closely related to the size and shape of the defined structural element. Structural element in accordance with the convex function is usually considered because the algorithm is simple and easy to implement. Usually, it is assumed that b is a square and $b(s,t) = 0$, where $s,t \in D_b$.

In essence, the result of the operation that A is opened by B is that set A is corroded by structural element B and then expanded by B. What can be known from the dilating and eroding definition is that open operation can smooth image contour, remove the part of the image which cannot contain structural element, which is to remove the tiny prominent part of image, some narrow part of the image or Bridges to the connection between the two objects. The boundary of the operation that A is opened by B is the collection of the boundaries that translations of B within A can achieve. Assume A is the input image, B is the image of structural element. The operation set A is opened by set B is as follows:

$$A \circ B = (A \otimes B) \oplus B \tag{7.3}$$

In essence, the result of the operation that A is closed by B is that set A is expanded by structural element B and then corroded by B. Close operation can also smooth image, remove the small holes in the original image, fill the small gap on the contour, and fuse the narrow gap and elongated curved mouth in the image. Assume A the input image, B the image of structural element. The operation set A is closed by set B is as follows:

$$A \bullet B = (A \oplus B) \otimes B \tag{7.4}$$

Due to dilating and eroding of duality, open operation and close operation can also get the corresponding duality theorem. The boundary that A is closed by B is the boundary of B when B is translated within A.

Next, mathematical morphology is to be extended to hyperspectral image processing.

In grayscale morphology, assume the numerical size of pixel as ordering relation to make calculate the maximum or minimum value, but in hyperspectral imagery, each pixel is multidimensional and comparing their sizes cannot be simple and intuitive. Therefore, the biggest challenge to extending the mathematical morphology to hyperspectral imagery is to define a proper ordering relation of the elements in the N dimensional vector space to sort the elements to determine the maximum and minimum elements. At present, extended mathematical morphology methods have been highly focused by the field of endmember extraction and classification of hyperspectral data.

In hyperspectral image processing, the introduction of a multidimensional vector measurement operator is employed to determine the multidimensional vector ordering relation of differences between the target and the background. The measurement operator is calculated by the cumulative distances from the structural elements within each pixel.

The mathematical expression is shown as formula (7.5):

$$D(f(x,y),b) = \sum_{s}\sum_{t} \mathrm{Dist}\{f(x,y),b(s,t)\} \tag{7.5}$$

with $x, y \in D_f$; $s, t \in D_b$, Dist is the pointwise linear distance measurement of N dimension vector. In order to effectively use the spectral and spatial information provided by hyperspectral data, cumulative distance D can sort the vector of structural elements according to the differences between the target and the background. According to the definition and narrative above, the eroding and dilating operation in hyperspectral data are defined as follows:

$$f \oplus b = \underset{s,t \in D_f}{\arg\_\mathrm{Max}} \{ D[f(x-s, y-t), b] \} \tag{7.6}$$

$$f \otimes b = \underset{s,t \in D_f}{\arg\_\mathrm{Min}} \{ D[f(x+s, y+t), b] \} \tag{7.7}$$

where arg _Max, arg _Min, respectively, denote pixel vector operator that make cumulative distance D achieve maximum and minimum. Through the above analysis, it is shown that the pixels with large differences from background in Structural elements are obtained from the extension to the dilating of the hyperspectral image, and the pixels similar with background in Structural elements are obtained from the eroding results, as shown in Fig. 7.1.

Now extended mathematical morphology is employed to build band selection method.

For computer processing and analysis of hyperspectral remote sensing image data, if all the original bands are used directly, the dimension of the spectral characteristics is quite high and AVIRIS data with the band numbers up to 224 are high-dimensional data of 224 dimensions.

For high-dimensional data processing and analysis, the method of linear transformation to decrease the dimension is adopted by the traditional method of multivariate data analysis (such as principal component analysis). This approach generally does not need all the band data, but first of all process all original band data through a mathematical transformation method. On the one hand, although this method reduces the dimension of hyperspectral remote sensing data, the band data after reducing dimensions of all the original data by some mathematical transformation is a linear or nonlinear combination of the original band data, and the physical meaning of the original band is changed. On the other hand, when the high-dimensional data dimension is higher, even if the measured sample size is very big, but compared with the dimension, sample size is always very little, the data points in high-dimensional space are often very sparse, high-dimensional space of some area is still empty and the "dimension disaster" problem arises. The stability

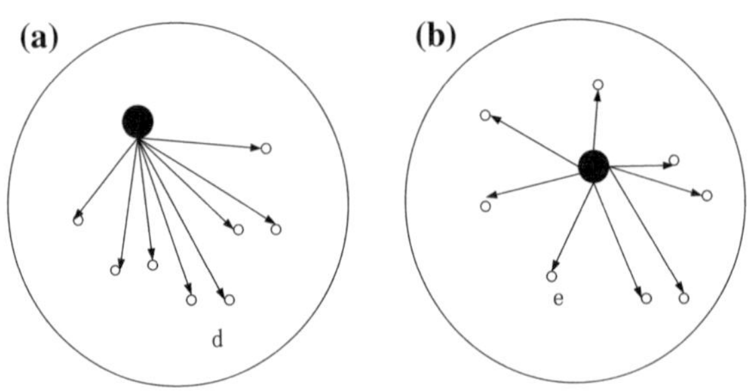

Fig. 7.1 **a** Expansion of dilating and **b** expansion of eroding in structural elements

of some statistical methods that make well in low-dimensional space becomes poor when extended to high-dimension space, and good effect cannot be obtained as in low-dimensional space processing and analysis.

Traditional optimal band selection methods of hyperspectral remote sensing are mainly based on the amount of information and classified separability. These methods often compare the relevance of the bands, the joint entropy of the bands and the best index. That is easy to cause the band information being discontinuous, effecting the target detection performance. Mathematical morphological close operation can smooth data, remove the small hole in the original image, fill the gap on the profile, fuse energy narrow opening, and gap, and a slender curved opening in the original image. Therefore, combining close operation to select the best bands can smooth data, and also reflect the whole data by sample space effectively.

Cross-correlation is widely used to describe the relevancy between two vectors. The correlation of spectral vector $\boldsymbol{x}$ and $\boldsymbol{y}$ is expressed as:

$$\rho = \frac{(x - \mu_x)^T \left(y - \mu_y\right)}{\left[(x - \mu_x)^T (x - \mu_x)\left(y - \mu_y\right)^T \left(y - \mu_y\right)\right]^{-\frac{1}{2}}} \tag{7.8}$$

where μ_x, μ_y are, respectively, the mean of spectral vector $\boldsymbol{x}$ and $\boldsymbol{y}$. Where $\rho \in [-1, 1]$, if ρ is closed to 1, the linear correlation of two spectral vectors is higher, and the shape of two spectral vector is more similar. Morphological close operation using the cross-correlation coefficient ρ instead of accumulated distance D, can effectively remove the redundancy, smooth data, and fill up data. Specific steps are as follows:

1. Map the cross-correlation curve of the hyperspectral image.
2. In order to expand the bands, make the minimum point of cross-correlation coefficient and the point whose ρ is the threshold T as the boundary point for partition processing.
3. Calculate the cross-correlation of the mean of two adjacent bands. Merge the two adjacent bands if the cross-correlation coefficient is greater than the threshold, and then corrode the band areas.
4. Calculate the cross-correlation of each band divided renewedly, and then select the maximum average correlation coefficient of the bands as a representative to be extracted.

7.1.2 *Kernel RX Algorithm Based on Morphology*

RX algorithm is one of the most widely used in anomaly detection algorithms. If the original sampling data are mapped to high-dimensional (possibly infinite dimensional) feature space by nonlinear mapping function ϕ, the KRX algorithm in kernel space is formed. The operator is expressed as:

$$\mathrm{RX}(\phi(\boldsymbol{x})) = \left(\phi(\boldsymbol{x}) - \boldsymbol{\mu}_{b\phi}\right)^T \boldsymbol{C}_{bf}^{-1} \left(\phi(\boldsymbol{x}) - \boldsymbol{\mu}_{b\phi}\right) \tag{7.9}$$

where $\boldsymbol{C}_{b\phi}$ is covariance matrix, and $\boldsymbol{\mu}_{b\phi}$ is estimate of the mean in the feature space. After the eigenvalue decomposition, RX algorithm in kernel space is expressed as follows:

$$\mathrm{RX}(\phi(\boldsymbol{r})) = \left(\phi(\boldsymbol{r}) - \boldsymbol{\mu}_{b\phi}\right)^T \boldsymbol{X}_{b\phi} \boldsymbol{\beta} \boldsymbol{\Lambda}_{\phi}^{-1} \boldsymbol{\beta}^T \boldsymbol{X}_{b\phi}^T \left(\phi(\boldsymbol{r}) - \boldsymbol{\mu}_{b\phi}\right) \tag{7.10}$$

where $\boldsymbol{\beta} = (\boldsymbol{\beta}^1, \boldsymbol{\beta}^2, \ldots, \boldsymbol{\beta}^N)^T$ is the feature vector normalized by square root eigenvalue of the kernel matrix $\boldsymbol{K}$. But because of the high-dimension data (or even infinite dimensional), the algorithm cannot be realized by mapping the original data into a high-dimensional feature space directly through a nonlinear mapping function. In order to avoid a direct calculation for formula (7.9), it is necessary to adopt kernel learning method theory, using kernel function in the original data space to indirectly realize the inner product in high-dimensional feature space, which is indirectly calculated by $K(\boldsymbol{x}_i, \boldsymbol{x}_j) = \langle \phi(\boldsymbol{x}_i) \cdot \phi(\boldsymbol{x}_j) \rangle$ for formula (7.10). Thus, KRX algorithm detection operator finally is as follows:

$$\mathrm{RX}_K(\boldsymbol{r}) = \left(\boldsymbol{K}_{\boldsymbol{r}}^T - \boldsymbol{K}_{\mu\boldsymbol{b}}^T\right)^T \boldsymbol{K}_b^{-1} \left(\boldsymbol{K}_{\boldsymbol{r}}^T - \boldsymbol{K}_{\mu\boldsymbol{b}}^T\right) \tag{7.11}$$

where,

$$\begin{cases} \boldsymbol{K}_{\boldsymbol{r}}^T \equiv K(\boldsymbol{X}_b, \boldsymbol{r}) - \dfrac{1}{N} \displaystyle\sum_{i=1}^{N} K(\boldsymbol{x}_i, \boldsymbol{r}) \\ \boldsymbol{K}_{\mu\boldsymbol{b}}^T \equiv \dfrac{1}{N} \displaystyle\sum_{i=1}^{N} K(\boldsymbol{x}_i, \boldsymbol{X}_b) - \dfrac{1}{N^2} \displaystyle\sum_{i=1}^{N} \sum_{j=1}^{N} K(\boldsymbol{x}_i, \boldsymbol{x}_j) \end{cases} \tag{7.12}$$

KRX operator maps original hyperspectral data to high-dimensional feature space to detect anomaly pixels and mine the nonlinear statistical characteristics of bands in hyperspectral images, which can improve the detection performance and achieve good results. But the algorithm makes only from the spectral information and spatial analysis, ignoring the spatial correlation between pixels. Therefore, morphological opening and closing operation of gray area for the gray-scale image after KRX operator detection (AOC), can take advantage of the spectral and spatial information, and analyze hyperspectral remote sensing image accurately.

Area morphology is based on two algebra operators: area morphological open operation and close operation, which are based on gray set theory. Compared with traditional morphological filter, area open and close filter have no any structure elements. The operations of the operators to image are not to change the shape. So that, the area open operation is to get rid of the targets with high brightness which does not meet the minimum area. Similarly, the area close operation is to remove the targets with low brightness that does not meet the minimum area. The targets here refer to the image gray connected region, and a gray set is a binary image that carried out through the gray decomposition in gray image. In the discrete domain D and image pixel p, we define a gray set S_l, where the gray $l \in [0, L-1]$. L is the largest gray level in gray image

$$S_l = \begin{cases} 1, & \text{if } l(p) \geq l \\ 0, & \text{otherwise} \end{cases} \quad (7.13)$$

In gray set S_l, connected regions is defined as follows:

$$C_{S_l}(p) = \{q : \exists P_{l \geq 1}(p, q)\} \quad (7.14)$$

where, $P_{l \geq 1}(p, q)$ is a connected path from p to q. For $S_l(\cdot) = 1$, can use four neighborhoods or eight neighborhoods. For the gray set S_l, area open operation is expressed as follows:

$$S_l \circ (a) = \{p : \exists |C_{S_l}(p)| \geq a\} \quad (7.15)$$

where a is the minimum area parameters (the number of pixels related), and $C_{S_l}(p)$ is the connected area determined by the point p. $|C_{S_l}(p)|$ shows the area of the connected regions. Area close operation is a similar operation to the complement of S_l.

For gray image, area open and close operation for all gray sets can be carried out by the stack operation. The definition of area open operations to image I with the scale a is as follows:

$$I \circ (a) = \sum_{l=1}^{L-1} S_l \circ (a) \quad (7.16)$$

The definition of area close operation is as follows:

$$I \bullet (a) = \sum_{l=1}^{L-1} S_l \bullet (a) \quad (7.17)$$

A morphological filter is a cascading of the open and close operations, which can filter out the noise block that is smaller than structural elements in the target image. It can cut long lap joint firstly with open operation, eliminating peaks, then connect short break with close operation, filling the holes. The filtering process of the morphological close–open filter is in contrast, but can attain the similar results. based on area open operation and area close operation, a series of area operators can be derived, such as area open–close operation (AOC) and area close and open operation (ACO). Obviously, AOC and ACO are the area mathematical morphology filter.

The application of ACO algorithm to the hyperspectral anomaly detection can effectively remove the influence of white noise, and reduce the false alarm probability in order to achieve good detection performance. The combination of area morphology ACO operation and KRX operator is proposed as a new algorithm of a hyperspectral target detection based on mathematical morphology, namely the ACO–KRX algorithm, and specific steps are as follows:

1. Reduce the dimension of the original hyperspectral images by extended mathematical morphology close operation.
2. Make anomaly detection for the reduced image data by KRX operator to get gray image after detection.
3. Make filter processing for the gray image that is the results of detection with AOC operation, getting gray image which has been removed the noise.
4. Set the threshold to binarize gray image, getting the final results.

7.2 Adaptive Kernel Anomaly Detection Algorithm

The introduction of kernel method could fully mine the nonlinear information in hyperspectral bands, and greatly improve the performance of the original linear detection algorithm. In a wide range of hyperspectral anomaly detection algorithms based on kernel methods (such as KRX algorithm, KPCA algorithm, Kernel Fisher separation algorithm, etc.), the kernel function generally chooses the Gaussian radial basis kernel function (RBF), because of the good generality of the Gaussian RBF and the shift invariance of its function. An important factor to determine the performance of the algorithm is the width of Gauss radial basis kernel function. The parameter selection artificially from a large number of experiments in most detection algorithms of kernel methods not only reduces the generality of the algorithm but also increases the detection work. And unified global detection parameters are difficult to adapt to the complex background for the utilization of local detection model in anomaly detection, leading to a decline in detection performance. This chapter will construct calculation model of the width factor of the Gaussian radial basis kernel function to form adaptive hyperspectral anomaly detection based on kernel method by the adaptive estimation for kernel parameters. The anomaly target detection algorithm based on support vector data description proposed by Banerjee is selected to avoid inversing covariance matrix in the process of detection, and achieve speed-up compared with kernel RX algorithm. According to the characteristics of hyperspectral data and local anomaly detection model, this chapter will analyze the change relationship between kernel parameters and the overall standard deviation of local background. The second-order statistic in different bands of the local background is used to construct the local kernel parameters, which makes it possible to adaptively adjust the kernel parameters according to different background distributions. It overcomes the problem that the traditional support vector description algorithm with constant kernel parameters leading to a low-performance in complicated background (Mei 2009).

7.2.1 The Method of Support Vector Data Description

For HSI, consider the spectra from a set of pixels with common characteristics in a hyperspectral image, denoted as $T = \{\boldsymbol{x}_i : i = 1, \ldots, M\}$. We seek the minimum enclosing hypersphere by the center $\boldsymbol{a}$ and radius $R > 0$. The hypersphere with the smallest volume is found by minimizing the R^2 to describe the data set and makes all the sample points (or as much as possible) are included in the sphere.

$$S = \left\{\boldsymbol{x} : \|\boldsymbol{x} - \boldsymbol{a}\|^2 < R^2\right\} \tag{7.18}$$

This is a constrained optimization problem stated as:

$$\min(R) \quad \text{subject to} \quad x_i \in S, \quad i = 1, 2, \ldots, M \tag{7.19}$$

The center $\boldsymbol{a}$ and radius R of the minimum enclosing hypersphere are found by optimizing the following Lagrangian:

$$L(R, \boldsymbol{a}, \alpha_i) = R^2 - \sum_i \alpha_i \left(R^2 - \langle \boldsymbol{x}_i, \boldsymbol{x}_i \rangle + 2\langle \boldsymbol{a}, \boldsymbol{x}_i \rangle - \langle \boldsymbol{a}, \boldsymbol{a} \rangle\right) \tag{7.20}$$

Taking the partial derivatives of L with respect to R and $\boldsymbol{a}$ setting them to 0 yields.

$$\frac{\partial L}{\partial R} = 0 \Rightarrow \sum_i \alpha_i = 1 \tag{7.21}$$

$$\frac{\partial L}{\partial \boldsymbol{a}} = 0 \Rightarrow \boldsymbol{a} = \frac{\sum_i \alpha_i \boldsymbol{x}_i}{\sum_i \alpha_i} \tag{7.22}$$

By combining (7.21) and (7.22), a simple expression for the center of the sphere is found

$$\boldsymbol{a} = \sum_i \alpha_i \boldsymbol{x}_i \tag{7.23}$$

Substituting (7.21) and (7.23) into (7.20) yields another expression

$$L = \sum_{i} \alpha_i \langle \boldsymbol{x}_i, \boldsymbol{x}_i \rangle - \sum_{i,j} \alpha_i \alpha_j \langle \boldsymbol{x}_i, \boldsymbol{x}_j \rangle \tag{7.24}$$

After optimizing L with respect to $\boldsymbol{a}$, it is typical to discover that a large amount of the α_i is zero. The training examples with nonzero α_i are called support objects and lie on the boundary of the hypersphere. Therefore, the SVDD yields a sparse representation of the support expressed entirely in terms of the support objects.

Once the minimum enclosing hypersphere has been found, outliers are identified by testing whether the test examples lie in or outside the hypersphere. The corresponding decision rule to detect an anomaly for a test pixel $\boldsymbol{y}$ is

$$\begin{aligned} \mathrm{SVDD}(\boldsymbol{y}) &= \|\boldsymbol{y} - \boldsymbol{a}\|^2 \\ &= \left(\boldsymbol{y} - \sum_{i} \alpha_i \boldsymbol{x}_i\right)^T \left(\boldsymbol{y} - \sum_{i} \alpha_i \boldsymbol{x}_i\right) \\ &= \langle \boldsymbol{y}, \boldsymbol{y} \rangle - 2 \sum_{i} \alpha_i \langle \boldsymbol{y}, \boldsymbol{x}_i \rangle + \sum_{i,j} \alpha_i \alpha_j \langle \boldsymbol{x}_i, \boldsymbol{x}_j \rangle \underset{H_0}{\overset{H_1}{\gtrless}} R^2 \end{aligned} \tag{7.25}$$

H_0 holds for the similar, as H_1 for heterogeneous.

In most instances, a hypersphere does not provide a tight representation for the supported data in the original input space. Additional flexibility is needed to model arbitrarily complex distributions. To address this, the nonlinear SVDD maps the data from the input space to a higher dimensional feature space through the use of a mapping $\Phi(x)$, and models the support of the distribution as a minimum enclosing hypersphere in the feature space. This hypersphere corresponds to a tighter boundary for the support region in the original input space and then nonlinear SVDD yields.

We now seek the smallest hypersphere in the feature space including the entire set of mapped training examples $T_\phi = \{\phi(\boldsymbol{x}_i) : i = 1, \ldots, M\}$, what is as follows:

$$S_\phi = \left\{ \phi(\boldsymbol{x}) : \left\| \phi(\boldsymbol{x}) - \boldsymbol{a}_\phi \right\|^2 < R^2 \right\} \tag{7.26}$$

where $\boldsymbol{a}_\phi$ is the sphere center in feature space, we must solve the following constrained optimization problem:

$$\min(R) \quad \text{subject to} \quad x_i \in S, \quad i = 1, 2, \ldots, M \tag{7.27}$$

The corresponding Lagrangian is expressed as

$$L(R, \boldsymbol{a}_\phi, \alpha_i) = R^2 - \sum_{i} \alpha_i \left(R^2 - \langle \phi(\boldsymbol{x}_i), \phi(\boldsymbol{x}_i) \rangle + 2\langle \boldsymbol{a}_\phi, \phi(\boldsymbol{x}_i) \rangle - \langle \boldsymbol{a}_\phi, \boldsymbol{a}_\phi \rangle \right) \tag{7.28}$$

Setting the partial derivatives of L with respect to R and $\boldsymbol{a}$ to zero, and substituting the results into (7.24) yields

$$L = \sum_i \alpha_i \langle \phi(\boldsymbol{x}_i), \phi(\boldsymbol{x}_i) \rangle - \sum_{i,j} \alpha_i \alpha_j \langle \phi(\boldsymbol{x}_i), \phi(\boldsymbol{x}_i) \rangle \tag{7.29}$$

With $\alpha_i \geq 0$, $\sum_i \alpha_i = 1$. This also gives an expression for the center of the sphere

$$\boldsymbol{a}_\Phi = \sum_i \alpha_i \phi(\boldsymbol{x}_i) \tag{7.30}$$

Because of the non-support vector corresponding to $\alpha = 0$, and the support vector corresponding to $\alpha > 0$, the hypersphere center is the center of gravity of the support vectors given the optimal weights $\boldsymbol{a}$ of L. The corresponding decision rule to detect an anomaly for a test pixel y with the mapping $\phi(\boldsymbol{y})$ in the feature space is

$$\begin{aligned} \mathrm{SVDD}(\phi(\boldsymbol{y})) &= \left\| \phi(\boldsymbol{y}) - \boldsymbol{a}_\phi \right\|^2 \\ &= \left(\phi(\boldsymbol{y}) - \sum_i \alpha_i \phi(\boldsymbol{x}_i) \right)^T \left(\phi(\boldsymbol{y}) - \sum_i \alpha_i \phi(\boldsymbol{x}_i) \right) \\ &= \langle \phi(\boldsymbol{y}), \phi(\boldsymbol{y}) \rangle - 2 \sum_i \alpha_i \langle \phi(\boldsymbol{y}), \phi(\boldsymbol{x}_i) \rangle + \sum_{i,j} \alpha_i \alpha_j \langle \phi(\boldsymbol{x}_i), \phi(\boldsymbol{x}_j) \rangle \end{aligned} \tag{7.31}$$

Due to the high-dimension of the feature space, it is not easy to achieve a direct calculation formula in the feature space. Because of the inner product in formula (7.29), (7.31), Kernel functions provide a computationally efficient technique to implicitly map the data into the induced feature space and compute the inner product. Kernel expression of the dot product in feature space is as follows:

$$K(\boldsymbol{x}_i, \boldsymbol{x}_j) = \langle \phi(\boldsymbol{x}_i), \phi(\boldsymbol{x}_j) \rangle \tag{7.32}$$

Under the formula (7.32), the formula (7.29), (7.31) can be simply expressed as

$$L = \sum_i \alpha_i K(\boldsymbol{x}_i, \boldsymbol{x}_i) - \sum_{i,j} \alpha_i \alpha_j K(\boldsymbol{x}_i, \boldsymbol{x}_j) \tag{7.33}$$

$$\mathrm{SVDD}(\phi(\boldsymbol{y})) = k(\boldsymbol{y}, \boldsymbol{y}) - 2 \sum_i \alpha_i K(\boldsymbol{y}, \boldsymbol{x}_i) + \sum_{i,j} \alpha_i \alpha_j K(\boldsymbol{x}_i, \boldsymbol{x}_j) \tag{7.34}$$

As can be seen, we do not need to know the concrete nonlinear mapping function ϕ and carry on the corresponding dot product computation. By mapping the dot product of the high-dimension feature space into kernel function in low-dimensional input space, it is easier for linear SVDD expansion into the nonlinear SVDD, and the key is to find a suitable kernel function.

7.2.2 Adaptive Kernel Anomaly Detection Algorithm

The selection of kernel function is the key to the property of nonlinear SVDD. In a sense, the choice of kernel function determines the mapping function and the generation of high-dimensional feature space. The Gaussian Radial basis function (RBF) is employed as most anomaly detection operator based on kernel method. Because the RBF kernel is shift invariant kernel, which depends only on the change $\boldsymbol{x}-\boldsymbol{y}$ between $\boldsymbol{x}$ and $\boldsymbol{y}$, and has nothing to do with the absolute position of each spectral vector. The expression is as follows:

$$K(\boldsymbol{x},\boldsymbol{y}) = \exp\left(\frac{-\|\boldsymbol{x}-\boldsymbol{y}\|^2}{c}\right) \tag{7.35}$$

In the formula, the width c of the radial basis kernel function is critical. Choosing the parameters properly can obtain ideal interface. In the most algorithms based on kernel technique, the selection of kernel parameters c is determined by experience of a large amount of experiments, affecting the performance of detection operator. In addition, because most anomaly detection algorithms use local detection model with the global kernel parameter, the local adaptability is affected due to the complexity of the real hyperspectral data. The value of c is determined by minimizing the expectation that support vector accounted for the total number of the training set. The expression for the optimization is as follows:

$$\tilde{c} = \min_c P_{\text{fa}} \approx \min_c \mathrm{E}\left[\frac{\#\mathrm{SV}}{N_{\text{tr}}}\right] \tag{7.36}$$

where $E[\frac{\#\mathrm{SV}}{N_{\text{tr}}}]$ is the expectation, #SV is the number of support vectors, and N_{tr} is the number of samples.

The choice of the radial basis kernel function simplifies the SVDD to

$$\mathrm{SVDD}(\boldsymbol{y}) = 1 - 2\sum_i \alpha_i K(\boldsymbol{y},\boldsymbol{x}_i) + \sum_{i,j} \alpha_i\alpha_j K(\boldsymbol{x}_i,\boldsymbol{x}_j) \tag{7.37}$$

The experiment is carried out by using a set of simulated data. A total of 50 sample points are represented by the (o) points, which is clearly a data set of nonlinear distribution. The data are described with SVDD algorithm, where the nonlinear SVDD uses a Gaussian radial basis kernel function.

Figure 7.2a is the boundary curves given by linear SVDD based on Gauss distribution assumption. Obviously, the decision interface cannot effectively characterize the distribution characteristics of the data sets, but the nonlinear SVDD with kernel method can give the ideal decision interface as shown in Fig. 7.2b–d. But, the value of c greatly affects the shape of the decision interface. As shown in Fig. 7.2b, with the smaller value of c, over-segmentation situation makes similar

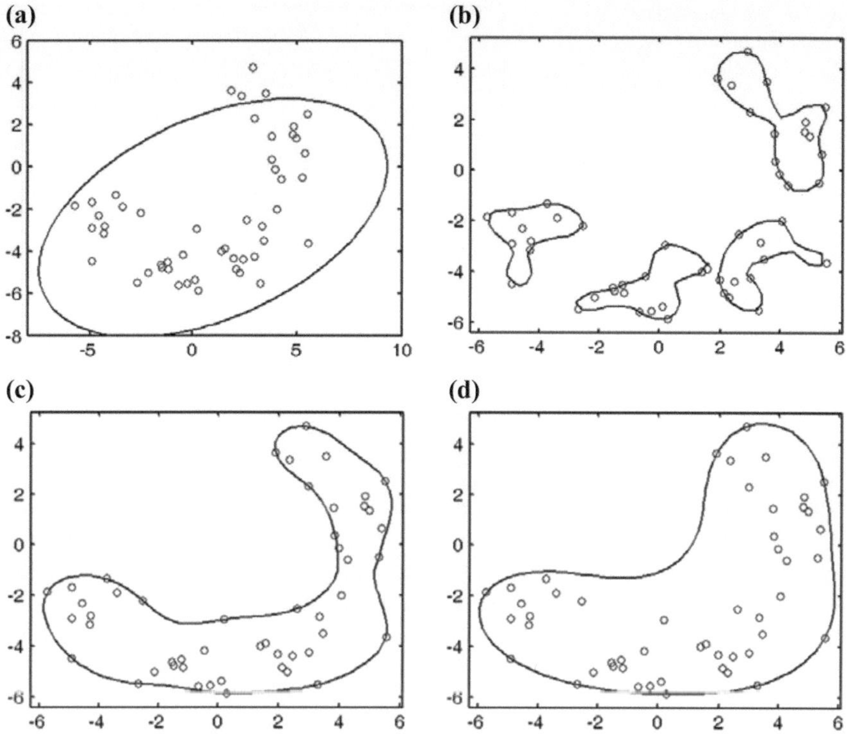

Fig. 7.2 The interface between the analog data given by SVDD. **a** Linear SVDD, **b** $c = 2$, **c** $c = 6$ and **d** $c = 20$

elements be sentenced to heterogeneities. While in Fig. 7.2d, as the value of c is relatively large to make the decision interface too loose, heterogeneities are easily sentenced to similar elements. An appropriate c is critical to the ideal decision interface as shown in Fig. 7.2c. It also shows that the selection of c is related with the distribution of the data itself.

According to the method of kernel parameter selection adopted by Banerjee, analysis of the simulated data in the different c is shown in Table 7.1. With the increase in c, the number of support vector is reduced, and the percentage of the support vector in the total training sample is reduced. However, in Fig. 7.2, it can be seen that the classifying boundary cannot describe the data distribution better with the increase of c. Therefore, minimizing the number of support vector cannot be simply used to the optimized estimation of kernel parameter c.

Table 7.1 The relationship between the value of c and support vector

c	2	6	10	15	20	30
#SV	28	15	14	12	11	9
#SV/N (%)	56	30	28	24	22	18

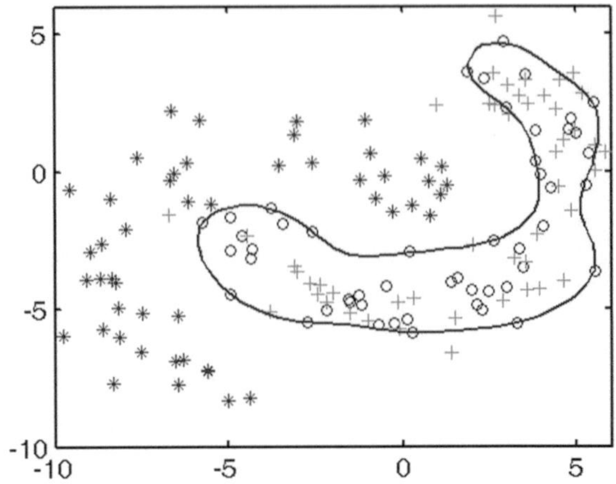

Fig. 7.3 Simulation data distribution of cross-test

The probability of false alarm detection is divided into two categories. One situation is to sentence the similar to be heterogeneous, referred to as the false negative rate (P_{fn}), the other situation is to sentence the heterogeneous to be similar, referred to as the false positive rate (P_{fp}). The optimal value of c is got by the cross-check. The simulation data is divided into three parts. Two hundred of the training samples are represented by the (o), 200 of the same similar test data with the same distribution as the training sample are represented by the (+) and 200 of the heterogeneous test data are represented by the (*). The distribution is shown in Fig. 7.3 (in order to facilitate observation, set the 50 sample points for each sample plotted on the graph).

Define the cost function P_{loss} as the weighted sum of the false negative rate and the false positive rate.

$$P_{\mathrm{loss}} = \beta P_{\mathrm{fn}} + (1 - \beta) P_{\mathrm{fp}} \tag{7.38}$$

where $0 < \beta \leq 1$, here $\beta = 0.5$. With different value of the kernel parameter c, firstly support vector description is obtained by training samples, and then tests them respectively with the similar data and heterogeneous data. Finally, the false negative rate curve (short dotted line), the false positive rate curve (dashed line), and the cost function curve (solid line) are shown in Fig. 7.4. It can be obviously seen that with the increase in c, the false negative rate is decreasing while false positive rate is increasing, and the cost function gets the smaller values in the interval range $c \in [6, 8]$. The decision surface obtained in this range is reasonable. The value of c is associated with the given sample data distribution. Of course, according to different requirements, you can weigh between the false negative rate and false positive rate, and adjust the value of β to obtain the cost function needed.

Local detection model is applied in hyperspectral anomaly detection. As the complexity of feature distribution determines the complexity of hyperspectral data distribution, in different spatial locations background distribution is different.

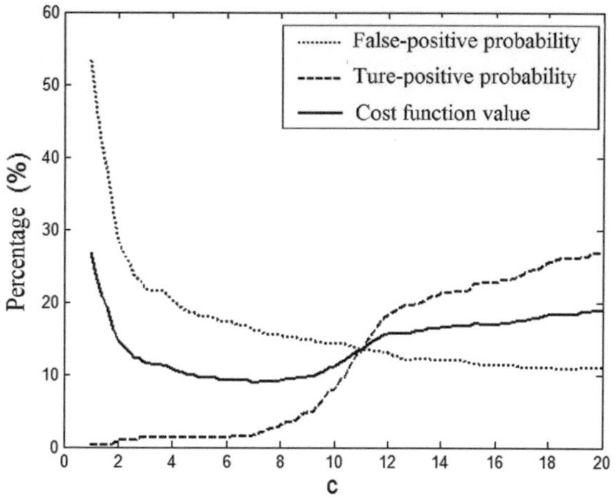

Fig. 7.4 The result curve of cross-check

Therefore, it is not suitable for the local detection operator using a global detection parameter. In order to detect the potential targets in different background, the kernel parameter c should be changed with background.

Gaussian radial basis kernel function is the vector relationship description based on the Euclidean distance of the two vectors $\boldsymbol{x} = [x_1, x_2, \ldots, x_J]$, $\boldsymbol{y} = [y_1, y_2, \ldots, y_J]$

$$\|\boldsymbol{x} - \boldsymbol{y}\| = \sqrt{\sum_{i=1}^{\mathrm{ND}} (x_i - y_i)^2} \tag{7.39}$$

M pixels in background window of hyperspectral data with ND bands are expressed as $\boldsymbol{P}_b = [\boldsymbol{p}_1, \boldsymbol{p}_2, \ldots, \boldsymbol{p}_M]$. The local background standard deviation of each band is defined as $\sigma_i\ (i = 1, 2, \ldots, \mathrm{ND})$

$$\sigma_i = \sqrt{\frac{\sum_{k=1}^{M} (p_{ik} - \mu_{bk})^2}{ND - 1}} \tag{7.40}$$

where ND is the number of the bands, $p_{ik} \in \boldsymbol{p}_k$ and μ_{bk} is the mean of the local background. And then, the standard deviation sum of each band in the background window is defined as the local total standard deviation of the window.

$$\sigma_{\mathrm{sum}} = \sum_{i=1}^{J} \sigma_i \tag{7.41}$$

As the sample data distribution is related to the value c, in a simulated data experiment, $\sigma_{\mathrm{sum}} = 6.7098$ in the simulation sample data obtained and ideal interval $c \in [6, 8]$ is experimentally obtained. Therefore, the sum of standard deviations of the samples can provide a more concise and reasonable measurement for the discrete degree of data distribution. Here, think c and σ_{sum} are positive correlation.

$$c = f(\sigma_{\text{sum}}) \propto \sigma_{\text{sum}} \tag{7.42}$$

Here, the linear function is used to describe this relationship

$$c = f(\sigma_{\text{sum}}) = a\sigma_{\text{sum}} + b \tag{7.43}$$

where $a > 0$, $b \geq 0$.

In the process of anomaly detection, because the priori information of the anomaly pixel cannot be got in advance, cross-check cannot be effectively carried out. But through derivation and experimental analysis, a better adaptive way to solve c is obtained. In order to prevent the rise of false negative rate, in the background where the range of the spectral curve and the distance between the spectrums are wide, the total standard deviation and kernel parameter of the local background are large. Relatively, in order to improve the detection probability and reduce false positive rate, in the background where the range of the spectral curve and the distance between the spectrums are narrow, the total standard deviation and kernel parameter of the local background are small. Thus, σ_{sum} provides an effective measure for the threshold distance in background spectrum, which makes results in different backgrounds better than that with constant kernel parameter method. When the detection window slides in hyperspectral image, the corresponding background total standard deviation is calculated to adjust kernel parameters adaptively. The process shows low computing complexity, in which the calculations can be performed in advance or in the process of detection, without massive calculations. The process of hyperspectral anomaly detection based on ASVDD is shown in Fig. 7.5.

7.3 Construction of Spectral Similarity Measurement Kernel in Kernel Anomaly Detection

Kernel method has obtained very good effect for nonlinear information of hyperspectral imagery, and how to select the kernel function is a main factor for algorithm performance. But for a long time, almost all the kernel-based hyperspectral processing algorithms choose Gaussian radial basis kernel function. The choice of kernel function is too monotonous. And from the characteristics of the kernel function itself, Gaussian radial basis kernel is based on L_2 distance, which has some limitations. In addition, in terms of hyperspectral data itself, the difference in the different materials is mainly reflected in the change of the spectrum. Although mixed pixels caused by poor spatial resolution make it different between real spectrum and pure spectrum, there is still a big difference between the spectrums of different materials.

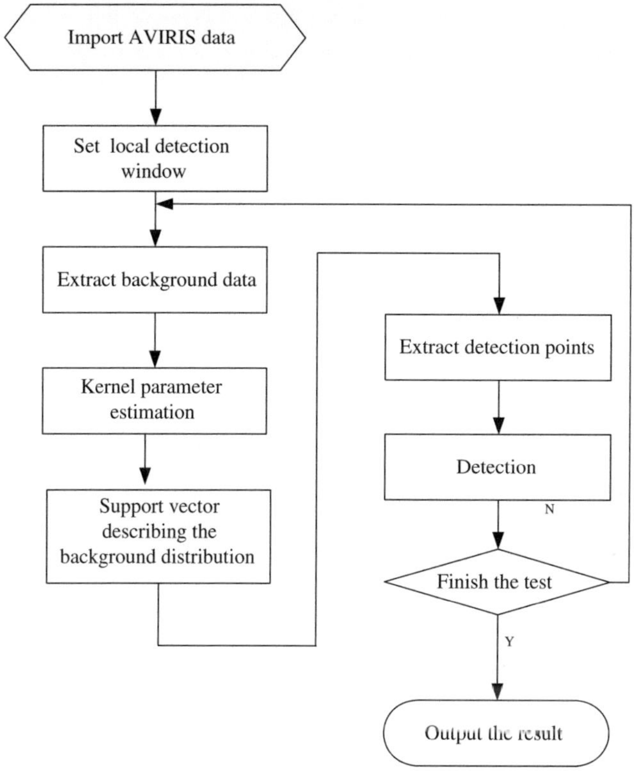

Fig. 7.5 Flow chart of hyperspectral anomaly detection algorithm based on ASVDD

Therefore, considering the distribution characteristics of HSI, especially the spectral curve, it is very important to construct a new kernel function suitable for HSI processing (Mei 2009).

7.3.1 The Limitations of Gaussian Radial Basis Kernel

The Gaussian radial basis kernel is a shifting constant kernel, shown as follows

$$K(\boldsymbol{x},\boldsymbol{y}) = \exp\left(\frac{-\|\boldsymbol{x}-\boldsymbol{y}\|^2}{c}\right) \tag{7.44}$$

The kernel only depends on the relative position between the two data points, so the expression is

$$K(\boldsymbol{x},\boldsymbol{y}) = K(\boldsymbol{x}+z,\boldsymbol{y}+z) \tag{7.45}$$

where $z \in \boldsymbol{X}$, the shifting constant kernel can be expressed in another form

$$K(\boldsymbol{x},\boldsymbol{y}) = f(d(\boldsymbol{x},\boldsymbol{y})) \tag{7.46}$$

where $d(\boldsymbol{x},\boldsymbol{y})$ is a distance metric, in Gaussian radial basis kernel function, the expression is

$$d(\boldsymbol{x},\boldsymbol{y}) = \|\boldsymbol{x}-\boldsymbol{y}\| = \sqrt{(\boldsymbol{x}-\boldsymbol{y})^T(\boldsymbol{x}-\boldsymbol{y})} \tag{7.47}$$

As a typical representative of the shifting constant kernel, the advantage of the Gaussian radial basis kernel mainly lies in the strong ability for grasping local information, on the other hand, it also has some limitations, mainly reflected in the following two aspects:

1. The global nature of the Gaussian radial basis kernel is poor. The ability of kernel mapping for transferring global information is poor from the low-dimensional space to the feature space, and the information transmission is lossy.
   For example: For 3 pixels, $\boldsymbol{x} = [a,b]^{\mathrm{T}}$, $\boldsymbol{y} = [a+c,b+d]^{\mathrm{T}}$, $z = [a-c,b-d]^{\mathrm{T}}$. since $d(\boldsymbol{x},\boldsymbol{y}) = d(\boldsymbol{x},z)$, so for Gaussian radial basis kernel $K(\boldsymbol{x},\boldsymbol{y}) = K(\boldsymbol{x},z)$, the information provided by $K(\boldsymbol{x},\boldsymbol{y})$ and $K(\boldsymbol{x},z)$ has nothing to do for distinguishing $\boldsymbol{y}$ and z, which means it is impossible to distinguish $\boldsymbol{y}$ and z only by $\boldsymbol{x}$ without other pixels. So, there is information loss and the Gaussian radial basis kernel is a local kernel.
   The integral kernel $K(\boldsymbol{x},\boldsymbol{y})$ can distinguish the three pixels above. The kernel is an inner product function of two vectors, so the absolute position information can be carried and the global nature is strong.
2. There are problems for the width of the Gaussian radial basis kernel. Because of the uneven spatial distribution, there exists less learning phenomenon in sparse regions, while more learning phenomenon in dense regions, which are the local risks of the constant kernel. The problem is better solved through the local selection of the adaptive kernel width as discussed in the previous chapter.

7.3.2 *Spectral Similarity Measurement Kernel Function*

Spectral similarity measure could describe the spectral shape similarity, making it widely used in hyperspectral classification, spectral unmixing, etc. But, because of its linear properties, the discerning ability is not ideal. The new kernel function is constructed through the spectral similarity measure to be employed in hyperspectral anomaly detection.

1. **Derivation of Spectral Similarity Measurement Kernel Function**

Spectral vector (or spectral curve) can be used to describe the characteristics of the pixel in hyperspectral data. In the view of classification, similarity measure can be used to describe the similarity degree between the two spectral curves. Two most commonly used similarity measure methods are presented in the following part.

Spectral angle measure makes the spectra as a multidimensional vector. It calculates the generalized angle of two spectral vectors, the smaller the angle is, the more similar the spectral vectors are.

$$\cos(\alpha) = \frac{\boldsymbol{x} \cdot \boldsymbol{y}}{||\boldsymbol{x}|| \cdot ||\boldsymbol{y}||}, \quad \alpha \in \left[0, \frac{\pi}{2}\right] \tag{7.48}$$

where α is a generalized spectral angle.

Spectral Angle Mapper has been applied in hyperspectral classification. But from the view of spectral discrimination ability, spectral angle has certain limitations.

Take three types of objects in Fig. 7.6 as an example. In order to facilitate the representation, only two-dimensional vector is carried out for the instructions. The calculation of spectral angle between samples of class A and B are $\cos(\alpha) = \boldsymbol{b}_1 \cdot \boldsymbol{b}_2/(\|\boldsymbol{b}_1\| \cdot \|\boldsymbol{b}_2\|)$, $\cos(\beta) = \boldsymbol{a} \cdot \boldsymbol{b}_2/(\|\boldsymbol{a}\| \cdot \|\boldsymbol{b}_2\|)$, as shown in figure. It is shown that $\alpha > \beta$ and the similar degree of spectral vector $\boldsymbol{a}$ and $\boldsymbol{b}_2$ is higher than that of and $\boldsymbol{b}_1$ and $\boldsymbol{b}_2$, which is clearly not appropriate. To solve this problem, we must centralize all spectral vectors, which is not easy. Such as in anomaly detection, only the local spectral vector can be obtained each time, while in the kernel-based algorithm, the original data need to be mapped into high-dimensional feature space, and it is difficult for centralizing.

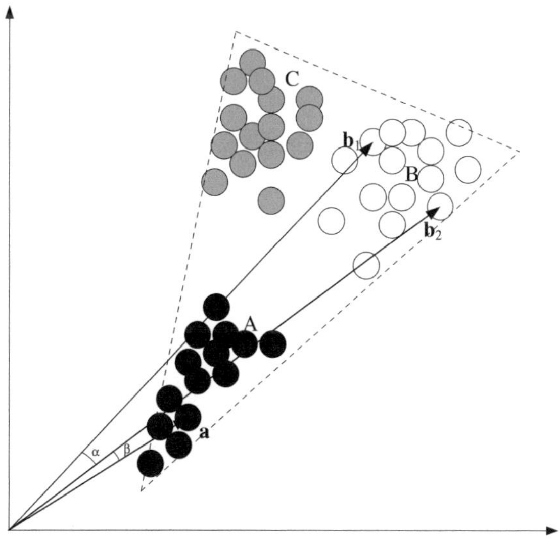

Fig. 7.6 The spectral vector angle

Correlation coefficient is widely used to describe the similarity degree between two vectors, the correlation coefficient of the spectral vectors $\boldsymbol{x}$ and $\boldsymbol{y}$ is expressed as

$$\rho(\boldsymbol{x},\boldsymbol{y}) = \frac{(\boldsymbol{x}-\overline{\boldsymbol{x}})^{\mathrm{T}}(\boldsymbol{y}-\overline{\boldsymbol{y}})}{\left[(\boldsymbol{x}-\overline{\boldsymbol{x}})^{\mathrm{T}}(\boldsymbol{x}-\overline{\boldsymbol{x}})(\boldsymbol{y}-\overline{\boldsymbol{y}})^{\mathrm{T}}(\boldsymbol{y}-\overline{\boldsymbol{y}})\right]^{\frac{1}{2}}} \tag{7.49}$$

where $\overline{\boldsymbol{x}}$ and $\overline{\boldsymbol{y}}$ are the mean of $\boldsymbol{x}$ and $\boldsymbol{y}$, respectively. $\rho \in [-1, 1]$, the closer to 1 the parameter ρ is, the higher linear correlation degree of the two spectral vectors is, and that is also to say the shapes of the two spectral vectors are more similar.

With the cross-correlation coefficient, spectral similarity kernel could be defined as follows:

$$K_s(x,y) = \begin{cases} \exp\left(\frac{-\cot[\pi(\rho+1)/4]}{\theta}\right), & \rho \in (-1,1] \\ 0, & \rho = -1. \end{cases} \tag{7.50}$$

where cot is the trigonometric cotangent function, θ is defined as the similar critical factor of the spectral similarity kernel, and formula (7.50) is a kernel function.

The definition of Gram matrices and the positive definite matrix is the necessary and sufficient condition of a function as a kernel function. In order to facilitate the proof, the definitions are described as follows:

Gram matrix: Set a function $\varphi : X^2 \rightarrow \boldsymbol{K}$, ($\boldsymbol{K}(i,j) \in R$ or C, R is the real set, C is the complex set) and a sample $\boldsymbol{x}_1,\ldots,\boldsymbol{x}_m \in X$, so the matrix $\boldsymbol{K} : \boldsymbol{K}(i,j) = K(\boldsymbol{x}_i,\boldsymbol{x}_j)$ with size of $m \times m$ is Gram matrix of the function φ with $\boldsymbol{x}_1,\ldots,\boldsymbol{x}_m$.

Positive definite matrix: In the complex matrix $\boldsymbol{K}$ of $m \times m$, if for any $c_i \in C$, $\sum_{i,j} c_i c_j \boldsymbol{K}(i,j) \geq 0$ is found, then the matrix $\boldsymbol{K}$ is a positive definite matrix.

Kernel function: Set X a non-empty set, a function k defined in $X \times X$. If they generate a positive definite Gram matrix to all $m \in N$ (N is natural numbers set) and $\boldsymbol{x}_1,\ldots,\boldsymbol{x}_m \in X$, k is defined as the positive definite kernel, simply as the kernel.

The proof process of spectral similarity measurement kernel is as follows:

Proposition 1 *K_s is a symmetric function.*

Proof From the definition of K_s, for any vector $\boldsymbol{x}$, $\boldsymbol{y}$, $\rho(\boldsymbol{x},\boldsymbol{y}right) = \rho(\boldsymbol{y},\boldsymbol{x})$. So $K_s(\boldsymbol{x},\boldsymbol{y}) = K_s(\boldsymbol{y},\boldsymbol{x})$ is found and K_s is a symmetric function. □

Proposition 2 *The matrix $\boldsymbol{K}(i,j) = K_s(\boldsymbol{x}_i,\boldsymbol{x}_j)$ generated by the function K_s is a positive definite matrix.*

Proof It can be learned from Proposition 1 that K_s is a symmetric function. Then $\boldsymbol{K}(i,j) = \boldsymbol{K}(j,i)$, so the matrix is symmetric.

Since $K_s \in R$ and $K_s(\boldsymbol{x},\boldsymbol{y}) \geq 0$, $\boldsymbol{K}$ is a real symmetric matrix and $\boldsymbol{K}$ is positive definite.

According to Definition 3, K_s can be determined as a positive definite kernel, which is called Spectral Similarity Measurement Kernel.

2. **The Nature of the Spectral Similarity Measure Kernel** □

A brief explanation of the spectral similarity measure kernel is as follows

Proposition 3 *The spectral similarity measure kernel K_s has the nature of translational invariance.*

Proof Set $\boldsymbol{x}, \boldsymbol{y} \in X$, $a, b \in R, \boldsymbol{I} \in X$ a unit vector, then the following formula is found.

$$K_s(\boldsymbol{x}, \boldsymbol{y}) = K_s(\boldsymbol{x} + a\boldsymbol{I}, \boldsymbol{y} + b\boldsymbol{I}) \tag{7.51}$$

According to the definition of spectral similarity measure kernel, it is expressed as

$$\rho(\boldsymbol{x}, \boldsymbol{y}) = \frac{(\boldsymbol{x} - \overline{\boldsymbol{x}})^{\mathrm{T}}(\boldsymbol{y} - \overline{\boldsymbol{y}})}{\left[(\boldsymbol{x} - \overline{\boldsymbol{x}})^{\mathrm{T}}(\boldsymbol{x} - \overline{\boldsymbol{x}})(\boldsymbol{y} - \overline{\boldsymbol{y}})^{\mathrm{T}}(\boldsymbol{y} - \overline{\boldsymbol{y}})\right]^{\frac{1}{2}}}$$

$$K_s(\boldsymbol{x}, \boldsymbol{y}) = \begin{cases} \exp\left(\frac{-\cot[\pi(\rho+1)/4]}{\theta}\right), & \rho \in (-1, 1] \\ 0, & \rho = -1. \end{cases}$$

Set $\boldsymbol{x}' = \boldsymbol{x} + a\boldsymbol{I}$, $\boldsymbol{y}' = \boldsymbol{y} + b\boldsymbol{I}$, then we could have $\overline{\boldsymbol{x}'} = \overline{\boldsymbol{x} + a\boldsymbol{I}} = \overline{\boldsymbol{x}} + a\boldsymbol{I}$. In the same way, $\overline{\boldsymbol{y}'} = \overline{\boldsymbol{y} + b\boldsymbol{I}} = \overline{\boldsymbol{y}} + b\boldsymbol{I}$. So $\boldsymbol{x}' - \overline{\boldsymbol{x}'} = \boldsymbol{x} + a\boldsymbol{I} - \overline{\boldsymbol{x}} + a\boldsymbol{I} = \boldsymbol{x} - \overline{\boldsymbol{x}}$ and $\boldsymbol{y}' - \overline{\boldsymbol{y}'} = \boldsymbol{y} - \overline{\boldsymbol{y}}$. Then,

$$\begin{aligned} \rho(\boldsymbol{x} + a\boldsymbol{I}, \boldsymbol{y} + b\boldsymbol{I}) &= \rho(\boldsymbol{x}', \boldsymbol{y}') \\ &= \frac{(\boldsymbol{x}' - \overline{\boldsymbol{x}'})^{\mathrm{T}}(\boldsymbol{y}' - \overline{\boldsymbol{y}'})}{\left[(\boldsymbol{x}' - \overline{\boldsymbol{x}'})^{\mathrm{T}}(\boldsymbol{x}' - \overline{\boldsymbol{x}'})(\boldsymbol{y}' - \overline{\boldsymbol{y}'})^{\mathrm{T}}(\boldsymbol{y}' - \overline{\boldsymbol{y}'})\right]^{\frac{1}{2}}} \\ &= \frac{(\boldsymbol{x} - \overline{\boldsymbol{x}})^{\mathrm{T}}(\boldsymbol{y} - \overline{\boldsymbol{y}})}{\left[(\boldsymbol{x} - \overline{\boldsymbol{x}})^{\mathrm{T}}(\boldsymbol{x} - \overline{\boldsymbol{x}})(\boldsymbol{y} - \overline{\boldsymbol{y}})^{\mathrm{T}}(\boldsymbol{y} - \overline{\boldsymbol{y}})\right]^{\frac{1}{2}}} \\ &= \rho(\boldsymbol{x}, \boldsymbol{y}) \end{aligned} \tag{7.52}$$

And we will have $K_s(\boldsymbol{x}, \boldsymbol{y}) = K_s(\boldsymbol{x}', \boldsymbol{y}') = K_s(\boldsymbol{x} + a\boldsymbol{I}, \boldsymbol{y} + b\boldsymbol{I})$. □

The translational invariance of the spectral similarity measure kernel illustrates that the value of the kernel function has nothing to do with the change of the relative distance but is closely associated with the shape of the spectrum, which enhances the hyperspectral data recognition ability for the change of the amplitude level of the same spectral curve caused by shadow, amplitude levels and so on.

7.4 Performance Assessment

7.4.1 Effect Testing of Morphology-Based Kernel Detection Algorithm

In the experiment, a part of real AVIRIS hyperspectral Santiago Airport data is used for simulation experiment to verify the validity of the ACO–KRX algorithm. The size of the sub-image is 100 × 100. The target number is large (four planes) and each target has small pixels. Select one of the planes as a target for testing in order to clearer show the effectiveness of the ACO–KRX algorithm. The size of the selected image is 30 × 30. The tenth band of the 100 × 100 image and the selected 30 × 30 sub-image are shown in Fig. 7.7.

In the experiment, adaptive subspace derivation of the data source is firstly adopted and formula (7.8) is used to calculate the correlation coefficients between each adjacent band. The correlation coefficient curves of each band are shown in Fig. 7.8. There are strong correlations between different bands and the HSI data are redundant. In order to eliminate redundancy, band selection is firstly addressed using the above mentioned band selection algorithm based on extended mathematical morphology. Set the minimum and the selected correlation coefficient threshold as the boundary values, which is somehow equivalent to data expansion of the HSI. The hyperspectral data space with the entire 126 bands is divided into 12 continuous subspaces, as shown in Table 7.2. Calculate the correlation coefficient between the mean of two adjacent band regions. If the correlation coefficient is greater than the threshold value, merge the two adjacent band regions. It is the equivalent to eroding operation of the band regions. At last, the spectral space is divided into nine continuous band areas, shown in Table 7.3. The entire process is equivalent to close operation for the HSI data, which can smooth data and avoid discontinuous band information. Then, calculate the correlation coefficient of each band and select the band with the largest average correlation coefficient as the representative of the group. Finally, the obtained HSI data after dimension reduction is used for the subsequent detection.

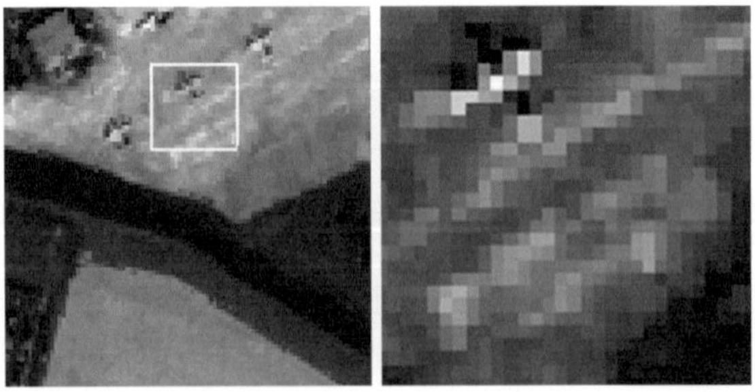

Fig. 7.7 Tenth band and the ground target distribution

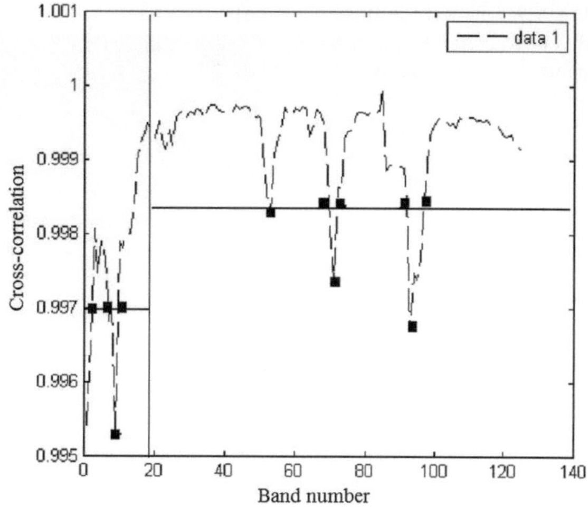

Fig. 7.8 The curve of correlation coefficients between adjacent bands

Table 7.2 Morphology-dilating-based band grouping

Group	1	2	3	4	5	6
Number of bands	1–2	8–7	8–9	10–14	18–53	58–69
Group	7	8	9	10	11	12
Number of bands	70–71	72–73	78–91	92–93	98–97	98–126

Table 7.3 Morphology-eroding-based band grouping

Group	1	2	3	4	5	6	7	8	9
Number of bands	1–2	8–7	8–9	10–14	18–53	58–73	78–93	98–97	98–126

According to image size, spatial resolution and the size of the abnormal target, the outer/ middle/ inner window sizes of ACO–KRX algorithm are set as 11 × 11, 9 × 9, and 3 × 3, respectively. The kernel function used in the algorithm is Gaussian radial basis kernel function with radial basis kernel width σ as its only one parameter. The experiment is very sensitive to the value σ. The optimal parameter is obtained through a large number of simulation experiments, and the parameter is set as 40.

The final ACO–KRX target detection results of real HSI data are shown in Fig. 7.9a, b. To facilitate analysis and comparison, the traditional RX algorithm and Kernel-based RX algorithm (KRX) are used for abnormal target detection, respectively. Final binary results with optimal thresholds are shown in Fig. 7.9c, d. It is obvious that the detection effect of the traditional RX algorithm is not ideal since it is originated from multispectral image anomaly detection. If it is directly used in HSI and the strong correlation between bands in hyperspectral image is ignored, more false alarms will be produced. As a result, dimension reduction is

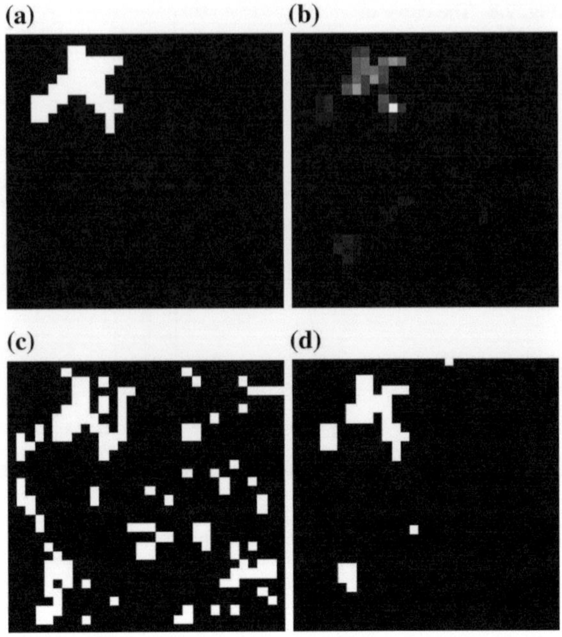

Fig. 7.9 The test results on different algorithms under the optimal threshold value. **a** The detection of gray image based on ACO–KRX. **b** The detection of binary image based on ACO–KRX. **c** The detection of binary image based on RX. **d** The detection of binary image based on KRX

needed in the original RX algorithm for the HSI detection. From the comparison between Fig. 7.9b, d, it can be seen that the detection result obtained from ACO–KRX algorithm is better than KRX algorithm. Although both the KRX algorithm and ACO–KRX algorithm use the nonlinear information between spectral bands, the ACO–KRX algorithm takes into account both the spectral and the spatial characteristics of HSI data, resulting in a lower false alarm rate with the same detection rate. Filter based on area mathematical morphology is used to remove the noise block smaller than structural elements. The gray closing operation is used to connect the short breaks and fill the holes. Then the opening operation is used to cut elongated lap and eliminate the spikes. Through these steps, the noise is removed and the result image is smoothed.

In order to illustrate the superiority of the algorithm more specifically, target number, pixels number of the target and the pixels number of the false alarm under the same detection threshold value are given to compare the performance. The comparison results are shown in Table 7.4.

Table 7.4 Comparison of the algorithm performance

Detection method	The pixel number in the target	The pixel number in the false alarm
ACO–KRX	39	1
KRX	31	9
RX	28	12

From Table 7.4, it can be clearly seen that the proposed algorithm can detect more targets with more pixels in the target region and less pixels in the false alarm region, and it fully proves the superior performance of the proposed algorithm.

7.4.2 Effect Testing of Adaptive Kernel Anomaly Detection Algorithm

In order to verify the effectiveness of the proposed ASVDD algorithm, the AVIRIS Santiago Airport HSI data source is used for simulation experiments. The image used in the experiment and the true target distribution are shown in Fig. 7.10. There are 38 abnormal targets in the image. Cross-validation experiment is performed before detection. Two different distribution regions A and B are respectively used as shown in Fig. 7.10, where the region A is complex with a large discrete degree $\sigma_{sum} = 21.372$ and the region B is relatively single with a small discrete degree $\sigma_{sum} = 9.775$. The spectrum curves for the two random regions A and B region are shown in Fig. 7.11. It can be easily seen from the figure that the spectrum discrete degree of region A is larger than that of the region B.

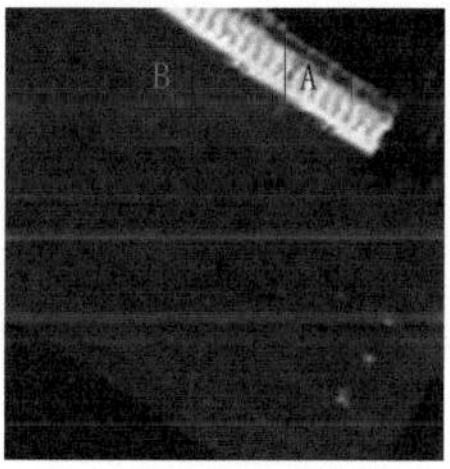

Fig. 7.10 The first band image of HSI in the experiment

Fig. 7.11 Spectrum comparison between region A and B

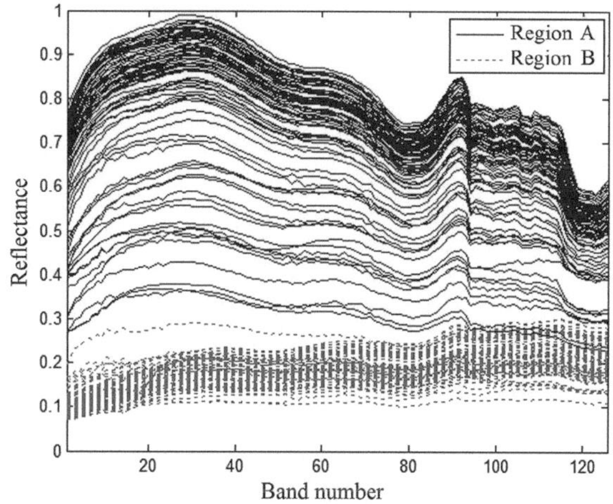

In the cross-validation experiment, 300 samples are randomly selected from region A and B, respectively. And then, 150 samples are respectively and randomly selected from the above 300 selected samples in every region as the training samples of the two regions A and B. The rest samples in each region are combined as one data set for testing. By setting different kernel parameters, support vector optimizations are respectively made for A and B, and then test data sets are used for the cross-check. The cost function P_{loss} is set as the mean of the negative false rate and positive false rate. The testing results are shown in Figs. 7.12 and 7.13, respectively.

From the cross-validation results of Figs. 7.12 and 7.13, it can be seen that region A has a larger spectral discrete degree and a wider spectral vector distance and the negative false rate is high when the kernel parameter c is small. As we can see that, for region A, if the kernel parameter c is smaller, a lot of background will

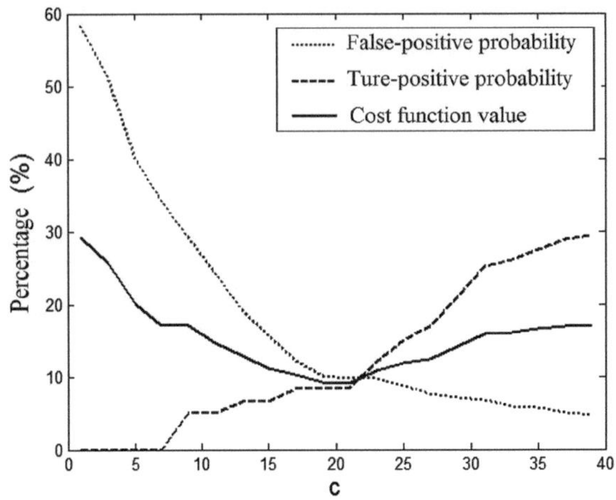

Fig. 7.12 The cross-validation results of region A as the training sample

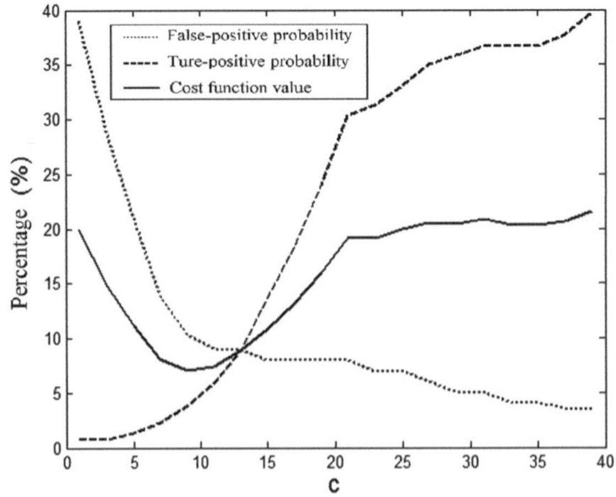

Fig. 7.13 The cross-validation results of region B data as the training sample

be detected as abnormal points. But for region B, because the spectral discrete degree is small and the spectral vector distance is narrow, the positive false rate is high when the kernel parameter c is small, which means if a larger kernel parameter c is used in the region B, some abnormal points will be seen as background and won't be detected. It can also be seen that the region A has a smaller cost function value when $c \in [18, 22]$, while the region B has a smaller cost function value when $c \in [9, 11]$. The total standard deviation of the two regions both lay in their optimal regions of the kernel parameter c, respectively.

The proposed adaptive detection algorithm based on support vector data description is used in HSI anomaly targets detection and the SVDD detection algorithm with the global kernel parameters, marked as SVDD, is used for comparison. As shown in Fig. 7.14, the concentric double window for detection is sliding on the HSI data, the background is in the outside window and the current pixel to be detected is in the inner window. In the experiment, after the normalization, according to the spatial resolution and the target size, the outside window is set as 11 × 11 and the inner window is set as 3 × 3. The value of c is set as σ_{sum}. ($c = a\sigma_{\text{sum}} + b$, subject to $a = 1, b = 0$). In SVDD algorithm, the value of the kernel parameter c is set as 40 after large trials of experiments. Three-dimensional representation of the ASVDD detection result is shown in Fig. 7.15. From the detection results in Figs. 7.14 and 7.15, it can be clearly seen that the ASVDD algorithm is better than the SVDD algorithm on the inhibition of false alarm. This is because by using adaptive kernel parameters, different kernel parameters are generated according to the backgrounds within the moving of the moving window, resulting in a stronger adaptability for complex hyperspectral backgrounds. So, the detection performance is better than the traditional SVDD algorithm.

Receiver operating characteristics (ROC) curve is used to describe the relations between the detection probability and the false alarm probability, providing a quantitative analysis of the algorithm performance. The Pf and Pd under different conditions are obtained by different threshold segmentation. The comparison ROC of the ASVDD algorithm and SVDD algorithm is shown in Fig. 7.16. It can be seen

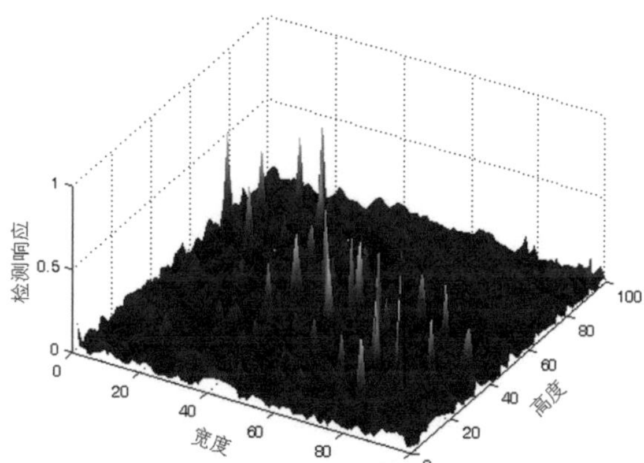

Fig. 7.14 The detection result of SVDD

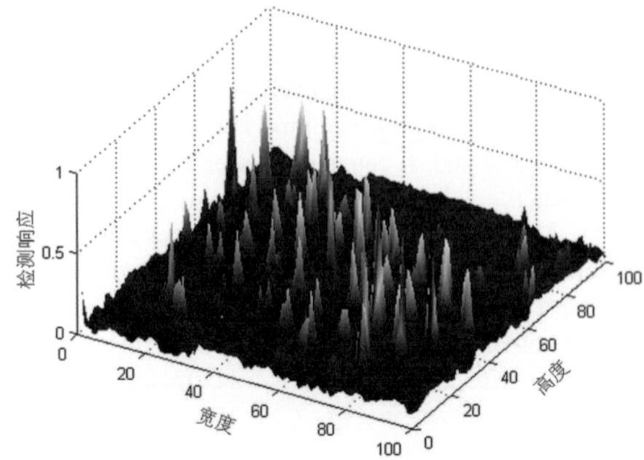

Fig. 7.15 The detection result of ASVDD

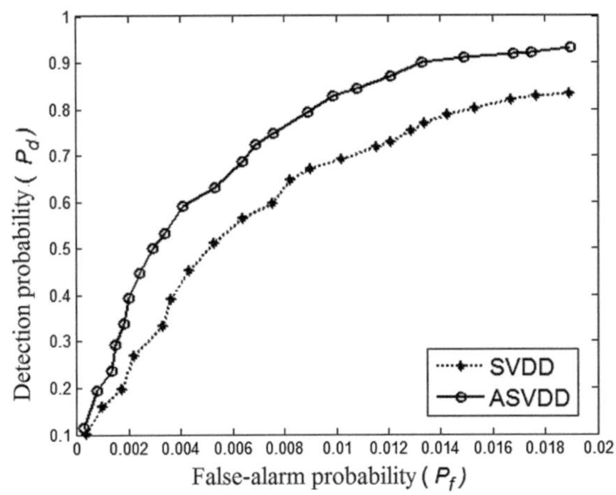

Fig. 7.16 The ROC curves of SVDD and ASVDD

that through using the selection of adaptive kernel parameters, the ASVDD algorithm could improve local adaptability under the different background and the detection ability in changing background. It has a strong adaptability for complex HSI data distribution with a reduced false alarm probability. Under the same false alarm probability, the detection probability in ASVDD is 10 % higher than commonly used SVDD algorithm.

7.4.3 *Effect Testing of Spectral Similarity Measurement Kernel-Based Anomaly Detection Algorithm*

In order to illustrate the comparison characteristics of the proposed spectral similarity measurement kernel (SSM-Kernel, denoted as SSM, K_s) and the Gaussian radial basis kernel (RBF K_r), real HSI data is used to conduct the experiments. The Gaussian radial basis kernel could be expressed as follows:

$$K_r(\boldsymbol{x},\boldsymbol{y}) = \exp\left(-\|\boldsymbol{x}-\boldsymbol{y}\|^2/c\right) \tag{7.53}$$

As shown in Fig. 7.17, three spectral vectors are selected from two kinds of ground objects, denote as spectral vectors $\boldsymbol{a}_1$, $\boldsymbol{a}_2$, and $\boldsymbol{b}$, respectively. Among them, $\boldsymbol{a}_1$ and $\boldsymbol{a}_2$ belong to the same class A. Due to the light intensity change and shadow effects, the absolute value of $\boldsymbol{a}_2$ is lower than $\boldsymbol{a}_1$, but their spectral shape resemblance can be clearly seen; The spectral $\boldsymbol{b}$ is taken from class B, it has the obvious difference in the aspects of spectral shape compared with spectral vector $\boldsymbol{a}_1$ and $\boldsymbol{a}_2$. Calculate kernel function values of spectrum $\boldsymbol{a}_1$ and $\boldsymbol{b}$, $\boldsymbol{a}_2$ and $\boldsymbol{b}$, $\boldsymbol{a}_1$ and $\boldsymbol{a}_2$ under RBF kernel and SSM kernel, respectively, where the RBF width factor c is set to 40, and the SSM similar critical factor θ is set to 0.1, the results as shown in Table 7.5.

The closer to 1 of the RBF or SSM kernel values, the higher similar degree of the spectral vectors, and the greater probability of the two vectors to be classified as the same class. As it can be seen from Table 7.5, the RBF kernel (K_r) of vector $\boldsymbol{a}_1$, $\boldsymbol{a}_2$, and $\boldsymbol{b}$ is greater than that of $\boldsymbol{a}_1$ and $\boldsymbol{a}_2$. Taking $\boldsymbol{a}_1$ as the reference spectral, the results show that $\boldsymbol{b}$ has more similarity than $\boldsymbol{a}_2$ and $\boldsymbol{b}$ is more likely to be judged as the same class of $\boldsymbol{a}_1$, while the actual situation is not the same. Instead, results of SSM kernel (K_s) give a higher similarity degree of $\boldsymbol{a}_1$ and $\boldsymbol{a}_2$, showing $\boldsymbol{a}_1$, $\boldsymbol{a}_2$ and $\boldsymbol{b}$

Fig. 7.17 spectral curves

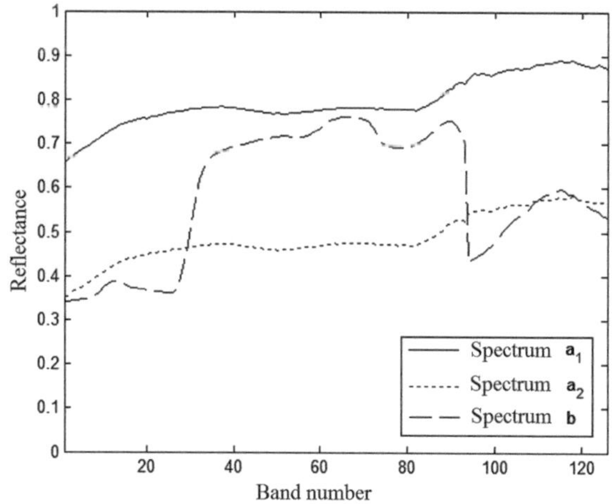

Table 7.5 Comparison of RBF kernel and SSM kernel

Spectral vector	$(\boldsymbol{a}_1, \boldsymbol{b})$	$(\boldsymbol{a}_2, \boldsymbol{b})$	$(\boldsymbol{a}_1, \boldsymbol{a}_2)$
K_r (RBF)	0.7616	0.8801	0.6691
K_s (SSM)	0.0008	0.0009	0.9954

are not the same class ($K_s(\boldsymbol{a}_1, \boldsymbol{a}_2) \gg K_s(\boldsymbol{a}_1, \boldsymbol{b})$, $K_s(\boldsymbol{a}_1, \boldsymbol{a}_2) \gg K_s(\boldsymbol{a}_2, \boldsymbol{b})$). Compared with the results of RBF kernel (K_r), we can also find that SSM kernel (K_s) results are more close to the actual situation from Table 7.5, but since the two parameters are not in the unified framework, the horizontal comparison between K_s and K_r is not expounded.

The adaptability SSM kernel is better than RBF kernel. The experiment is conducted by a specific hyperspectral anomaly detection method.

Choose KRX as the anomaly detection algorithm, it can be expressed as

$$\mathrm{KRX}(\phi(\boldsymbol{x})) = (\phi(\boldsymbol{x}) - \hat{\boldsymbol{\mu}}_{b\phi})^{\mathrm{T}} \hat{\boldsymbol{C}}_{b\phi}^{-1} (\phi(\boldsymbol{x}) - \hat{\boldsymbol{\mu}}_{b\phi}) \tag{7.54}$$

After the nucleation, the expression can be rewritten as

$$\mathrm{KRX}(\phi(\boldsymbol{x})) = (K_{\boldsymbol{r}}^T - K_{\hat{\mu}_b}^T)\boldsymbol{K}_c^{-1}(K_{\boldsymbol{r}} - K_{\hat{\mu}_b}) \tag{7.55}$$

In order to test the effectiveness of SSM kernel, a real AVIRIS HSI data is used to conduct simulation experiments. The image and its ground truth distribution are shown in Fig. 7.18, and 38 anomaly targets are contained in the image.

The target size in region A is large with very little influence caused by spectral mixture and shadow, as shown in Fig. 7.18 of region A, while the target size in region B is generally smaller, with very serious influence caused by spectral mixture and shadow, as shown in Fig. 7.18 of region B.

The background spectral and target spectral of A and B are shown in Figs. 7.19 and 7.20, respectively. It could be clearly seen from the two figures that, targets within region A can be relatively easy to detected since the target size is larger and its spectral is less affected by surrounding background and shadow, what is more, target and the background spectral characteristics are very different in terms of distance and shape. While on the other side, it is very hard to distinguish background and targets within region B since the target size is smaller, and its spectral is severely affected by the shadow and mixed pixels is more and the target is submerged in the background.

The outer window size is set to 11 × 11 and the inner window size is set to 3 × 3 for local detection. The selection of Gaussian radial basis kernel width c is crucial

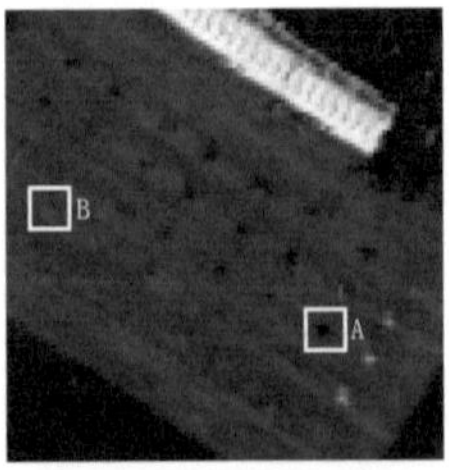

Fig. 7.18 Targets in different regions of real HSI data for experiment

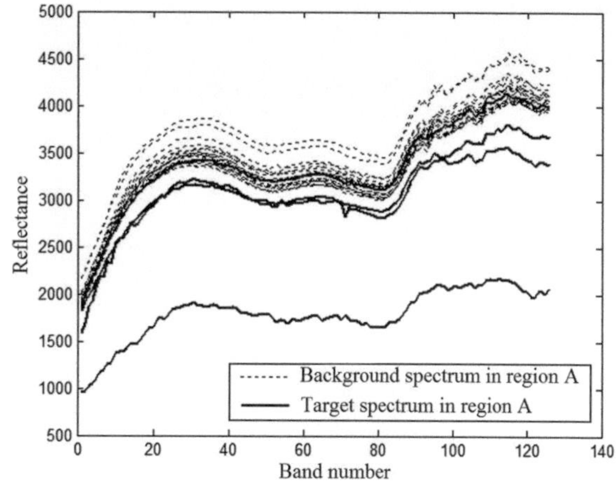

Fig. 7.19 Background and target spectrum in region A

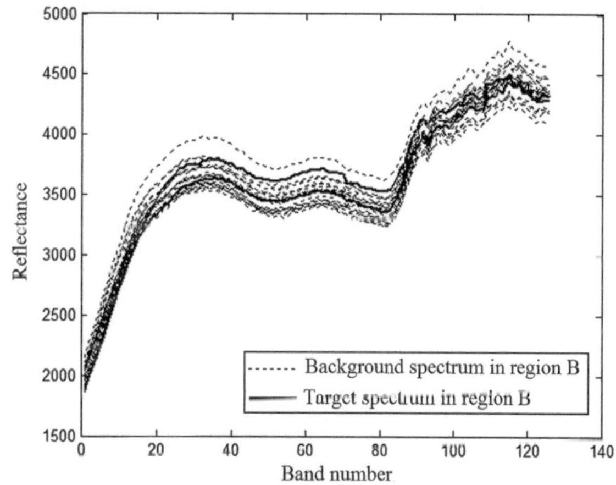

Fig. 7.20 Background and target spectrum in region B

and it is set to 40 through repeated experiments. KRX algorithm based on the Gaussian radial basis kernel function is denoted as RBF-KRX, while KRX algorithm based on spectral similarity measurement kernel is denoted as SSM-KRX. Through experimental comparison, similar critical factor of the spectral SSM kernel is set to 0.08, and the detailed discuss is given later.

Results as shown in Figs. 7.21 and 7.22, it can be seen from the figures that, RBF-KRX performs poorly in the regions severely affected by mixed pixel and shadow. This is because the RBF kernel is based on distance measure and cannot effectively distinguish subtle shape difference. However, the detection performance of SSM–KRX algorithm is better than RBF–KRX due to the enhanced ability for distinguishing spectral shape.

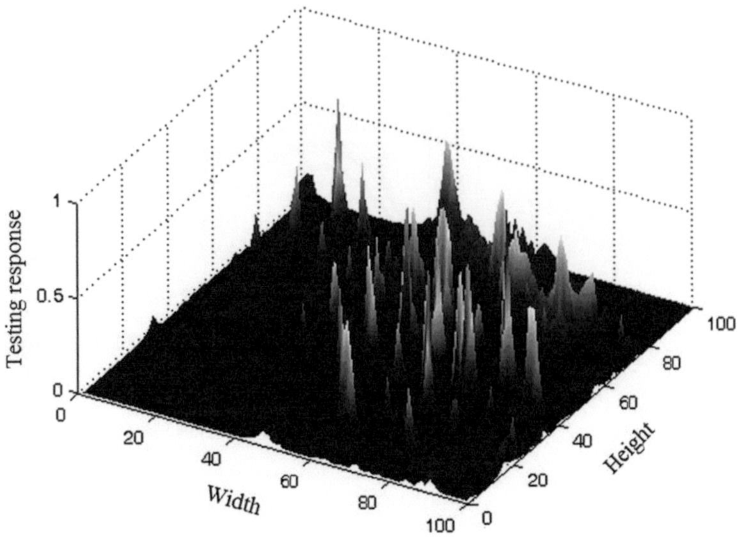

Fig. 7.21 3D plot of RBF-KRX algorithm detection result

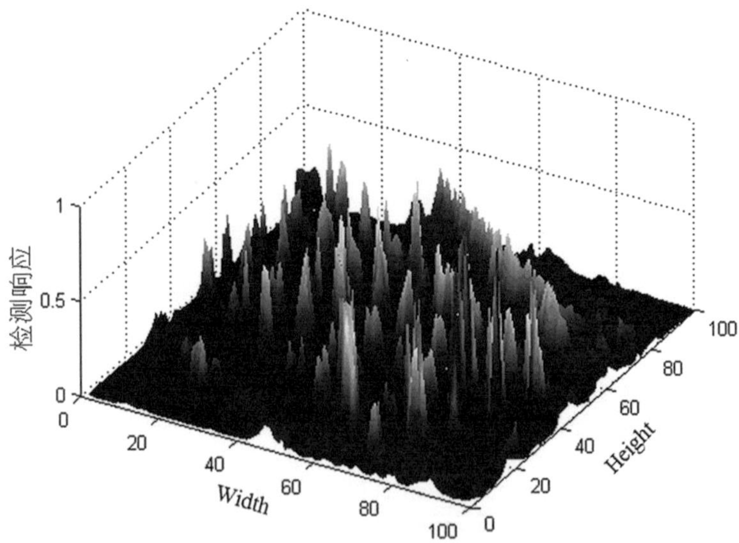

Fig. 7.22 3D plot of SSM-KRX algorithm detection result

7.5 Introduction of Other Anomaly Detection Algorithms

7.5.1 Spatial Filtering-Based Kernel RX Anomaly Detection Algorithm

KRX algorithm adopts local dual window model, as shown in Fig. 7.23, background is located in the outer window, and pixel under test is located in the inner window. The pixel is detected as anomaly target when the spectral has notable difference from the background. It can easily realize nonlinear anomaly detection by converting the dot product of high-dimensional feature space into the kernel function of input space (Mei 2009).

Simulated experiments are conducted to illustrate the nonlinear effect. Figure 7.24a is a set of simulated data, where red circle represents targets to be detected, and blue circle represents the background. We can see clearly from the figure that, a hyperplane (here is a straight line) is not existed to linearly separate the two groups. Figure 7.24b is the detection results and decision surface contour by RX algorithm. It can be seen that RX algorithm cannot effectively separate the two groups of data, which is to say it cannot effectively detect the target. Figure 7.24c is KRX results, which can distinguish the target and background, and generated decision surface coincides with the data distribution.

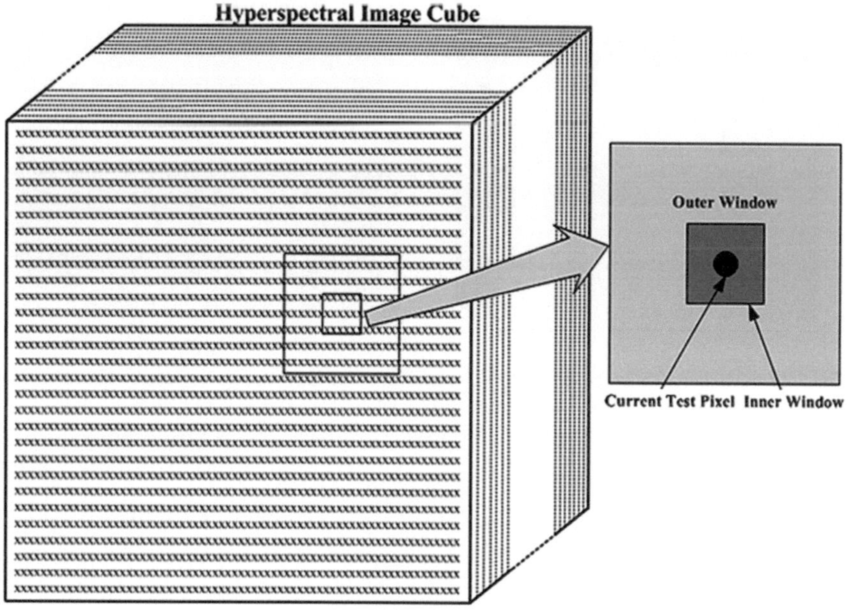

Fig. 7.23 Hyperspectral anomaly detection concentric double window model

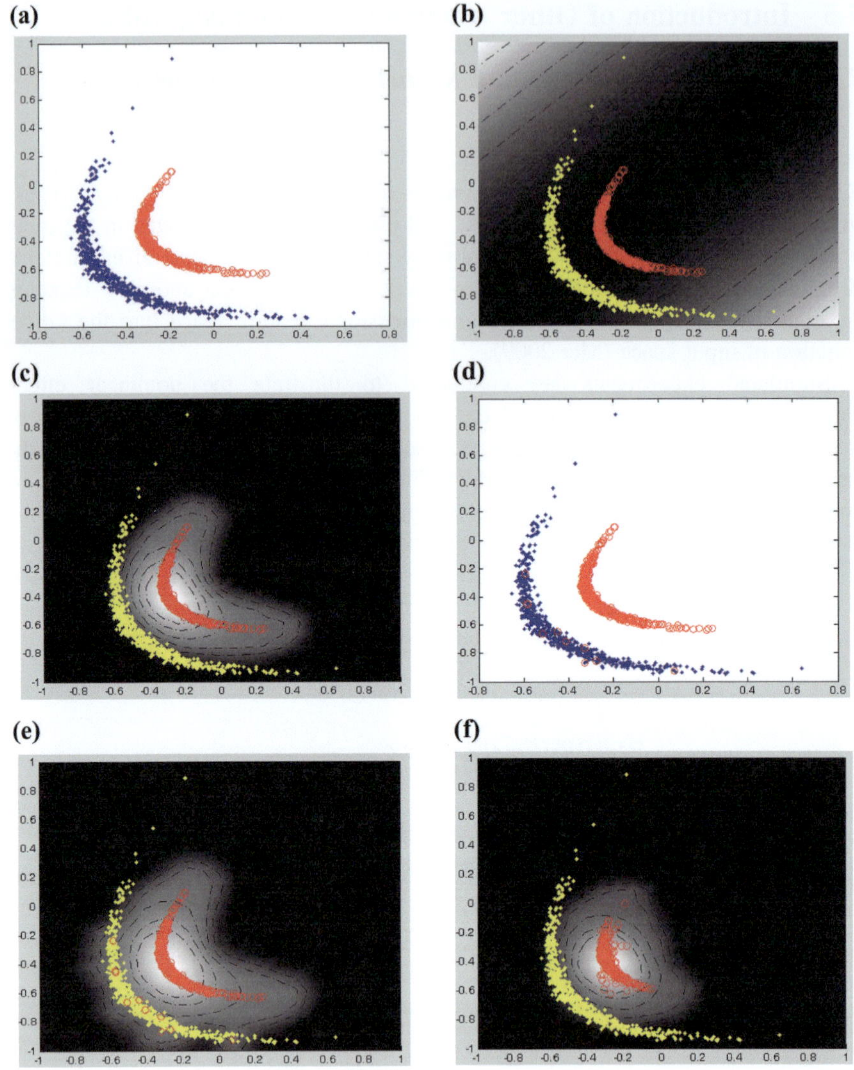

Fig. 7.24 Simulated data and experimental results. **a** Simulate the nonlinear distribution data. **b** RX algorithm contour line. **c** KRX algorithm contour line. **d** Simulation background doping data. **e** KRX contour line after doping. **f** KRX contour line after doping data filtering

The abnormal pixels in HSI data are very small and the background is statistically uniform, therefore the covariance matrix $\hat{C}_{b\Phi}$ of outer window pixels is approximated as the background covariance matrix. In actual situation, due to the unpredictability of target distribution, the complexity of the ground distribution, and the window size selection, the outer window does not necessarily reflect the actual background distribution, it is likely to be mixed with abnormal pixels, especially in the dense region of targets. The KRX algorithm has a strong nonlinear resolution, as a result it is sensitive to abnormal data. The miss rate rises when the background data is mixed with abnormal pixels. The experimental results are given based on simulation data in which background data is mixed with the target pixels (noise). Figure 7.24d is the simulation data in which background data is mixed with the target pixels. We can be seen from result in Fig. 7.24e that, although the mixed target is very few, the decision surface shape has changed a lot, in this case, the final miss ratio and false alarm ratio will increase.

Hyperspectral data is the combination of graph and spectrum; it not only exists the spectral neighboring correlation but also has a strong spatial correlation. Due to detect anomaly target in the whole testing data set is very few, so in the process of construct background data, it can reduce or even eliminate abnormal data generated the impact for background data. Here, use hyperspectral data space correlation to conduct sub-band spatial filtering, in order to reduce abnormal data generated impact for background data. Here, choose median filter to finish the optimization of background data.

Median filter is a nonlinear filtering, under certain conditions, it can overcome the blurring of detail caused by the linear filter (such as mean filter, etc.), and it is the most effective way to filter out pulse interference and image scanning noise. As a result, it has been widely used in digital image processing. Median filtering first determines one neighborhood by one certain pixel as the center point, then sorts the gray value of each pixel in epsilon neighborhood, and finally take the middle value as the new gray scale value of the center pixel. When the window slides over the image, the median filter can commendably conduct smooth processing for the image.

Two-dimensional median filtering algorithm is defined as follows:

$$Y(m,n) = \text{median}[X(m \pm i, n \pm j);\ (i,j) \in W] \quad (m,n) \in Z^2 \tag{7.56}$$

where W is the window, $2i + 1$ is horizontal size of the window, $2j + 1$ is vertical size of the window, $X(m,n)$ is a pixel to be processed image with coordinates (m,n), $Y(m,n)$ is median gray scale value within the $(2i + 1) \times (2j + 1)$ window centered at $X(m,n)$, and it is the output value of median filter.

HSI data is a kind of high-dimensional data, it can be seen as "a stack of" different wavelengths radiation image in view of the spatial space. Use matrix $\boldsymbol{D}^i_{W\times H}$ to present ith band of hyperspectral image (W and H present the width and height of HSI respectively), $i = 1, 2, \ldots,$ ND, A is the filter window, pixel $d^i_{l,k} \in \boldsymbol{D}^i_{W\times H}$ $(l = 1, 2, \ldots, W, k = 1, 2, \ldots, H)$, $y^i_{l,k}$ is filtering result of window A in $d^i_{l,k}$.

$$y^i_{l,k} = \text{med}\{d^i_{l+r,k+s}; d^i_{l,k} \in \boldsymbol{D}^i_{W\times H}, (r,s) \in A\} \tag{7.57}$$

Background data were optimized after filtering, the corresponding background matrix is expressed as $\boldsymbol{X}^m_{\phi c}$, $\hat{\boldsymbol{\mu}}^m_{b\phi}$, and $\boldsymbol{K}^m_c$ are used to present background mean vector and background nonlinear matrix after filtering respectively, and for testing point still use the original data, KRX algorithm can be rewritten as (7.58) after filtering, the algorithm is simplified as MKRX algorithm:

$$\text{MKRX}(\phi(\boldsymbol{x})) = \left(\phi(\boldsymbol{x}) - \hat{\boldsymbol{\mu}}_{b\phi}^{m}\right)^{\mathrm{T}} \boldsymbol{X}_{\phi c}^{m} \boldsymbol{K}_{c}^{m-1} \boldsymbol{X}_{\phi c}^{m\mathrm{T}} \left(\phi(\boldsymbol{x}) - \hat{\boldsymbol{\mu}}_{b\phi}^{m}\right) \tag{7.58}$$

The feature space dot product is converted into the input space kernel function by kernel function properties, the type (7.63) is simplified.

$$\text{MKRX}(\phi(\boldsymbol{x})) = \left(\boldsymbol{K}_{\boldsymbol{r}}^{m} - \boldsymbol{K}_{\hat{\boldsymbol{\mu}}_b}^{m}\right)^{\mathrm{T}} \boldsymbol{K}_{c}^{m-1} \left(\boldsymbol{K}_{\boldsymbol{r}}^{m} - \boldsymbol{K}_{\hat{\boldsymbol{\mu}}_b}^{m}\right) \tag{7.59}$$

Now let us make a simple analysis of filter selection. Firstly, the abnormal point will also show an anomaly spatial distribution in most bands, so just look from one band, abnormal points is similar to salt and pepper noise in the image, median filter has very good effect in the aspect of filter out the this kind of noise; Secondly, the filter result will be considered as a background data to conduct anomaly detection, so the background data change will have a great influence on detection results. Spatial resolution of a single band hyperspectral data is very low, and it has strong spatial correlation, median filter only removes the abnormal "noise" without changing the distribution of other data, so the background data distribution will keep the consistency of the original data. Figure 7.24f shows the result of simulation background mixed with target data after median filter, despite that the decision surface shape changes a bit, but it is already a big improvement compared with the non-filtering results. What's more, since the simulated data is randomly generated at the same time, there does not exists spatial correlation of adjacent points as the HSI data do, therefore, it fails to achieve the ideal effect. The following real hyperspectral experiment will give a big improvement using median filtering.

7.5.2 Multiple Window Analysis-Based Kernel Detection Algorithm

The influence of noise is often ignored in a double window detection model and the detection results are severely affected by noise, so it cannot achieve good results. In order to improve the detection efficiency, this section describes a multiple window analysis-based KRX algorithm (You 2011; Zhang 2009).

Traditional KRX algorithm is based on the dual window detection theory. As shown in Fig. 7.25, background is located in the outer window, and detection point is located in inner window, others in the inner window are buffer zones. This method is largely affected by white noise. In order to reduce the influence of white noise, three-layer window-based model is used for detection. Pixel under detect is located in the inner window, others in the inner window are buffer zones. Background is located in the middle window, while the outer window is used to

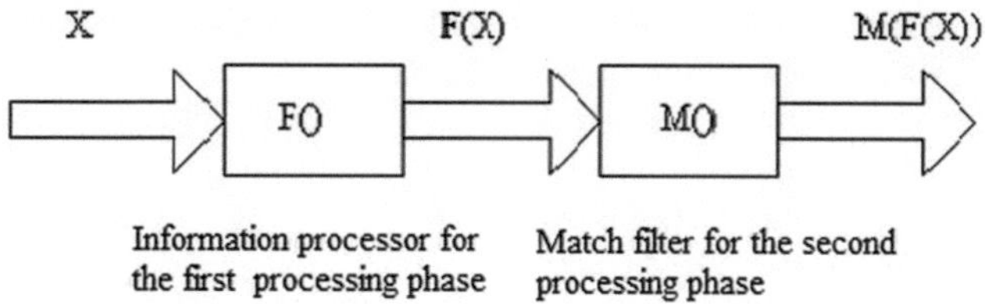

Fig. 7.25 Algorithm structure

eliminate the white noise. The algorithm is designed based on matched filter, and the processing procedure can be divided into two steps, as shown in Fig. 7.25.

The first phase is an information processor, which makes use of a priori or a posteriori spectrum information to suppress interference or white noise caused by unknown or not interested signals. The second phase is a match filter to extract interested target information. It uses linear mixed model-based anomaly and background separation. Let $\boldsymbol{x}$ is m-dimensional hyperspectral pixel vector $\boldsymbol{S} = [\boldsymbol{s}_1, \boldsymbol{s}_2, \ldots, \boldsymbol{s}_k]$ is the $m \times k$ endmember spectral matrix, $\boldsymbol{\alpha} = [\alpha_1, \alpha_2, \ldots, \alpha_m]^T$ is m-dimensional vector, in which each component element corresponding to pixel components, $\boldsymbol{n}$ is m-dimensional random noise. The linear pixel mixed model is:

$$\boldsymbol{x} = \boldsymbol{S\alpha} + \boldsymbol{n} \tag{7.60}$$

When decompose the mixed pixels, $\boldsymbol{S}$ is divided into interested signal $\boldsymbol{d}$ and not interested signals $\boldsymbol{U}$. Set $\boldsymbol{d} = \boldsymbol{s}_1$ as first endmember spectral, $\boldsymbol{U} = [\boldsymbol{s}_2, \boldsymbol{s}_3, \ldots, \boldsymbol{s}_k]$ is other endmember spectrums, $\boldsymbol{S} = [\boldsymbol{d}, \boldsymbol{U}]$, The formula (7.60) can be rewritten as

$$\boldsymbol{x} = \boldsymbol{d}\alpha_d + \boldsymbol{U}\alpha_U + \boldsymbol{n} \tag{7.61}$$

where α_d is percentage of the first endmember, α_U is components of the other endmembers, $\boldsymbol{\alpha} = (\alpha_{\mathrm{d}}, \alpha_U)$. $\boldsymbol{d}$ is abnormal endmember, $\boldsymbol{U}$ is the background endmember matrix, $\boldsymbol{n}$ is white noise.

Different from the traditional dual window-based detection where the noise interference is ignored, three layers window-based detection makes use of the OSP operator to eliminate background interference in the inner window and middle window, and removes the white noise effectively. Orthogonal Subspace Projection (OSP) algorithm is a commonly used anomaly detection algorithm. The biggest problems of statistic-based anomaly detection algorithms are the uncertainty of statistical distribution and estimation of relevant parameters, while multiple windows feature analysis could effectively solve this problem. It directly use the subspace of outer window pixels instead of endmember spectrum matrix **S**. Background pixels from the middle window and pixel under detection from the inner window s are projected onto the subspace, respectively, which could eliminate background noise interference.

Under the assumption that noise and signal are uncorrelated, OSP operator could get the projection matrix expression of signal $\boldsymbol{d}$:

$$\boldsymbol{P}_{\mathrm{OSP}} = \mu \boldsymbol{d}^T \boldsymbol{P}_{1/U} \tag{7.62}$$

where, $\boldsymbol{P}_{1/U} = \boldsymbol{I} - \boldsymbol{U}\boldsymbol{U}^{\#}$ is the projection matrix, its role is to make the signal projection onto elements formed subspace of the signal $\boldsymbol{U}$, ($\boldsymbol{U}^{\#}$ is the generalized inverse of $\boldsymbol{U}$, $\boldsymbol{U}^{\#} = (\boldsymbol{U}^T\boldsymbol{U})^{-1}\boldsymbol{U}^T$), $\boldsymbol{I}$ is $m \times m$ unit matrix, $\mu = (\boldsymbol{d}^T \boldsymbol{P}_{1/U}\boldsymbol{d})^{-1}$ is a constant term of indicate decomposition precision. The distribution of anomaly target in the image can be represented as (assuming constant term μ is 1):

$$\delta_g(\boldsymbol{x}) = \boldsymbol{d}^T \boldsymbol{P}^{\boldsymbol{x}}_{1/U} \tag{7.63}$$

When use OSP operator, we can know that it must to extract the interested signal $\boldsymbol{d}$ by the formula (7.63), so need to know image prior spectral information $\boldsymbol{S} = [\boldsymbol{d}, \boldsymbol{U}]$, that is endmember spectral matrix $\boldsymbol{S}$,and usually $\boldsymbol{S}$ is difficult to be obtained completely, thus brought difficulties for small target detection.

In view of the OSP algorithm exists endmember spectrum matrix $\boldsymbol{S}$ unknown problem in the actual detection, multiple window characteristics analysis algorithm clever apply the three layers window detection mode, use the two layers local background pixel windows, remove noise interference for the hyperspectral data and then to conduct anomaly detection. In the external window, the algorithm use the OSP operator to eliminate not interested in interference or white noise from the signal source in the inner window and middle window, in order to reduce the false alarm probability, it has better detection effect.

The principle of three layers window noise suppression is applied to the KRX method, in order to solve the influence of white noise for traditional double window detection. Expression form is as follows:

$$\mathrm{KRX}^{\mathrm{MW}} = \mathrm{KRX}\left(\mu_{\mathrm{inner}}\left(\boldsymbol{P}_{\frac{\perp}{\mathrm{outer}}}\boldsymbol{x}\right), \mu_{\mathrm{middle}}\left(\boldsymbol{P}_{\frac{\perp}{\mathrm{outer}}}\boldsymbol{x}\right)\right) \tag{7.64}$$

Among them, $\boldsymbol{P}_{\frac{\perp}{\mathrm{outer}}} = \boldsymbol{I} - \boldsymbol{U}_{\mathrm{outer}}\boldsymbol{U}^{\#}_{\mathrm{outer}}$ is the projection matrix of outer window pixels. Its role is to projection the pixel signal of the inner window and the middle window onto the subspace from outer window pixel ($\boldsymbol{U}^{\#}_{\mathrm{outer}}$ is $\boldsymbol{U}_{\mathrm{outer}}$ generalized inverse matrix $\boldsymbol{U}^{\#}_{\mathrm{outer}} = \left(\boldsymbol{U}^T_{\mathrm{outer}}\boldsymbol{U}_{\mathrm{outer}}\right)^{-1}\boldsymbol{U}^T_{\mathrm{outer}}$), $\boldsymbol{I}$ is $m \times m$ unit matrix.

Detecting point is located in the inner window, the other pixels in inner window is a buffer zone. The size of the inner window is decided by the size of anomaly target, and the size of middle window and outer window is decided by the number of selected background pixels. Under the condition of selecting the least background pixel, assume that the number of selected pixel in the external window is $n \times n - (n-2) \times (n-2)$, the number of selected pixel in the middle window N is

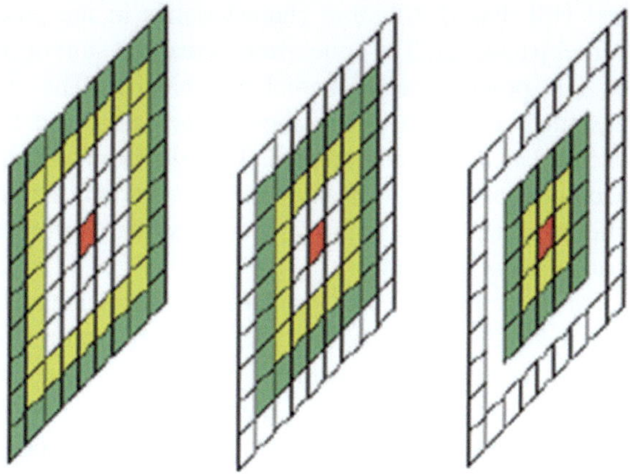

Fig. 7.26 Three layers window detection legend

$$N = (n-2) \times (n-2) - (n-4) \times (n-4) \tag{7.65}$$

As shown in Fig. 7.26, it shows an example of a three layer windows distribution, dark pixel area is pixel to be detected in the inner window. The white area in the middle is the buffer area, and it is adjusted according to the size of anomaly target. The light green color is the middle window area with background pixels. Darker area is the outer window, it is used to eliminate uninterested interference or white noise in the inner window and middle window. As shown in Fig. 7.26, three different sizes window models are given. The first window model has 32 pixels in outer window and 24 pixels in inner window. The second window model has 24 pixels in outer window and 16 pixels in inner window. The last window model has 16 pixels in outer window and 3 pixels in inner window.

7.6 Summary

The traditional detection algorithms ignore the spatial correlation between pixels. In order to solve this problem, this chapter presents a morphology-based anomaly detection algorithm. This algorithm can better consider the spatial characteristics of hyperspectral image, which could help remove the background noise and smooth detection result. AVIRIS HSI data is used to conduct simulation experiment. Compared with other algorithms, the results show that the detection performance of the proposed algorithm is superior to the traditional algorithms.

In view of kernel parameter estimation problem in kernel-based anomaly detection algorithms, this chapter presents an adaptive SVDD-based anomaly detection algorithm. Firstly, introduce the support vector data description method, and then conduct in-depth discussions on optimization of kernel parameter selection. Cross-validation is used to discuss the relationship between kernel parameter and HSI data distribution characteristics in the process of constructing adaptive kernel parameter. The conclusion is that the sum of partial band standard deviation σ_{sum} is proportional to kernel parameter c. The adaptive estimation formula is designed according to the above conclusion, and finally combined with support vector data description method. This algorithm adaptively adjusts kernel parameter according to different background distribution, making a stronger local adaptation ability, which will improve the detection performance under the condition of complex background.

In order to solve the problem that Gaussian radial basis kernel function has strong local information ability but its global information capacity is very weak, this chapter combines the characteristics of HSI data and makes use of the spectral similarity measure to form a new SSM kernel. The construction process of SSM kernel is expounded and the certainty of this kernel is proved. The SSM kernel has the property of invariability. Through a deep comparison between RBF kernel and SSM kernel, we can draw the conclusion that the shape distinguish ability of SSM kernel better than Gaussian RBF kernel, and SSM kernel has a stronger adaptability of spectral changes caused by the shade, shadows, and spectral mixing. This chapter also conducts a brief discussion for the parameter selection problem. As the Gaussian RBF kernel has very good local superiority and the SSM kernel has very good resolution of spectral similarity, through the inference of Mercer kernel theorem, this chapter finally construct the mixture kernel function of these two kernels and apply to the hyperspectral anomaly detection. Simulation results show that the combination of these two kernels can show their respective advantages, improve the detection probability and reduce the false alarm probability.

References

Mei F (2009) Hyperspectral anomaly target detection algorithm research based on nuclear machine learning. Harbin Engineering University master's degree thesis, Harbin

You J (2011) The hyperspectral image anomaly detection algorithm research based on kernel method. Harbin Engineering University master's degree thesis, Harbin

Zhang WS (2009) Hyperspectral images small target detection algorithm research based on the kernel method. Harbin Engineering University master's degree thesis, Harbin

Chapter 8
Dimensionality Reduction and Compression Technique of HSI

Hyperspectral imagery (HSI) suffers from extremely large data volumes for storage, transmission, and processing. Band selection and feature extraction are the data reduction techniques for data analysis and processing, and compression is data reduction technique mainly for storage and transmission. Dimensionality reduction and compression are usually related to each other.

8.1 Dimensionality Reduction Technique

Band selection is presented for the introduction of dimensionality reduction techniques in this section.

8.1.1 SVM-Based Band Selection

Sensitivity analysis is a popular approach for structure determination method in neural network. It is applied in the decision function of support vector machine (SVM) for band selection.

Let the total number of sample be Np, and the dimension of sample be ND, which means $\boldsymbol{x}_i = [x_i^1, x_i^2, \ldots, x_i^{\mathrm{ND}}]^T$, $i = 1, 2, \ldots, \mathrm{Np}$. The corresponding fuzzy classification decision function is defined as:

$$f(\boldsymbol{x}) = \sum_{i=1}^{\mathrm{Np}} \alpha_i^* K(\boldsymbol{x}_i, \boldsymbol{x}) + b^* \tag{8.1}$$

Firstly, consider the band selection method based on a single sample point $\boldsymbol{x}$. If the classification result of the sample $\boldsymbol{x}$ is changed after getting rid of a certain band, the band would be relatively important. If the decision function for each band is additive separable or multiplicative separable as

L. Wang and C. Zhao, *Hyperspectral Image Processing*,
DOI 10.1007/978-3-662-47456-3_8

$$f(\boldsymbol{x}) = \sum_{d=1}^{\mathrm{ND}} f_d(\boldsymbol{x}) \quad \text{or} \quad f(\boldsymbol{x}) = \prod_{d=1}^{\mathrm{ND}} f_d(\boldsymbol{x}) \tag{8.2}$$

where the value of $f_d(\boldsymbol{x})$ is entirely depends on the d th band of training sample. So, the degree of effects on classification for each band could be directly obtained by the component of adding formula or multiplier formula, then the trade-offs of the bands could be decided. However, only some simple kernel functions (such as polynomial kernel function) have the separable character for decision function, and most of the kernel functions do not have this character. For this case, the fuzzy classification value is calculated after removing the corresponding band by

$$f_{\mathrm{ND}/d} = \sum_{i=1}^{\mathrm{Np}} \alpha_i^* K_{\mathrm{ND}/d}(\boldsymbol{x}_i, \boldsymbol{x}) + b^* \ , \ d = 1,\ 2,\ \ldots,\ \mathrm{ND} \tag{8.3}$$

where $K_{\mathrm{ND}/d}$ is a new kernel function matrix which does not contain the data of the dth band. The difference value between the above result and the fuzzy classification results in (8.1) shows the influence of the removed bands for sample $\boldsymbol{x}$ on classification results:

$$\begin{aligned} \mathrm{order}(d) &= \left| f(\boldsymbol{x}) - f_{\mathrm{ND}/d}(\boldsymbol{x}) \right| \\ &= \sum_{i=1}^{\mathrm{Np}} \alpha_i^* \left(K_{\mathrm{ND}/d}(\boldsymbol{x}_i, \boldsymbol{x}) - K(\boldsymbol{x}_i, \boldsymbol{x}) \right) \\ d &= 1,\ 2,\ \ldots,\ \mathrm{ND} \end{aligned} \tag{8.4}$$

The above band selection is only for a sample points, classification accuracy is often a census of all the samples. Therefore, band selection should also from the perspective of global. However, classification influence exists a very large amount of calculation for statistics of each band for all the sample points. Because the support vector is located in the boundaries of the classification hyperplane, the classification results relative to the location of the samples are very sensitive. To this purpose, we use the automatic choosing support vector function of SVM, only to test the overall impact of each band for support vector classification as a global statistical approximation. So, after selecting support vector, calculate the overall impact of all support vector classification results after removing a band, further to determine the influence of the band for correct classification. Using support vector label set, at this point, corresponding formula (8.4) is written in the form below:

$$\begin{aligned} \mathrm{order}(d) &= \sum_{j \in SV} \sum_{i=1}^{\mathrm{Np}} \left| \alpha_i^* \left(K_{ND/d}(\boldsymbol{x}_i, \boldsymbol{x}_j) - K(\boldsymbol{x}_i, \boldsymbol{x}_j) \right) \right| \\ d &= 1,\ 2,\ \ldots,\ \mathrm{ND} \end{aligned} \tag{8.5}$$

We can use the original kernel function matrix $\boldsymbol{K}$ obtained directly from the new kernel function matrix $\boldsymbol{K}_{\mathrm{ND}/d}$ by simple operation rather than by the original data delete some bands, and the method can reduce the amount of calculation. For example, when the kernel function is a gauss model, it is not difficult to launch the relation of both as:

$$K_{\mathrm{ND}/d}(\boldsymbol{x}_i, \boldsymbol{x}_j) = \boldsymbol{K}(i,j) / \exp\left(-\frac{(x_i^d - x_j^d)^2}{2\sigma^2}\right) \quad d = 1, 2, \ldots, \mathrm{ND} \tag{8.6}$$

After calculating the sorting of the fuzzy vector order, it can retain the corresponding band of large numerical in the component according to the predetermined proportion. The above band selection method is built for two kinds of target samples. As for multiple target samples, we can train multiple SVM decision functions, then apply each decision function to band selection process to get more weight vectors. The average vector is taken as the final weight vector, which decides whether a band is remained or deleted.

In this part, the necessity of the subspace division and the choosing reason of dividing method will be described.

1. **The necessity of subspace division**

There exists a strong correlation between adjacent bands, and the relationship between the spectrums better than spatial correlation. This situation led to the gray scale change of each pixel in image, and it shows the tendency of roughly continuous with the change of band that the gray scale can be the approximation for continuous function of the band. So, we can draw the following conclusion by arithmetic property of continuous function that the dependent variable order(d) in formula (8.5) is continuous function of wave number d. And then we can launch the fact by local number theorem of continuous function: The chose band from the formula (8.5) tends to focus on one or several continuous wave jokes space. This kind of problem also exists in the article of Webb (1999), Hermes and Buhmann (2000) and other scholars. The experiment verified our above inference. Figure 8.1 is a group of 200d hyperspectral data and shows the relationship function between the calculated band and the sorting fuzzy value by the methods and formulas (8.4) derived from Webb (1999) and Hermes and Buhmann (2000) scholars. Abscissa shows the band number, and ordinate shows the corresponding sort fuzzy value. When we choose five bands according to the size of sorting fuzzy factor, only sorting fuzzy factor corresponding to bands of fuzzy value more than dotted line in the picture is selected. In this way, the selected five bands by three methods are all concentrated in narrow bands (38–48). Since sorting fuzzy factors are continuous to band numbers, the final selected bands are often concentrated in a continuous subspace, leading to information redundancy. The result is not only very bad for data classification, but also greatly influenced other subsequent applications of band selection, such as information fusion. Although the hyperspectral image is actually

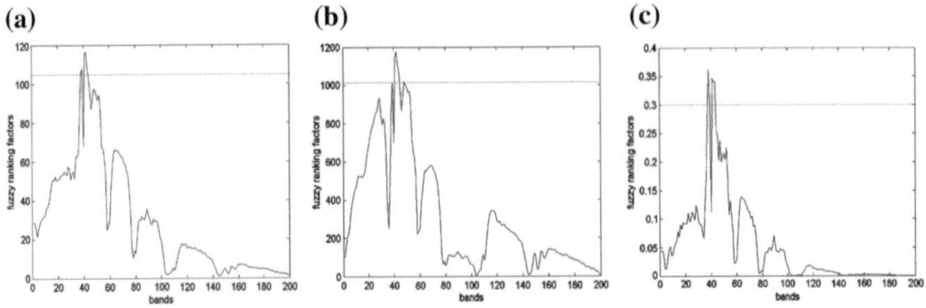

Fig. 8.1 The approximation continuity of sorting fuzzy value relative band based on different methods **a** Hermes L etc (2000) **b** Webb A etc (1999) **c** This paper method

a data source, the reflective spectral characteristics of different regions is different, and the correlation difference of the whole data is also bigger, only selected band correlation difference is bigger, to more fully utilized hyperspectral image information. In addition, global statistical characteristic and local characteristic of hyperspectral image data exist differences, thus it is not the best for band selection in the whole space. Therefore, we divide all the bands into several continuous band subspaces before band selection.

2. **The selection of subspace classification method**

The simplest way of subspace division is to divide all the bands into some predefined band groups evenly, but the way is not effective for lacking in theoretical rationale. Using the differences of spectral characteristics can also classify the subspace. For example, according to the wavelength of electromagnetic wave, the whole high spectral coverage area can be divided according to the visible light (0.4–0.7 μm) near infrared (0.7–1.3 μm) and short-wave infrared (1.8–2.5 μm). This classification method has its reasonable rationale, but it just divides bands according to the spectral range, without giving sufficient consideration to the property of hyperspectral data. The method does not reflect the local characteristics of data itself. Gu and Zhang (2003) put forward a automatic subspace division method, hyperspectral data space is divided into appropriate data subspace by defining band correlation coefficient matrix, and its neighbors can transfer related vector. This classification method reflects local characteristics of hyperspectral data. The obtained band subspaces generally have different dimensions with each other. The image data in each subspace have similar spectral characteristics. After finishing the subspace division, band selection will be completed within each subspace. In this way, the conditions of a band can be chosen are as follows: first, the band corresponds to the larger sort fuzzy values; second, no band is elected within the band is located in the subspace, or the correlation coefficient of band and this selected band more than a set threshold. Our band selection algorithm process is shown in Fig. 8.2.

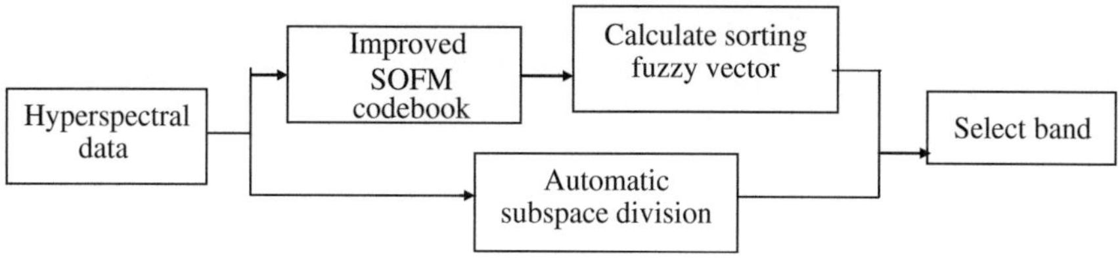

Fig. 8.2 Band selection process schematic diagram

To verify the effectiveness of the proposed method, we use Indian AVIRIS agriculture and forestry remote sensing data for simulation experiment. Keep 200 valid bands of the data, select two kinds of samples among them and each type of sample is 300. Positive and negative samples are evenly distributed for training sample set and test sample set. Using least squares, SVM, and Gaussian kernel function to train the training sample in the experiment, among them, select 30 samples (accounting for 10 % in all the training samples) as support vector set. We choose the methods from Webb (1999) and Hermes and Buhmann (2000) as the effective reference of this paper method. First of all, three band selection methods without subspace divisions are directly implemented on experimental data, and then the selected bands are applied to SVM-based classification to verify the selection effect. Table 8.1 shows band selection results and classification results based on different methods. The comparison results show that the proposed method based on SVM has achieved a better classification results. Experiment classification accuracy of application of all 200 bands is 95.33 %. Then, we carried out subspace division of the experimental data, and the data source is divided into five space: 1–35, 36–79, 80-103, 104–144, and 145–200. Table 8.2 shows the classification results for the bands selected by three band selection methods combined with subspace division. We can see the subspace division processing that overcomes the situation of the selected band is too concentrated and greatly improves the accuracy of classification. We can know from the comparison result of Tables 8.1 and 8.2 that classification accuracy of combining subspace division band selection is not in the combining subspace division band selection, and the classification accuracy is more than 10 %. In case of only 5 band is retained (accounting for 2.5 % of the total number of bands) and the classification accuracy reached 94.33 % based on this

Table 8.1 The band selection and classification results of not in combination with the subspace division method

Method	The selected band	Precision
Webb (1999)	38 39 41 42 43	0.7967
Hermes and Buhmann (2000)	41 42 43 44 48	0.7900
New method	38 39 41 42 43	0.7967

Table 8.2 The band selection and classification results of combination with subspace division method

Method	The selected band	Precision
Webb (1999)	34 42 89 116 161	0.9133
Hermes and Buhmann (2000)	29 42 89 116 160	0.9300
New method	28 38 89 117 160	0.9433

paper method, the result is close to the classification accuracy of all the bands. Subspace division successfully applied to the two reference methods, which fully shows the effectiveness and generality of this division method.

8.1.2 Application of Typical Endmember Methods-based Band Selection

It is worth noting that whether band selection or endmember selection, some typical vector to meet the demand, is selected in numerous vectors. As a result, in principle, there exists an algorithm portability between endmember selection and band selection. If this algorithm is transplanted successfully, we probably get band selection method and it is more suitable for specific tasks, which will promote the further development of hyperspectral image processing. In this section, the relationship between endmember selection and band selection is explored. Three typical endmember selection algorithms, such as IEA, PPI, and N-FINDR, are applied to the band selection, to solve the problem from interoperability and amount of calculation. For the detailed information of three endmember selection algorithms, readers can refer to the articles of Boardman (1995) and Winter (1999) that are in the references of Chap. 1, and the limit of the length, no longer described it again. The endmember selection and band selection in all applicable occasions are referred as vector selection. It can be seen that endmember selection and band selection are, respectively, carried out in pixel space and band space, and endmember selection and band selection can be referred as vector selection in vector space.

1. **The application of PPI algorithm in the band selection**

Original PPI algorithm conducts dimension reduction pretreatment by the maximum noise fraction (MNF) transformation. PCA transform is easy to implement and has a similar effect with the MNF transform (the former is a special form of the latter), so MNF is often replaced by the PCA. There exists a good linear relationship between the selected vectors by the PPI algorithm and other vectors. If we are sure that there is no strong correlation between the selected vectors, this algorithm can be used as a band selection. We know that PPI algorithm is not able to obtain directly the target vector. If the algorithm combined with automatic subspace classification methods, on the one hand, it can finish automatic selection of the target vector, and on the other hand, it can also guarantee the weak correlation between the selected vectors. At this point, it can construct band selection method based on PPI algorithm in accordance with the following main process. First of all, conduct dimension reduction processing by PCA transform. And then, L straight lines with stochastic directions are randomly generated in the transformed Nd dimension data space (as shown in Fig. 3.1). All the points in this data space are then projected onto the lines, and the points corresponding to the projection extremities are counted. After a lot of statistics, obtain the corresponding statistics score of each band. And then, the original feature space is divided into Nd consecutive subspace of the specified number by automatic subspace algorithm. In the end, combined with the results of the first two steps, select the band of highest statistics number and integral number to constitute all Nd target bands in each subspace.

The number of pixels Np is often larger than the number of band Ndgreatly. When PCA transform is applied in band space, an $Np \times Np$ matrix needs to be decomposed. When Np is bigger, the calculated amount of feature decomposition will be very big and even difficult to complete. And in view of band selection of the whole hyperspectral image, Np value is usually bigger (its value is generally more than 20,000). To reduce the amount of calculation, this decomposition calculation (such as Hubert et al. 2005) can be converted into characteristics decomposition of a matrix by nuclear method, to achieve the purpose of reduced amount of calculation. This technique will now be described as follows.

Hypothesis data $\boldsymbol{s}_i = [S_i^1, S_i^2, \ldots, S_i^{\mathrm{Np}}]^{\mathrm{T}}$ $(i = 1, 2, \ldots, \mathrm{Nd})$ have been centralized processing, that is the average result of all the data is 0. First of all, using transformation ϕ $(\phi : R^{\mathrm{Np}} \rightarrow R^{\chi})$. The sample set $\{x_k\}$ is map into feature space R^{χ}; in the feature space R^{χ}, the computation formula of covariance matrix $\boldsymbol{\Sigma}_\phi$ is as follows:

$$\boldsymbol{\Sigma}_\phi = \frac{1}{\mathrm{Nd}} \sum_{i=1}^{\mathrm{Nd}} \phi(x_i)\phi(x_i)^{\mathrm{T}} \tag{8.7}$$

$\boldsymbol{V}(\boldsymbol{V} \neq 0)$ expression eigenvector of $\boldsymbol{\Sigma}_\phi$ corresponding to the eigenvalue λ, $\boldsymbol{V}$ can be generated linearly by the feature space vector, that is $\boldsymbol{V} \in \mathrm{span}\{\phi(x_1), \phi(x_2), \ldots, \phi(x_{\mathrm{Nd}})\}$. Assume that

$$\boldsymbol{V} = \sum_{i=1}^{\mathrm{Nd}} \beta_i \phi(x_i) \tag{8.8}$$

$\boldsymbol{\beta} = (\beta_1, \beta_2, \ldots \beta_{\mathrm{Nd}})^{\mathrm{T}}$, it is known as the dual vector of $\boldsymbol{V}$. The basic formula based on the eigenvalue and eigenvector is as follows:

$$\mathrm{Np}\lambda \cdot \boldsymbol{V} = \boldsymbol{\Sigma}_{\phi} \cdot \boldsymbol{V} \tag{8.9}$$

If both sides are multiplied by $\phi(x)$ at the same time, we obtain the following equation:

$$\mathrm{Np}\lambda\beta = \boldsymbol{K}\beta \tag{8.10}$$

Here, $\boldsymbol{K}$ is the nuclear matrix of Nd × Nd:

$$\boldsymbol{K}(i,j) = \langle \phi(x_i), \phi(x_j) \rangle \tag{8.11}$$

When using linear kernel function (that is the arbitrary $x \in R^{\mathrm{Np}}$, $\phi(x) = x$), feature space is equivalent to the original space, the obtained characteristics vector is the characteristics vector of the original data, and matrix characteristics decomposition is aimed at the nuclear matrix $\boldsymbol{K}$ of Nd × Nd.

2. **The application of IEA algorithm in band selection**

In principle, the IEA algorithm can select some vector; the rest of bands can use them to conduct constrained or unconstrained linear synthesis with the error as small as possible. Easy to see, the selected band is suitable for linear analysis processing of hyperspectral images. In accordance with the subprime standards, this algorithm will be carried out by iterative way. At the beginning of the iteration, first initialize a vector, usually choose the farthest point of distance from origin point in band vector, then apply the selected band linear fitting the rest of the bands, and then choose the band of fitting error is largest as the second selected band. Further, the two selected bands are used to linearly fit the rest bands, and the band with the largest fitting error is selected as the third band.. Repeat this process until the specified number of bands is obtained. As to applicate the constraint condition, it can be determined according to actual needs. As we can see, the principle and method of IEA application to band selection and endmember selection is basically same.

3. **The application of N-FINDR algorithm in band selection**

The content has been described in detail in Sect. 3.6.2, and it is not repeat again.

8.1.3 Simulation Experiments

The effectiveness of the proposed band selection method is verified by the following experiments. In this experiment, a rapid and efficient band selection method (Referred to as the FS method, the specific process and performance advantages refer to the article of Mitra (2002) in Chap. 3) is used to conduct the comparison with the proposed method. The experiment images from a part of Indian hyperspectral remote sensing agriculture and forestry experimental area. In the first set of experiments, the detailed comparison result based on the number of transform

Table 8.3 MSE comparison results of four band selection methods

Number of samples	Method	MSE of selected band number is Nd			
		Nd = 5	Nd = 10	Nd = 15	Nd = 20
Np = 200	FS	0.0141	0.0085	0.0069	0.0058
	IEA	0.0175	0.0092	0.0073	0.0050
	PPI	0.0147	0.0089	0.0075	0.0056
	N-FINDR	0.0087	0.0057	0.0049	0.0042
Np = 2000	FS	0.0132	0.0088	0.0072	0.0061
	IEA	0.0170	0.0086	0.0073	0.0060
	PPI	0.0129	0.0090	0.0075	0.0058
	N-FINDR	0.0093	0.0059	0.0052	0.0045
Np = 20, 000	FS	0.0220	0.0096	0.0074	0.0062
	IEA	0.0188	0.0120	0.0089	0.0081
	PPI	0.0141	0.0085	0.0069	0.0058
	N-FINDR	0.0115	0.0065	0.0055	0.0049

samples and the number of selection band (selected, respectively, 200, 2000, 20,000 pixels points and selected, respectively, 5, 10, 15, 20 bands) method performance is shown. The rest of band to make use of the selected band conducts unconstrained linear fitting and the fitting mean square error (MSE) as evaluation indexes. The experimental results are shown in Table 8.3.

As shown in the table, N FINDR method have achieved the best results in most cases; PPI methods and FS methods have similar effects; The effect of IEA method is relatively poor when the number of selected bands is small, because the initially selected bands of the method have weak rationale. When the number of selected bands is large, the effect of IEA is slightly better than FS method.

Select 5 kinds of crops (corn, soybean, grass, forest, and hay) in the second experiment and a total of 2000 (400 samples in each category) samples. Traditional maximum likelihood (ML) classification method is used for the evaluation of various band selection methods. Uniformly selected half samples are treated as the training data and the other half as test samples. The classification results are shown in Table 8.4. As shown in table, this experiment has achieved similar comparison results with the first set of experiment.

Table 8.4 The classification accuracy comparison of four band selection methods

Band selection method	The classification accuracy (%) of selected band number is Nd			
	Nd = 3	Nd = 5	Nd = 7	Nd = 9
FS	86.2	88.6	92.1	92.2
IEA	85.0	89.9	92.8	92.8
PPI	86.9	88.7	92.3	92.5
N-FINDR	88.4	90.1	92.8	93.7

From the point of view operating speed, the unconstrained IEA method and FS method are relatively close to the fastest, PPI method takes the second place, and N-FINDR method is relatively slower. These methods have close running speeds, but their running speeds are faster than traditional greedy optimization search method (Devijver and Kittler 1982) in dozens of times and even more than one hundred times.

8.2 Compression Technique

8.2.1 Vector Quantization-based Compression Algorithm

1. **Vector quantization**

The vector quantizer of K dimension and size is N can be defined as the mapping of finite set Y, and Y is constructed from K dimension euclidean space R^k that contains N outputs(reconstruction) points, as shown in formula (8.12):

$$Q(\boldsymbol{x}|\boldsymbol{x} \in R^k) = y_p, \text{ in which, } y_p \in Y = \{y_0, y_1, \ldots, y_{N-1}\} \tag{8.12}$$

Among them, the set Y is called code book, and its size is N; the N elements of $\{y_0, y_1, \ldots, y_{N-1}\}$ code book are called code word, and it is the vector in R^k; since $\boldsymbol{x}$ as the input vector, the corresponding code word of $\boldsymbol{x}$ is y_p in the code book Y.

The mapping should meet the distortion error of the input vector $\boldsymbol{x}$ and y_p is the minimum distortion error of $\boldsymbol{x}$ with other code word in code book

$$d(\boldsymbol{x}, y_p) = \min_{1 \le j \le N} (d(\boldsymbol{x}, y_j)) \tag{8.13}$$

Among them, $d(x, y_j)$ acts as distortion error between vector $\boldsymbol{x}$ and code word y_j.

After input vector space R^k gets through the quantizer Q, and it is divided into N non-overlapping cells, and this process is known as the division of the input vector space. Cell R_i, $i = 0, 1, \ldots, N-1$ is defined as in formula (8.14):

$$R_i = \{\boldsymbol{x} \in R^k : Q(\boldsymbol{x}) = y_i\} \tag{8.14}$$

$$\bigcup_i R_i = R^k, \quad \text{when } i \neq j,\ R_i \cap R_j = \Phi \tag{8.15}$$

The distortion error of the whole system is expressed by the mean square of error between all the input vectors and its corresponding code word, as shown in formula (8.16):

$$\text{MSE} = \frac{1}{M} \sum_{i=1}^{M} d(\boldsymbol{x}_i, Q(\boldsymbol{x}_i)) \tag{8.16}$$

where M is the number for training vector.

Vector quantization process is the input vector space R^k that is divided into several small space, and each small space is called a cluster area; it contains all the vectors are expressed by the centroid (code word). The essence of the vector quantization problem is to find a best space division and the centroid of each cluster area, so that the distortion error is minimum.

Vector quantization has been able to compress data because it can remove redundancy and can effectively use the four interrelated properties between each component in the vector: linear dependence, nonlinear dependence, the shape of probability density function, and vector dimension. And scalar quantization can only use linear dependence and the shape of probability density function to eliminate the redundancy. Under the same coding rate, vector quantization distortion is significantly smaller than scalar quantization distortion; and under the same distortion condition, vector quantization code rate is much lower than the scalar quantization code rate. But due to the complexity of the vector quantization in the form of index increased with vector dimension, its complexity is higher than the complexity of scalar quantization.

2. **The basic SOFM algorithm**

It is rephrased as "The classical self-organizing feature map (SOFM) neural network has unsupervised function. It is a double-layer structure network with lateral associative ability, and includes input layer and output layer (competition layer). The number of network input layer neurons is equal to N, the size of code book. Input layer neurons connected to output layer neurons by variable weights. All the output layer neurons adjust connection weights by competition with each other and adaptive learning algorithm. After training, the weights of all output layer neurons constitute the code book.

Suppose the number of training vector is M, and training vector set is $X = \{\boldsymbol{X}_0, \boldsymbol{X}_1, \ldots, \boldsymbol{X}_{M-1}\}$; network includes K input nodes (vector dimension) and N output nodes (codebook size), and the weight of each input node to output node is W_{ij}, $i \in [0, N-1], j \in [0, K-1]$; that is y_i subscript is j component in code book $Y = \{y_0, y_1, \ldots, y_{N-1}\}$, and the basic SOFM algorithm steps are as follows:

1. The connection weight value of initialize input neurons to the output neurons is $\boldsymbol{W}_{ij}(0)$, $i \in [0, N-1], j \in [0, K-1]$, and it can be realized by random select vector in the training set.
2. Select the training vector $\boldsymbol{X}_k$ from the training set $k \in [0, M-1]$, and it is input to each neuron by parallel style.
3. Calculate the distortion between $\boldsymbol{X}_k$ and each code word (that is, the weight vector of output node) in code book and select the neuron with the smallest distortion $\hat{i}$ as winning neuron, adjust the code word $\hat{i}$ and weight value of each code word in $\hat{i}$ topology field according to formula (8.18); other weight value remains the same, that is

$$d_{\hat{i}} = \min_{0 \le i \le N-1} \left| \sum_{j=0}^{K-1} [\boldsymbol{x}_{kj} - \boldsymbol{W}_{ij}(t)]^2 \right| \tag{8.17}$$

$$\boldsymbol{W}_{ij}(t+1) = \boldsymbol{W}_{ij}(t) + \phi(t)[\boldsymbol{x}_{kj} - \boldsymbol{W}_{ij}(t)] \tag{8.18}$$

Among them, t is the number of iterations. $\phi(t)$ is the learning rate factor, generally select $0 < \phi(t) < 1$, to ensure the convergence of the algorithm.

4. For all the training vectors, repeat steps (2) and (3), until the algorithm convergence or reach the initial setting maximum number of iterations.

3. **The improved SOFM algorithm**

Basic SOFM algorithm needs N distortion calculation when selecting the winning neuron, and M training samples complete one iteration need MN distortion calculation, every time distortion calculation needs K multiplication and $2K - 1$ addition, operand is larger, algorithm speed and convergence performance are all affected. On the other hand, because of the influence of training vector statistical properties, some output neurons competition winning probability is big, weights need to be adjusted regularly, and some weight values of output neurons are rarely adjusted. Therefore, improve the basic SOFM algorithm by using distance inequality criterion and introducing the frequency sensitive factor, to improve the performance of code book.

Each component sum of a vector is defined as a vector sum value, set sum value of input vector $\boldsymbol{X}_k$ is $\boldsymbol{S}_{X_k}$, $k \in [0, M-1]$, and the sum value of code word $\boldsymbol{W}_i$ is $\boldsymbol{S}_{W_i}$, $i \in [0, N-1]$, that is $\boldsymbol{S}_{X_k} = \sum_{j=1}^{K} \boldsymbol{x}_{kj}$, $\boldsymbol{S}_{W_i} = \sum_{j=1}^{K} \boldsymbol{W}_{ij}$. It can be proved that

$$d(\boldsymbol{X}_k, \boldsymbol{W}_i) = \sum_{j=1}^{K} (\boldsymbol{x}_{kj} - \boldsymbol{W}_{ij})^2 \ge \frac{(\boldsymbol{S}_{X_k} - \boldsymbol{S}_{W_{ij}})^2}{K} \tag{8.19}$$

Assume that the current minimum distortion is $d_{\min}$, $\mathrm{MD} = Kd_{\min}$, if meet the formula

$$(\boldsymbol{S}_{X_k} - \boldsymbol{S}_{W_i})^2 \ge \mathrm{MD} \tag{8.20}$$

We can conclude the following formula according to formula (8.19)

$$d(\boldsymbol{X}_k, \boldsymbol{W}_i) \ge d_{\min} \tag{8.21}$$

As a result, before searching the winning neuron, the sum value $\boldsymbol{S}_{W_i}$, $i \in [0, N-1]$ of N code word can be calculated in advance and stored in a code book. Meanwhile, MD is precalculated in the process of search winning neuron, and then whether the sum value S_{W_i} of code word W_i satisfies the formula (8.20) is judged. If yes, the code word W_i can be ruled out from the calculation of distortion error.

To make full use of each code word vector, we introduce the frequency sensitive factor μ_i in algorithm; the initial value of μ_i is 1, once output neuron i wins one time in the competition process, μ_i plus 1, the frequency sensitive factor as the number of neuron winning. The distortion error correction is shown below:

$$\hat{d}_i = \mu_i d_i \tag{8.22}$$

Therefore, with the increase of μ_i value, the chance of output neuron i once again become the winning neuron is decrease; increasing the winning chance of other neurons further improves the convergence speed and performance of the algorithm.

Improved SOFM algorithm steps are shown below:

1. Initialize the connection weights between input neuron and output neuron, $W_{ij}(0),\ i \in [0, N-1], j \in [0, K-1]$, and it can select randomly the vector to achieve from the training set.
2. Compute the sum value of each code word W_i in the code book, and its sum value is $S_{W_i},\ i \in [0, N-1]$.
3. Calculate the sum value of input training vector X_k, and its sum value is S_{X_k}, $k \in [0, M-1]$, X_k is input to each neuron in parallel.
4. According to the given inequality criterion from formulas (8.19) and (8.20), search minimum distortion neurons $\hat{i}$ with training vector X_k will be the winning neuron, according to the formula (8.24) to adjust the code word $\hat{i}$ and the weight of each code word $\hat{i}$ in topology field; other weight remains the same, that is

$$d_{\hat{i}} = \min_{0 \le i \le N-1} \left| \sum_{j=0}^{K-1} [x_{kj} - W_{ij}(t)]^2 \right| \tag{8.23}$$

$$W_{ij}(t+1) = W_{ij}(t) + \phi(t)[x_{in} - W_{mn}(t)] \tag{8.24}$$

   Among them, t is the number of iterations. $\phi(t)$ is the learning rate factor, generally select $0 < \phi(t) < 1$, to ensure the convergence of the algorithm, the adjusted distortion error is adjusted according to the formula (8.22).
5. For all the training vectors, repeat steps (2) and (4), until the algorithm convergence or achieve maximum number of iterations.

4. **The experimental simulation results and its analysis**

Indian AVIRIS agriculture and forestry remote sensing data are still adopted in the following experiments. The block diagram of the compression algorithm is shown in Fig. 8.3.

This algorithm firstly adopts the adaptive band selection to conduct dimension reduction process for hyperspectral image and effectively focuses on the energy distribution by spectrum compression, and it can realize larger compression ratio

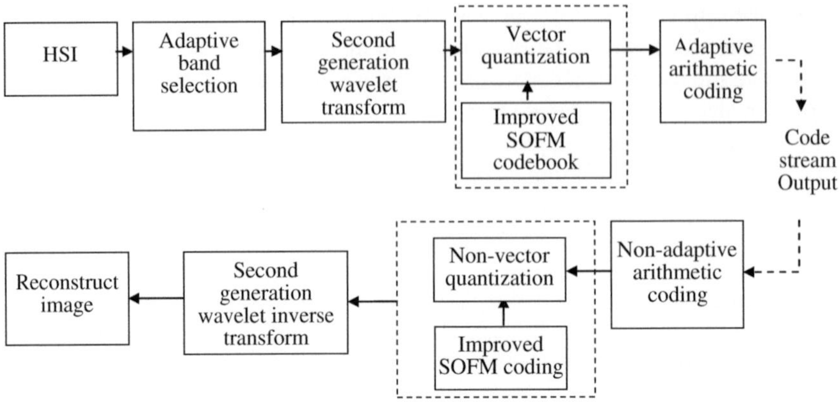

Fig. 8.3 The block diagram of hyperspectral remote sensing image compression algorithm

(CR) and adopt the spectrum compression method that has significantly improve the computing speed than directly adopt vector quantization method. But on the other hand, dimension reduction process is a non-reversible process, so it adopts spectrum compression that leads to the part of detail information loss and never recovers and cannot realize lossless compression. Then, algorithm for dimension reduction image conducts second generation wavelet transform vector quantization, respectively, to adopt basic SOFM algorithm and the improved SOFM algorithm and compared with the traditional LBG codebook and uniform scalar quantization with the center "dead zone."

With the increase of codebook size, the number of vector in training set is increased. If the vector quantization distortion error decreases, then residual image information entropy should also reduces accordingly. But on the other hand, with the increase of codebook size, the occupied storage space will enlarge, design time will also go up, and as a result, the choice of codebook size should be determined according to actual situation; the each band of selected image in the experiment is 145×145 pixels, so the codebook size is 128.

Figure 8.4 shows the compression effect comparison of basic SOFM algorithm, improved SOFM algorithm, and traditional LBG algorithm. Figure 8.4a shows the120th band of original image, and Fig. 8.4b, c, d shows the compressed images of traditional LBG algorithm, basic SOFM algorithm, and improved SOFM, respectively. The subjective compression effect of the three methods is similar. From the reconstruction compressed image, we can seen that this article method generate noise performance is a blurring effect based on an overall image. This is because the image conducts wavelet transform before taking the vector quantization, and this process is conducted for all images as a whole and adopts a periodic

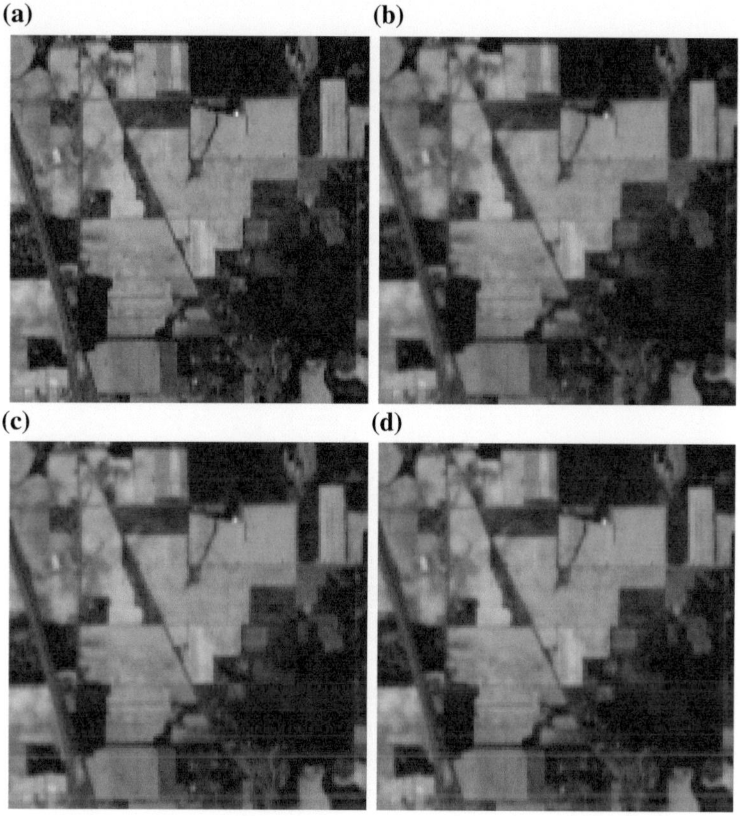

Fig. 8.4 Hyperspectral remote sensing image compression effect picture (take 120th band image as an example), **a** The original image **b** The compressed reconstruction image based on traditional LBG algorithm **c** The compressed reconstruction image based on basic SOFM algorithm; **d** The compressed reconstruction image based on improved SOFM algorithm

boundary continuation way when compressing the images; the image matches well with mirror filter group, eliminates the edge distortion, and improves effectively the recovery image signal-to-noise ratio.

As can be seen from the technical statistics data in Table 8.5, the improved SOFM algorithm can reduce effectively the design time of code book and improve codebook performance; it can achieve a better compression effect, compared with traditional LBG algorithm and basic SOFM algorithm; the improved SOFM algorithm increases by 82.41 and 73.14 % in the aspect of codebook design time; PSNR value, respectively, increases by 1.3 and 0.7 dB.

Compared with uniform scalar quantization with the center "dead zone," all kinds of vector quantization methods have increased significantly in the aspect of CR; the improved SOFM algorithm increases more than 1.5 times; peak signal-to-noise ratio is also much better than the effect of scalar quantization, only

Table 8.5 The statistical properties of different hyperspectral image compression methods

Scalar quantization		Code book size	Traditional LBG algorithm			Basic SOFM algorithm			Improved SOFM algorithm		
CR	PSNR (dB)		CR	PSNR (dB)	Design time (s)	CR	PSNR (dB)	Design time (s)	CR	PSNR (dB)	Design time (s)
		64	2.91	33.922	1052	3.01	34.524	689	3.24	35.258	185
2.03	33.231	128	2.88	35.236	1522	3.11	36.582	805	3.22	37.215	267
		256	2.89	35.650	1793	3.18	36.612	912	3.19	37.229	356

because the vector quantization codebook design needs take some time, so it makes a greater sacrifice in terms of compression time.

8.2.2 Lifting Scheme-based Compression Algorithm

1. **Ascension format**

Classical wavelet lifting scheme is a kind of simple and effective method of construction compactly supported by biorthogonal wavelet; It does not depend on Fourier transform, and completes the construction of biorthogonal wavelet filter in 2-D space. Its main idea is conduct simple multi-resolution analysis (Split) for a basic wavelet, then alternately using dual lifting (Predict) and the original lifting (Update) to improve its performance, toward a certain characteristics to gradually approaching (Lifting).

The process of construct wavelet positive transform and wavelet inverse transform by lifting scheme is shown in Fig. 8.5.

The standard lifting algorithm consists of three steps: (Split), (Predict)和□, and (Update).

1. Split. The purpose of split is that the original signal $x(i)$ is divided into two subsets $x_o(i)$ and $x_e(i)$, which are interconnected but not intersected. In general, it does not need to specify split rules and the size of subset after splitting, while it just needs to determine how to reconstruct original signal $x(i)$ from $x_o(i)$ and

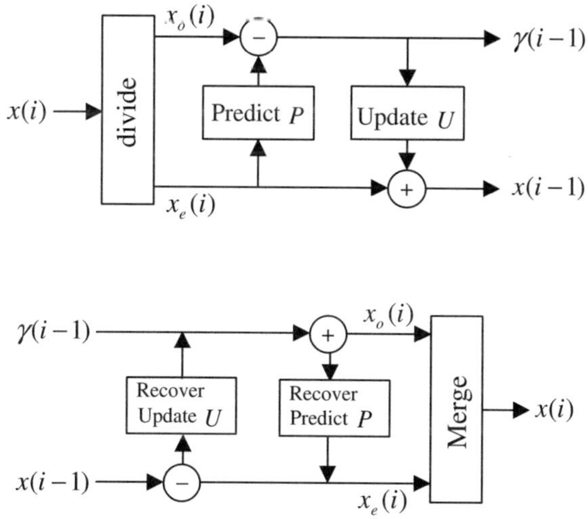

Fig. 8.5 The positive transformation and inverse transformation of lifting scheme

$x_e(i)$. The stronger of the correlation between $x_o(i)$ and $x_e(i)$ is, the better the split effect is.

The simplest split adopts usually the lazy wavelet method, assuming that there is a maximum correlation between adjacent data (it tend to be so in practice), then interval sampling the original signal according to odd and even number of the data; the signal $x(i)$ is split into odd–even two sets, $x_o(i) = x(2i + 1)$, $x_e(i) = x(2i)$.

2. Predict. It is also known as dual lifting link that is used $x_e(i)$ to predict $x_o(i)$; the prediction error is defined as shown in the formula (8.25):

$$\gamma(i - 1) = x_o(i) - P[x_e(i)] \tag{8.25}$$

Among them, $P(\cdot)$ is the prediction operator. This prediction is a reversible process, that is as long as select a prediction operator $P(\cdot)$, $x_o(i)$ can fully recovered by $x_e(i)$ and $\gamma(i - 1)$ [As shown in formula (8.26)], and then restore signal $x(i)$.

$$x_o(i) = \gamma(i - 1) + P[x_e(i)] \tag{8.26}$$

Predict link has two main functions:

The original data should be represented in compact form. The original signal $x(i)$ generally has local relevance, so the prediction error $\gamma(i - 1)$ is always much lower than $x_o(i)$. In other words, the representation of $x(i)$ by $x_e(i)$ and $\gamma(i - 1)$ is much more compact than by $x_o(i)$ is much more compact than use $x_e(i)$.

The isolated high-frequency component of signal $x(i)$ is used as a prediction in the spatial domain, due to use a smooth curve based on $x(2i + 1)$ point and $x(2i)$ point to predict the middle point $x(2i + 1)$ (that is, low-order polynomial interpolation); the smooth signifies low frequency at this moment, and the prediction error $\gamma(i - 1)$ signifies high-frequency component of signal $x(i)$.

3. Update. Update is also known as the original lifting link, that is use $\gamma(i - 1)$ to deal with $x_e(i)$, make the processed $x_e(i)$ (it is marked as $x(i - 1)$) contains only the low-frequency component of signal $x(i)$, as shown in formula (8.27):

$$x(i - 1) = x_e(i) + U[\gamma(i - 1)] \tag{8.27}$$

$U(\cdot)$ represents update operator; in this process, making the envelope line of $x(i - 1)$ becomes a smooth fitting curve of signal $x(i)$. Mathematically, $x(i - 1)$ and $x(i)$ should have the same low-order disappear moment (i.e., the same dc component).

Whether it is predict or update, it can be called as lifting link, and the whole process is completely reversible.

2. **The construction method of lifting wavelet**

Red and black transform and quincunx grid. Uytterhoeven and Bultheel first proposed the use of quincunx grid to signify red and black wavelet transform based on lifting algorithm. First, take an image as a two-dimensional signal, as shown in Fig. 8.6, and the interval sampling is divided into such as quincunx grid; the original signal can be divided into two parts, and it includes red pixel point and black pixel point; this segmentation method can also be referred to as the "chessboard" or "red and black" segmentation, and it is more simple and very close to split step of lifting format. The red pixel point is used to predict black pixel point, and black pixel point carries details information that is used to update red pixel point at the same time.

The second-level predict and update filter is given by the formulas (8.28.) and (8.29):

$$P_x(i,j) = [x(i-1,j) + x(i,j-1) + x(i+1,j) + x(i,j+1)]/4, \quad i \bmod 2 \neq j \bmod 2 \tag{8.28}$$

$$U_x(i,j) = [x(i-1,j) + x(i,j-1) + x(i+1,j) + x(i,j+1)]/8, \quad i \bmod 2 = j \bmod 2 \tag{8.29}$$

Neville filter. Neville filter is a kind of classical filter structure; the principle is that polynomial sampling is applied to a grid and generates the same polynomial structure; it is very important for the filter group and the structure of wavelet analysis based on lifting algorithm, some famous filter such as Coiflets and Deslauriers Dubuc filter are all applicable to Neville filter structure.

Predict filter and update filter are used for quincunx grid, and it usually can be represented by formulas (8.30) and (8.31):

$$P_x(i,j) = \sum_{(n,m)\in S_{\bar{N}}} a_{\bar{N}}(n,m)x(i+n,j+m), \quad i \bmod 2 \neq j \bmod 2 \tag{8.30}$$

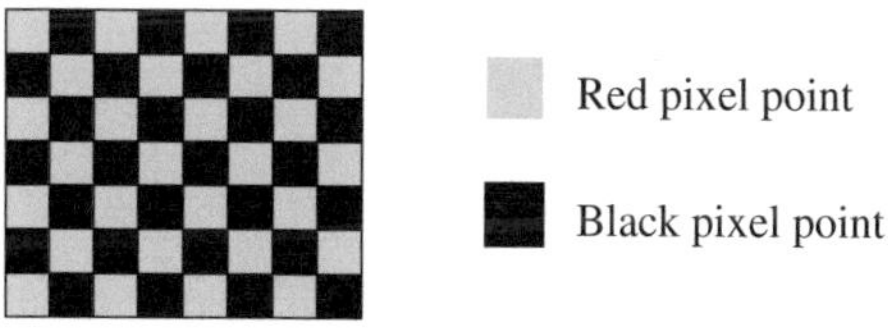

Fig. 8.6 The matrix grid is decomposed into quincunx grid

$$U_x(i,j) = \sum_{(n,m)\in S_N} a_N(n,m)x(i+n,j+m)/2, \quad i \bmod 2 = j \bmod 2 \tag{8.31}$$

Among them, $S_{\bar{N}}$ and S_N are all a subset of $\{(n,m) \in Z^2 | (n+m) \bmod 2 = 1\}$, $a_{\bar{N}}(s)$, $s \in S_{\bar{N}}$ is a set of wavelet coefficients in real number domain. N value corresponding to original N order disappears moment of the filter in S_N, then several elements contain same coefficient in S_N are put together as its subset, as shown in formula (8.32):

$$\begin{aligned}
V_1 &= \{(+1,\ 0), (0,\ +1), (-1,\ 0), (0,\ -1)\} \\
V_2 &= \{(+1,\ +2), (-1,\ +2), (-2,\ +1), (-2,\ -1), (-1,\ -2), (+1,\ -2), (+2,\ -1), (+2,\ +1)\} \\
V_3 &= \{(+3,\ 0), (0,\ +3), (-3,\ 0), (0,\ -3)\} \\
V_4 &= \{(+2,\ +3), (-2,\ +3), (-3,\ +2), (-3,\ -2), (-2,\ -3), (+2,\ -3), (+3,\ -2), (+3,\ +2)\} \\
V_5 &= \{(+1,\ +4), (-1,\ +4), (-4,\ +1), (-4,\ -1), (-1,\ -4), (+1,\ -4), (+4,\ -1), (+4,\ +1)\} \\
V_6 &= \{(+5,\ 0), (0,\ +5), (-5,\ 0), (0,\ -5)\} \\
V_7 &= \{(+3,\ +4), (-3,\ +4), (-4,\ +3), (-4,\ -3), (-3,\ -4), (+3,\ -4), (+4,\ -3), (+4,\ +3)\}
\end{aligned} \tag{8.32}$$

Neville filter $N = 4$ is shown in Fig. 8.7., in which, the numbers 1, 2 correspond to the filter coefficient values of V_1 and V_2 in corresponding positions, respectively. As shown in Fig. 8.7a, quincunx grid signal is converted into rectangular grid signal, while in Fig. 8.7b, which is obtained by rotating the filter in Fig. 8.7a, rectangular grid signal is converted into quincunx grid signal. This kind of transformation is symmetric in horizontal and vertical directions.

As shown in Table 8.6, the all $a_N(s)$ values of quincunx Neville filter correspond to different N values, ($s \in V_k$). When $N = 2$, it is simplified as mentioned earlier red–black conversion. When $N = 8$, $S_8 = V_1 + \ldots + V_7$.

3. **The experimental simulation results and analysis**

The algorithm adopts, respectively, lifting format method and traditional two-dimensional wavelet method, conduct the comparison of lossless compression and lossy compression under different CR. As shown in Fig. 8.8, the comparison

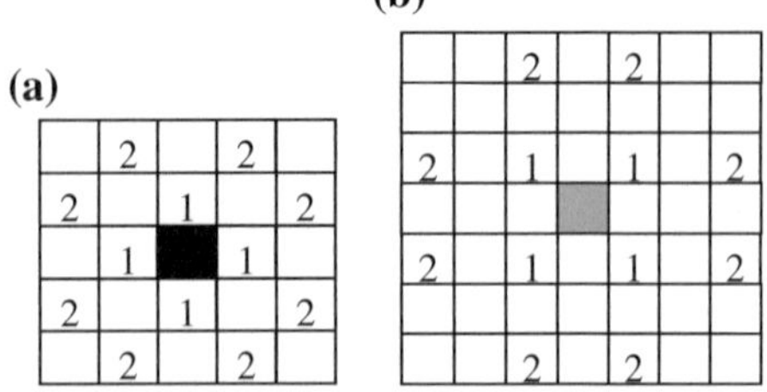

Fig. 8.7 Neville4 filter decomposition **a** Rectangular **b** Quincunx grid

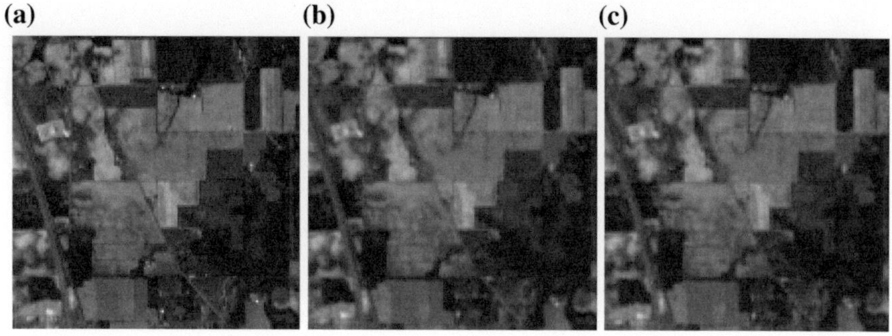

Fig. 8.8 Hyperspectral remote sensing image compression effect picture (CR = 4 as a example, take the 120th band), **a** Original image **b** Traditional two-dimensional wavelet compression **c** Lifting wavelet compression

Table 8.6 Neville filter coefficient of quincunx grid

N value	V_1	V_2	V_3	V_4	V_5	V_6	V_7
2	1/4	0	0	0	0	0	0
4	10/32	−1/32	0	0	0	0	0
6	$87/2^8$	$-27/2^9$	2^{-8}	$3/2^9$	0	0	0
8	$5825/2^{14}$	$-2235/2^{15}$	$625/2^{16}$	$425/2^{15}$	$-75/2^{16}$	$9/2^{16}$	$-5/2^{12}$

results of lifting format compression effect and traditional two-dimensional wavelet compression effect. Figure 8.8a shows the original image 120th band, and Fig. 8.8b shows the traditional two-dimensional wavelet compressed 120th band image, and Fig. 8.8c shows compressed 120th band image based on lifting format. The subjective effect of compressed image based on two methods is similar, and it can all reflect the general picture of original image. As can be seen from the technical statistics data in Table 8.7, when CR is equal, peak signal-to-noise ratio (PSNR) of lifting algorithm is all higher than the obtained value from traditional two-dimensional wavelet, and its value increased by 6 % on average, and the compression performance is higher and computing speed is fast; constructor method is obviously superior to traditional two-dimensional wavelet method. On the other hand, it concentrates effectively the energy distribution by spectral compression, and it can realize larger CR; the computing speed based on spectrum compression method has improved obviously than conduct directly lifting format compression method, but adopts spectrum compression method that makes part of detail information loss.

Table 8.7 The statistical characteristics of hyperspectral image based on different compression methods

Predetermine CR		1		4		10		20	
	Original image	2D wavelets	Lifting scheme	2D wavelets	Lifting scheme	2D wavelets	Lifting scheme	2D wavelets	Lifting scheme
M(bit)	2522.8	3222.7	3250.3	3520.9	3490.8	3871.2	3672.7	4023.7	4023.1
practical CR	–	1.21	1.03	4.4	4.12	10.33	9.98	20.379	20.325
PSNR (dB)	–	40.562	43.231	36.120	38.520	35.879	37.460	34.755	36.254

8.3 Chapter Conclusions

1. SVM-based band selection method. In this method, SVM is used for training decision function and selecting support vector, and then sensitivity analysis is progressed for the decision function of SVM. The bands are divided into several subspaces. According to the combination results from sensitivity analysis and subspace division, effective band selection could be achieved. The bands selected by this band selection method would not be too concentrated, to reduce the redundancy within a wide range of the spatial band and keep more image information. So, it has the ability to improve the classification accuracy, substantially. And, it could reduce the influence on accuracy of subsequent data analysis. Of course, the subspace division method can also be applied in other band selection algorithms. Because of its universality, it has significance and application value.
2. The band selection method based on endmember selection. After revealing the relationship between endmember selection and band selection, the three typical endmember selection methods are applied to construct band selection methods. Theory and experiment show that this transformation method is effective, especially the wavelength selection method based on N-FINDR algorithm achieved satisfactory effect. In this way, we can make full use of existing endmember selection methods to construct the effective band selection methods, thus provide more and more effective theoretical tools for hyperspectral image analysis. It is important to note that due to restriction on the principle or constraint on the complexity, not all the endmember selection can be applied to band selection, in the practical application, it need be earnest analysis from the perspective of theory and experiment. In addition, the established several kinds of band selection methods are susceptible to noise interference in this paper, as a result, before conducting the algorithm, it is very necessary to make the corresponding pretreatment on hyperspectral data. In addition to the transplantation of several typical algorithms mentioned in this paper, the existing other endmember selection algorithms and prospective new endmember selection algorithms are all need to conduct this aspect inquiry.
3. The compression method based on the combination of vector quantization and adaptive band selection.

   The selected spectrum compression algorithm based on adaptive band method can obviously increase the CR, but the spectrum between compressions leads to lose some important characteristics and some detail information of original image, and it is non-reversible process, unable to realize lossless compression. After adopting the second generation wavelet transform, conduct vector quantization, and adaptive arithmetic coding in space compression, the improved SOFM algorithm is used for designing codebook based on vector quantization, due to neural network has stronger fault tolerance, so it can solve the matching problem of non-typical vector in vector quantization. The improved SOFM algorithm improves the training speed and performance of the codebook. Compared with traditional LBG algorithm that is used to design codebook, the paper method has better compression effect.

4. Hyperspectral remote sensing image compression method based on lifting format. The first generation wavelet transform is a kind of image compression method, and it does not bound by the bandwidth and its time–frequency characteristics is excellent; therefore, since its inception, it has always been one of the hot spot image compression methods. However, it is limited by its complex structure and heavy calculation burden. Compared with the first generation wavelet transform, the second generation wavelet transform based on lifting algorithm is easier to be constructed, and the relationship between inverse transformation and forward transformation is simple. It just needs to reverse the transformation order, and so is very convenient to be implemented in computer. In addition, lifting format can be easily implemented in terms of integers.. So it can more easily realize lossless compression of medical image, military remote sensing images, and so on. Adoption of spectrum between compression methods can obviously increase the CR, but spectrum between compression methods leads to lose of some important characteristics and some detail information of original image. Quincunx grid is adopted to construct lifting algorithm for space compression. The lifting algorithm not only has the characteristics of simple structure and fast speed, but also is better than the first generation wavelet transform in terms of actual compression effect. It is superior to the traditional two-dimensional wavelet transform in terms of peak signal-to-noise ratio under the same CR condition.

References

Chang CI, Du Q, Tu LG et al (1999) A joint band prioritization and band-deccorelation approach to band selection for hyperspectral image classification. IEEE Trans Geosci Remote Sens. 37:2631–2640

Davis GM, Nosratini A (1999) Wavelet-based image coding: an overview. Appl Comput Control Signals Circ 1999(1):6

Devijver PA, Kittler J (1982) Pattern recognition: a statistical approach. Prentice-Hall, Englewood Cliffs

Foerster H, Stüben K, Trottenberg U (1981) Nonstandard multigrid techniques using checkered relaxations and intermediate grids. In: Schultz M (ed) Elliptic problem solvers. Academic Press, New York, pp 285–300

Fry TW, Hauck S (2002) Hyperspectral image compression on reconfigurable platforms. The 10th annual IEEE symposium on field-programmable custom computing machines, 251–260

Gu YF, Zhang Y (2003) Unsupervised subspace linear spectral mixture analysis for hyperspectral images. Image Process 1:801–804

Hubert M, Rousseeuw PJ, Vanden BK (2005) ROBPCA: a new approach to robust principal component analysis. Technometrics 47(1):64–79

Hermes L, Buhmann JM (2000) Feature selection for support vector machines. Pattern Recog 2(3–7):712–715

Jelena K, Wim S (2000) Wavelet families of increasing order in arbitrary dimensions. IEEE Trans Image Process 9(3):480–496

Marin JA, BrpckhausJ, Schafer J (1999) A assessing band selection and image classification techniques on HYDICE hyperspectral data. IEEE Trans Syst Man Cybernetics 1:1067–1072

Pabitra M, Murthy CA, Sankar KP (2002) Unsupervised feature selection using feature similarity. IEEE Trans Pattern Anal Mach Intell 3:301–302

Suykens JAK, Brabanter JD, Lukas L, Vandewalle J (2002) Weighted least squares support vector machines: robustness and sparse approximation. Neurocomputing 48(1–4):85–105

Uytterhoeven G, Bultheel A (1998) The red-black wavelet transform. Proc IEEE Benelux Signal Process Symp 3:191–194

Vapnik VN (1995) The nature of statistical learning theory. Springer Press, New York

Wang Q, Guo L (2003) Multispectral image compression based on 1+2 dimension wavelet transform. J Photons 32(9):1126–1129

Webb A (1999) Statistical pattern recognition. Wiley, New York

Wim S (1995) The lifting scheme: a new philosophy in biorthogonal wavelet constructions. In: Proceeding of SPIE, wavelet applications in signal and image procession III (c): 68–79

Wim S (1997) The lifting scheme: a construction of second generation wavelets. SIAM Math Anal 29(2):511–546

Yan JW (2002) Digital image processing technology and image graphics basic tutorial. Sci Press, Beijing

Zhang LY (2005) The research of hyperspectral remote sensing image compression algorithm. Harbin engineering university master's degree thesis

Zhang Y, Zhang JP (2001) Remote sensing spectrum (Hyperspectral) image processing technology. J Chin Image Graph 6(1):6–13

Chapter 9
Introduction of Hyperspectral Remote Sensing Applications

9.1 Agriculture

Currently, hyperspectral remote sensing has been fully applied to wheat, paddy, soybean, corn and other crops (Anguo et al. 2009). As a simple, fast, low-cost and non-invasive spectral analysis technique, hyperspectral technique gets much concern in agricultural production applications, mainly used in crop growth monitoring and yield estimation, nutrition diagnosis and fertilizer, agricultural product quality and safety testing and many other aspects. In recent years, the application research in the field of agriculture of hyperspectrum focuses on the following areas: food crops, including wheat, paddy, soybean, maize, etc.; agricultural products, including fruits and livestock products; economic crops, including cotton, tea and tobacco and so on.

9.1.1 Wheat

Make use of remote sensing technology to establish the contact between the hyperspectrum and wheat biochemical indices. In 2000, Daughtry C et al. made use of crop canopy reflectance spectra to evaluate the canopy chlorophyll content, and in 2004, Huang Wenjiang et al. made use of structure insensitive vegetation index $[\mathrm{SIPI} = (R_{800} - R_{445})/(R_{800} - R_{680})]$ to invert the content of crop canopy pigment ratio. With further research, more factors influencing hyperspectral changes are gradually taken into account. In 2008, Feng Wei et al. was based on the field experiments with different years, different species and types and nitrogen levels to comprehensively analyze the quantitative relationship between pigment content of wheat leaves and canopy hyperspectral parameters, and comparatively analyze the effect of a variety of hyperspectral parameter estimate leaf pigment content, and put forward that red edge position REPLE can better monitor leaf chlorophyll a and

L. Wang and C. Zhao, *Hyperspectral Image Processing*,
DOI 10.1007/978-3-662-47456-3_9

chlorophyll a + b content as well as red edge position REPIG can relatively better monitor chlorophyll b content. Under the experimental conditions, Feng Wei et al. also used characteristic spectral parameters relevant with pigment characteristic index $[(R_{750-800}/R_{695-740}) - 1$ and VOG2 and VOG2 $= (R_{734} - R_{747})/(R_{715} + R_{726})]$ as well as moisture index (e.g. FWBI, Areal190) to effectively evaluate sugar to nitrogen ratio changes in the wheat leaves. In addition to studying the relationship between hyperspectrum and traditional indicators of wheat, some scholars also attempted to only apply hyperspectral classification, matching and other techniques. In 2006, Wang Changyao et al. carried out feature selection and classification study on Hebei Luancheng wheat variety identification with the MAIS imaging spectrometer, classifying imaging spectrometer data by using Fuzzy-Artmap classifier and the selected best band and distinguishing four wheat varieties, with wheat overall classification accuracy over 97 %. In addition to studying different varieties, but in 2008, when Wang Xiaoping et al. studied spring wheat canopy and blade hyperspectrum in different densities at different growth stages (booting stage and milky ripe stage), they found that there were some differences for different varieties of spring wheat canopy and blade spectrum, with obvious differences in the near infrared region of the blade spectrum, but spring wheat canopy and blade spectral waveform of different planting densities was similar, which showed that differences at the near infrared spectrum and similarities of spectrum can be used to distinguish spring wheat varieties. Assessment of pests and freeze disaster extent of losses in wheat with the hyperspectral technique is also an important direction of the current applications of hyperspectral technique. In 2004, Qiao Hongbo measured winter wheat aphids and powdery mildew canopy spectra in different growth periods on different levels of harm, and found that the first derivative (first derivative) of different pest and disease reflectivity changed significantly but red edge displacement was not obvious. In 2007, Jiang Daohui used wheat hyperspectral data affected by stripe rust measured in Xiangfan region and the regression analysis method to establish the relationship model between the wheat chlorophyll content affected by stripe rust and the hyperspectrum. It was proven that the model had better accuracy. In 2008, Li Zhangcheng et al. found that frost stress caused significant damage on the blades, resulting in reduced blade chlorophyll content, forming hyperspectral differences, with significant negative correlation between red edge position and the extent of frost damage and obvious blue shift. Whether the research on the contact between the high spectral reflectance of wheat and physiological parameters or the separate study of different types of wheat hyperspectral characteristics in different stages of growth can provide a reference for future hyperspectral inversion of wheat biochemical indices, and can also provide a reference for intelligent identification of wheat species and other agricultural refinement operations in future large-scale aerial fertilization and dispensary. And the difference of wheat hyperspectrum from normal state under the disaster or other abnormal situations as well as change characteristics of hyperspectrum before the disaster have an important value for accurate forecasting and disaster prevention.

9.1.2 Paddy

In recent years, domestic hyperspectral researches on paddy have a lot, mainly taking paddy canopy, blades and spikes as the research objects for differences of species, nitrogen levels and growing period and other research conditions on the leaf area index (leaf area index, LAI) and vegetation index (vegetation index, VI). Plant leaf surface has a strong absorption feature in the visible red band and has a strong reflection feature in the near infrared band, which is the physical basis of remote sensing monitoring of vegetation. The different combination of two band measured values can obtain different vegetation indices and chlorophyll density (chlorophyll density, CH.D), which is as the research method to study the relationship between protein content, pigment and high spectral or hyperspectral relationship between various elements previously mentioned. These studies make hyperspectral technique possible in the application of paddy. The paddy high spectral characteristics under nitrogen stress, paddy soil leaded or white paddy fringe after inspect pest and other unusual circumstances have been reported, and these studies can observe whether the nitrogen is excess and the paddy field is contaminated as well as identify whether the paddy is damaged by pests in a very short period of time, which will have more practical significance for the production.

9.1.3 Soybean

When studying the relationship between plants and hyperspectrum, the chlorophyll (plant pigments with its function to absorb solar radiation energy in photosynthesis, and terrestrial plants contain mainly chlorophyll a and b) is often mentioned, mainly because that the chlorophyll can indirectly reflect the health of vegetation and photosynthetic capacity as well as the physiological state after a variety of factors stress in the environment. In 2006, Song Kaishan et al. measured the hyperspectral reflectivity and chlorophyll content data of soybean canopy and carried out the correlation analysis, and then made use of chlorophyll sensitive bands to establish vegetation index chlorophyll estimation model, and finally used the greater correlation coefficient band as the input variable of neural network model to estimate the chlorophyll content. The vegetation indices (VI) include ratio vegetation index (ratio vegetation index, RVI), modified second ratio vegetation index (modified second ratio index, MSRI), modified secondary soil adjusting vegetation index (MSAVI2), normalized difference vegetation index (normalized difference vegetation index, NDVI), renormalized difference vegetation index (renormalized difference vegetation index, RDVI), soil adjusted vegetation index (soil adjusted vegetation index, SAVI) and modified secondary soil adjusting vegetation index (MSAVI) and so on. There is good correlation between LAI and chlorophyll content, so the size of LAI can indirectly explain the chlorophyll content. In 2008, the correlation coefficients of soybean LAI estimation model (namely fitting equation) produced by the power function $[y = a \times x^b]$ constructed by RVI and the exponential function $[y = a \times \exp(b/x)]$ and logarithmic function $[y = a + b \ln(x)]$ constructed by MSAVI used by Huang Chunyan et al. were all above 0.9 and the root mean square error were all below 0.24.

9.1.4 Maize

When studying maize hyperspectrum, in addition to research hyperspectral inversion pigments, the researchers began to consider hyperspectrum from a practical point of view of maize. When the corn is used as silage, which is like alfalfa, we consider the crude protein content. In 2005, Shi Yunlu analyzed the change contact between the changes of crude protein content of maize silage and each growth period and established the crude protein content hyperspectral inversion method. In the same year, Yi Qiuxiang also carried out a more complete hyperspectral study on maize and established the hyperspectral optimal estimation models of main agronomic parameters of maize (including nitrogen content, fat content, crude fiber content and pigment content); the correlation coefficients of these models are high.

Apart from the above applications, hyperspectral technique is gradually replacing HPLC (high performance liquid chromatography, HPLC) and mass spectrometry (mass spectrometry, MS) and other traditional detection methods, playing an increasingly important role in food quality and safety. There are many application researches of hyperspectrum in terms of fruit detection, mainly focusing on fruit production, damage and firmness detection. Hyperspectral research mainly focuses on food safety and meat quality in terms of applications of livestock products. Hyperspectrum also has some applications in terms of tea, tobacco and cotton and other economic crops.

9.2 Forest

Applications of hyperspectral remote sensing in terms of forest include forest inventory, forest biochemical composition and forest health research, forest hazard analysis, invasive species monitoring, etc. (Tan et al. 2008). Hyperspectral remote sensing has been used in forest mapping, forest resource survey, forest area estimate, biochemical and physical factor estimation and so on. Forest information extracted from hyperspectral remote sensing data mainly includes forest biophysical and biochemical parameters, forest health factors and so on.

9.2.1 Forest Investigation

Forest investigation is typically completed by using conventional ground plot method or image interpretation and other means. In the past 30 years, large-scale of application remote sensing data (such as TM, SPOT) conducted the forest resource investigation practice, however, due to the limitations of multi-spectral remote sensing spectral resolution as well as the uncertainty of the used remote sensing data in space and time, the obtained forest information was limited by data accuracy, completeness and level of details. Hyperspectral remote sensing can provide better, blending and accurate measurement data of forest information, which is essential to investigating the production of high quality forest products. Hyperspectral data can be used to estimate some forest biophysical parameters with high precision, such as forest type, leaf area index, canopy density or degree of coverage, etc.

1. **Identification of forest types**

Hyperspectral remote sensing can improve the classification accuracy of forest species, and hyperspectral data classification can get a more accurate distribution map of forest species, which have been fully illustrated in many studies. In 1998, Martin M et al. combined unique biochemical characteristics of different forest species with the relationship established between the hyperspectral AVIRIS data and foliage chemical components to identify 11 kinds of forest types, and they considered that the hyperspectral remote sensing technique can divide forest types finer, which is mainly decided by vegetation reflected light nature controlled by biochemical substances, while the conventional wide-band remote sensing data is almost impossible to reflect these subtle spectral differences. In 1998, Gong Peng et al. made use of field measured spectral data to identify six kinds of major conifer species in California, proving that hyperspectral data have strong potential for tree species identification, and they considered that hyperspectral data simple transformation can effectively improve the recognition accuracy and the identification of conifer is better with spectral data with the band width of 20 nm or narrower. In 1999, Davison D et al. evaluated the ability of CASI hyperspectral data monitoring Canada Ontario forest parameters, and the results showed that CASI hyperspectral data had the ability to distinguish main tree species. In 2002 and 2003, Goodenough D et al. classified five kinds of forest types in Canada Victoria area by using Hyperion, ALI and ETM3 remote sensing data with the corresponding classification accuracy respectively as 92.9, 84.8 and 97.5 %, indicating that hyperspectral remote sensing data had a stronger ability to identify forest types. In 2007, Chen Erxue et al. made use of Hyperion hyperspectral data and ground observation data to carry out a comparative evaluation study on several advanced hyperspectral statistical pattern recognition methods of overseas development, showing the reduction in dimensionality of hyperspectral data, and used the second-order statistic estimation method, and then applied the classification algorithm combining the spatial contextual information and spectral information, such as ECHO, which can effectively improve the recognition accuracy of forest types. However, we must see that there exist vegetation types that can not be recognized by hyperspectral remote sensing, which requires field investigations of investigators.

2. **Leaf area index (LAI) estimation**

Previous studies on the forest LAI estimation in the remote sensing method are mainly confined to a few relatively wide-band multi-spectral data. Most researches have been devoted to find some simple statistical relationship between LAI and various vegetation indices extracted from remote sensing data is estimate LAI, but the accuracy is not high. One reason is that the wide-band remote sensing data is often mixed with a considerable proportion of non-plant spectrum, resulting in the relationship between a variety of vegetation indices and LAI not close; this non-vegetation spectra can be repressed by using the spectral differential technique in the hyperspectral remote sensing data so as to improve the correlation between remote sensing data and LAI. In 1994, Gong P et al. made use of CASI hyperspectral data to conduct experimental predictions on LAI of Oregon coniferous forest successfully. In 1997, Zhang Liangpei et al. carried out estimation study on biological variables by using hyperspectrum and considered that the first derivative operation by using spectrum signals could compress the soil spectral signal in mixed spectrum, resulting in more objectively reflecting the actual biological index, such as LAI and so on. In 2000, Pu Ruiliang et al. made use of CASI data to use the estimation method related to variables and based on vegetation indices and multiple regression prediction method to estimate the forest LAI, and found that the stepwise regression could be yet regarded as a method to predict with high accuracy; in the single variable regression analysis, the hyperbolic relationship between LAI and NDVI is the optimal method to estimate LAI. In 2004, Gong P et al. estimated forest vegetation LAI in Argentine Patagonia semi-arid areas by using three kinds of remote sensing data, used the stepwise regression method to select the spectral bands close to LAI and then created multiple regression equation of each band and LAI. The examination proved that LAI estimation accuracy of AVIRIS was the highest, followed by Hyperion, and ALI estimation accuracy was the lowest, and the wave band estimation LAI with the center wavelength as 820, 1040, 1200, 1250, 1650, 2100 and 2260 nm was the most potential.

3. **Canopy density information extraction**

Forest canopy density is very important on the forest ecosystem research and forest management. Conventional forest canopy density estimation is obtained by field survey and aerial photo interpretation technique. This method is labor-intensive, time-consuming and costly. The promotion and use of remote sensing technique, especially the emergence of hyperspectral remote sensing, provides a powerful tool for forest canopy density estimation in the region scale as well as large areas. In 2000, Pu Ruiliang et al. carried out spectral mixing pixel decomposition on the calibration AVIRIS image, showing that the extracted forest canopy density information component figure was 2–3 % higher than the value of infrared aerial photo interpretation and the canopy density distribution was more reasonable, so it is reliable to extract forest canopy density information from hyperspectral image data with the spectral mixing model method. In 2004, Pu R et al. made use of Hyperion image of wavelet transform to select a variable closely related to the forest canopy density by the stepwise regression method, and then created multiple regression relationship with canopy density, with the estimation accuracy reaching 85 %, which can meet the production needs. In 2004, Lee K et al. estimated the high canopy forest LAI through the establishment of relationship between LAI and spectral reflectance, and thought that shortwave infrared spectral reflectance might be an important factor to increase the optical remote sensing data to estimate the potential of high canopy forest LAI. Tan et al. (2006) carried out a comparative evaluation of multivariate statistical regression estimation method based on the spectral feature selection and spectral feature extraction by making use of forest resource survey data. Verification results show that the two kinds of methods can achieve more than 85 % of canopy density estimation accuracy and the precision of spectral feature extraction method is slightly higher than the selection method based on spectral feature.

9.2.2 Forest Biochemical Composition and Forest Health Status

The appearance of hyperspectral remote sensing technique makes it possible to extract biochemical parameters from remote sensing data. The extraction of biochemical information in the region as well as on a global scale is very important on the study and understanding of ecosystem processes such as photosynthesis, carbon and nitrogen cycles and forest litter decomposition rates as well as description and simulation of ecological systems. The canopy biochemical constituents can be extracted from hyperspectral remote sensing data, such as chlorophyll a and b, nitrogen, lignin, water content, etc., and these estimated chemical compositions are related to the concentration of special chemical elements in the tree body and are also related to the measured total forest health status. Some studies show that the narrow band carried by the airborne sensor can be used to monitor the early damage of coniferous species in the forest decline and the hyperopia field narrow band spectrometer can successfully monitor the invisible stress damage of plants caused by the invisible herbicide in the mixed conifer forest stand. The report indicates that the judge of narrow band reflection in critical spectral region is the basis of remote sensing applied to victim monitoring of the forest canopy. The growth and decline of specific biochemical elements may cause the measured forest health status and the generated earning precision not high, so the forest damage distribution diagram can provide valuable method and basis for sustainable forest management practices. For example, the estimation of particular chemical elements such as N element can be used in the actual operations of precision forestry, whereas the fertilizer only can be used in those forest areas lacking of N element; the forest land around the newly exploited mine can be carried out environmental monitoring and evaluation by using hyperspectral imaging (HYMAP), which is completed by monitoring the concentration of the chemical elements of local forests in a period of time. In 1989,

Wessman C et al. pointed out that the aviation imaging spectrometer AIS radiation data and pure conifer forest canopy lignin and effective N existed significant correlation. In 1994, Johnson L et al. analyzed the relationship between AVIRIS hyperspectral data in Oregon middle and western region forest stands and the corresponding forest stand canopy biochemical characteristics change, indicating that the changes in canopy N content and lignin and the selected AVIRIS band data change existed a corresponding relation and AVIRIS data and starch content had no significant relationship. In 1994, Matson P et al. used AVIRIS and CASI data to confirm that the chemical composition of the canopy carried a variety of ecological system change process information in climatic zones, and recommended that such information could be estimated from hyperspectral data, and found that the first derivative spectral data with the center wavelength at 1525–1564 nm could be used to describe the change of N amount in the canopy. In 1997, Pu Ruiliang et al. made use of CASI hyperspectral data to estimate the concentration of the chemical composition of the forest foliage; for the chlorophyll, the best value R^2 came from the second derivative spectral trinomial regression equation ($R^2 = 0.944$), with the center wavelength respectively as 748, 507 and 735 nm; the optimal value R^2 of full N came from the first derivative spectral trinomial regression equation ($R^2 = 0.933$), with the center wavelength respectively as 780, 764 and 566 nm; these results showed that the differential spectral technique could significantly improve the estimation accuracy of the chemical composition of the forest foliage. In 1994, Datt B et al. studied visible—near infrared reflectance spectrum characteristics of several eucalyptus leaves, better improving the chlorophyll content estimation error, and the results showed that the reflection spectrum at a wavelength of 710 nm had a maximum sensitivity for the chlorophyll content, followed by the reflection spectrum at a wavelength of 550 nm. The test of several reflectance spectral indices showed that as a remote sensing estimation parameter aiming at the higher plant chlorophyll content, the effect of the ratio $(R_{850}-R_{710})/(R_{850}+R_{710})$ is the best; the correlation between the ratio D_{1754}/D_{1704} of the first derivative spectrum and red edge position and chlorophyll content is the best; the second order reflection spectral derivative ratio is equally one of the best parameters for the chlorophyll content.

9.2.3 Forest Disaster

For managers who want to make a decision on reducing environmental degradation and loss of timber, the forest disaster monitoring is an imminent thing. Forest disasters include forest diseases, pest invasion and fire. Hyperspectral remote sensing images can detect the key invasive trees in the early stage of pest victims, so fighting the pest attack as soon as possible. However, the pests and pathogens have the physiological regulation ability, so multi-spectral remote sensing image is often lack of the sensitivity monitoring the canopy reflectance subtle changes. The forest canopy understory shaded part can be regarded as an indicator of the environmental change to affect the canopy health status, which is an early warning, and changes of the forest stand environment are related to the CO_2 loss of soil surface, acidification, N availability and soil water content and so on. Hyperspectral image can make forest classification more accurate, and the accuracy is related to the impact of the observed hyperspectral information on the reflection coefficient of the understory. The main problem of forest fire monitoring is to obtain real-time updates on the fire boundary data. In 2000, Jerred B pointed out that burning vegetation spectral signal had about 767 nm of small pulse, which was caused by the burning K element and could be used to monitor the burned area of standing trees. The wavelength of this pulse is very short, which can only be obtained by hyperspectral remote sensing image. Canopy water content is a key indicator and a determining factor of the outward spread ability of the forest fire through the canopy. The moisture content can also be measured by moisture absorption features in the near-infrared short wave region of hyperspectral imaging. In addition, the study found that the absorption peak of water vapor and carbon dioxide mixture was near 1,400 nm, which could be regarded as a credible direction index of the leading edge of fire.

9.2.4 Exotic Species Monitoring

Hyperspectral remote sensing image can be also used to draw the invasion figure of alien species. Invasion of non-domestic species poses a major threat for maintaining global biodiversity and ecosystem. Multispectral remote sensing can monitor only when the number of invasive species causes density rising and scope expansion. Relative to multi-spectral remote sensing, hyperspectral remote sensing can make full use of biochemical substances and provide a potentially enhanced of invasive species figure in the early stage of invasion with the intruder as the goal according to its invasion way, which greatly benefits the control of invasion species expansion and the reduction of the destruction. Many studies have indicated the support for this result.

9.3 Meadow

Hyperspectral remote sensing applications in the meadow include meadow biomass estimation, meadow species identification, estimation of meadow chemical compositions and other aspects (Zhou et al. 2009). Meadow can both protect the environment and maintain ecological balance, and the meadow ecosystem is the guarantee of the entire ecosystem safety and plays an important role in supporting and promoting agro-ecosystem and forest ecosystem. Physical and chemical properties of the vegetation canopy control the primary productivity of vegetation to some extent. The basic principle of the meadow vegetation remote sensing is to make use of the spectral characteristics of meadow vegetation canopy, that is, vegetation has a strong absorption peak in the visible light part and a strong reflection peak in the near-infrared part, and data from various treatments on these sensitive bands can be used to reflect all kinds of information of the vegetation. With the advent of hyperspectral remote sensing, more bands are combined to carry out more accurate estimation on meadow information.

9.3.1 Biomass Estimation in Meadow

In a broader scope, the biological parameters of vegetation mainly refer to some key variables used in the study of terrestrial ecosystem, including leaf area index, photosynthetically active radiation absorption rate, biomass, vegetation coverage and so on. Among them, the leaf area index is the most important structural parameter, not only as an important input parameter of earth surface evapotranspiration model, but also as the key factor to determine the biomass and yield, so it has been the focus of remote sensing estimation biophysical parameters. The leaf area index is generally defined as the area sum of all leaves per unit area of surface area or the area sum of downward projection of all leaves per unit area of surface area. Photosynthetically active radiation absorption rate is an important variable to determine the net primary productivity, dry matter accumulation and crop yield and is also an important parameter of the land surface energy budget and land—atmosphere exchange hydrological model. Biomass is one of the important parameters of ecosystem research and refers to the sum of all plant quality on the surface area, usually referring to the weight of dry matters. Coverage is one amount of distribution sparse degree and canopy spread degree, with its value between 0 and 1. The inversions of these biological parameters mainly focus on the leaf green inversion, leaf area index inversion and moisture content inversion. About the contents, Mr. Tong Qingxi discussed in more detail in his work, no longer talking about here.

Grassland vegetation biomass monitoring is an important basis for the rational use of grassland resources. The accurate and timely acquisition of regional forage yield and the feature with time and spatial variation is the key to achieve the reasonable, efficient and sustainable use of grassland resources and the judgment of grassland ecosystem integrity and sustainability as well as an important condition to protect and manage grassland. The frequently used parameters for grassland biomass measurement are NDVI and “red edge” (REP); with the advent of hyperspectral remote sensing, NDVI has more alternative bands to characterize plant information, but the change in soil and atmospheric environment is too sensitive, and “red edge” is more stable. Accurate estimation of the amount of grass edible forage can reasonably control the amount of livestock and maintain the dynamic equilibrium of livestock. Liu Zhanyu et al. used high spectral model method to estimate the biomass of grassland and confirmed the stepwise regression model with five original high spectral band variables as the optimal model by comparing the estimation accuracy of univariate linear and non-linear regression model and stepwise regression model, with the estimated standard deviation as $0.404\text{kg}/\text{m}^2$ and the estimate accuracy as 91.62 %. Moses et al. carried out hyperspectral data processing by comparing the partial least squares regression (PLS), NDVI and red edge position methods, and finally proved that the accuracy of partial least squares regression was the highest by comparing three methods of accuracy on the grassland biomass estimation. Although it is based on empirical models, relative to the high spectral index of univariates (NDVI, REP), PLS regression is a better method. When biomass is estimated by using NDVI and if the vegetation coverage is relatively high, there will be relatively large deviations. Onisimo et al. proposed a new approach to solve this problem. Under controlled experimental conditions, the plant spectral reflection curve at a wavelength of 550–750 nm is obtained by a spectrophotometer, and after the continuous removal method, 4 wavelength depth indices are obtained including wavelength depth (BD), wavelength depth ratio (BDR), normalized wavelength depth index (NBDI) and normalized wavelength depth area (BNA); the biomass estimation is also conducted, with correlation coefficients respectively as 0.81, 0.83, 0.86 and 0.85, but the estimated biomass correlation coefficient with a wide band of NDVI is 0.31 and 0.85, but the estimated biomass correlation coefficient with a wide band of NDVI is 0.31 and 0.32. Abdullah found that $\text{WBI} = R_{970}/R_{990}$ played a good indication role in 4 physicochemical properties (fresh weight, dry weight, water content and leaf area index) of natural grassland in the burned semi-arid areas, and had a high Pearson correlation coefficient with the grassland biomass. The above several examples with high-spectral data to estimate grassland biomass have achieved relatively high accuracy, but the used hyperspectral data are ground-based data, and the estimation of biomass is not achieved from hyperspectral images indeed.

9.3.2 Grassland Species Identification

Hyperspectral remote sensing data can greatly improve the identification and classification accuracy of vegetation. People can screen the significantly different spectral band of the plant species from the numerous narrow bands, and then carry out identification and classification of the screed several narrow bands of plant species. Some data compression techniques can also be used to recombine several comprehensive bands and make full use of the spectral information of vegetation, which will greatly improve the identification and classification accuracy of vegetation. The main purpose of hyperspectral remote sensing for grassland species identification is to monitor the extent of degradation of grassland, and the grassland degradation has become the most serious problem for the sustainable development of grassland animal husbandry. Grassland degradation will have a major impact on the climate, the environment and the local economy. Currently monitoring grassland degradation is mainly based on changes in vegetation index, but the vegetation index is vulnerable to environmental changes, which is not conducive to grassland monitoring. Therefore, a more effective way should be to find out the characteristic bands that characterize the indicator species of grassland degradation, thus achieving the identification on hyperspectral images. In 2001, Schmidt K et al. made use of the spectrophotometer, statistical analysis, envelope line removal analysis and distance analysis and other methods to verify that the Themeda triandra, Hyperthelia dissoluta, Panicum maximum, Brachiaria brizantha, Dichanthium insculpta and other three kinds of grass can be distinguished under laboratory conditions. In the same method, the spectral information extracted by the hyperspectral image can be analyzed and compared, and the results show that all eight kinds of grass have separability. These differences will provide a theoretical basis for remote sensing classification of grass species. Yamano et al. measured the spectral characteristics of Achnatherum splendens, Caragana microphylia, Stipa grandis and Leymus chinensis in the Inner Mongolia Xilingol region by using a spectrometer with the spectral resolution of 3 nm and then interpolated the spectral curve to 2 nm; carried out three times of smoothness to remove the noise with Savitsky-Golay least squares method, modified weighting function and normalization factor and found that Caragana microphylla had different characteristics from the other three grass in the peak district near 670 and 720 nm wavelength. These different characteristics can effectively differentiate Caragana microphylla from the other three grass, which has been verified in Mongolia region. It shows that monitoring grassland degradation by using these characteristics is possible. But the data is obtained from the ground spectral measurement data analysis, and whether it can be distinguished on hyperspectral images needs to be further explored.

9.3.3 *Chemical Constituent Estimation*

The hyperspectral remote sensing technique makes it possible to extract biochemical information parameters from remote sensing data. The application of remote sensing technique measurement and the analysis of changes of the leaf and canopy biochemical information in time and space can help to understand the productivity of plants, litter decomposition rate and nutrient availability. According to changes in the concentration of various chemical ingredients, the growing situation of grassland can be evaluated. Spectral characteristics of nitrogen in plant leaves are easily concealed by water features. Ofer et al. made use of aerial hyperspectral imaging to estimate the number $\left(\mathrm{kg/hm^2}\right)$ and quality of forage in two areas of interest (AOI); the quality was represented by the ratio of C:N; the relative error of the biomass obtained and C:N was respectively 18 and 8 %; these two AOI can be used to calculate the total biomass and C:N of two plains, with the total accuracy over 80 %. Then the obtained two values can be used to calculate the crude protein content of forage. The crude protein content of grassland vegetation is very important for pasture managers to decide whether the grassland can be continued to use. In 2007, Gianelle et al. made use of partial least squares regression to validate the data obtained by different measurement methods, which had a better correlation with grassland vegetation physicochemical parameters [biomass, nitrogen content (total amount, percentage), the nitrogen content of the ground part (total amount, percentage)]. The results show that the data obtained by vertical measurement, especially "red edge" location, had a good indication role in the total nitrogen content and nitrogen concentration. Compared with traditional broadband vegetation indices, the hyperspectral data obtained from whichever measurement method can describe the vegetation physical and chemical information well. Currently, remote sensing monitoring study on the grassland vegetation growth and health status is relatively few and remote sensing monitoring study on the crop growth is many, with mature method and worth learning from. Moreover, hyperspectral applications in other areas of grassland monitoring also made certain progress. In 2007, Black SC et al. attempted to use high spectral reflectance curves to calculate the CO_2 exchange rate of the grassland; by calculating several narrow-band vegetation indices and spectral depth analysis, the final result was that Pearson correlation coefficient can find high correlation between the exchange rate of CO_2 and optical reflection coefficient, but the linear regression method only had 46 % of correlation and did not get the same results as that in the laboratory. In 2006, Liu Zhanyu et al. made use of the model estimation method to research the coverage of natural grassland.

9.4 Ocean

Hyperspectral remote sensing applications in the ocean aspect mainly focus on marine and coastal resource environmental monitoring, and also include other related basic researches (Lou et al. 2008). Ocean remote sensing is one of the key technologies for significant progress in marine science in late 20th century, with its main purpose to understand the ocean, research the ocean, develop and utilize and protect marine resources, so it has a very important strategic significance. With the development of science and technology, hyperspectral remote sensing has become a frontier domain in ocean remote sensing. The moderate resolution imaging spectrometer is with a wide spectral coverage, high resolution and many bands and other advantages, so it has become an effective tool to detect ocean water color and water temperature. It not only can be used in detecting seawater chlorophyll concentration, suspended sediment content, certain contaminants and surface water temperature, but also can be used to detect sea ice and coastal zones and so on.

9.4.1 Basic Research on Ocean Remote Sensing

In the analysis and research of hyperspectral remote sensing information of ocean water color, in 1998 and 2001, Pan Delu et al. combined the radiative transfer mechanism and mathematical approximation theory to develop the ocean water color hyperspectral information multifactorial inversion algorithm to study the quantitative relationship of the chlorophyll, suspended sediment and yellow substance and other factors on the water leaving radiation rate and proposed multi-factor leaving radiation model. In 1991, Chinese Academy of Sciences Remote Sensing Application Institute made use of MAIS imaging spectral data to make the seawater chlorophyll concentration distribution diagram of Australian Darwin City. In 2000, Tang Junwu established ocean optics three-dimensional Monte Carlo model to simulate the impact of different solar zenith angles and water body composition and other parameters on the water leaving radiance direction characteristics. The simulation results show that under certain remote sensors, the sun and pixel geometrical conditions, the water leaving radiation change brought by the same water body light field dichroic feature may be greater than the water leaving radiation error of the existing operational water color atmospheric correction algorithm inversion, and the simulation results have a certain significance on the correct spot data acquisition and remote sensing and ground data contrast in the water color remote sensing; in 1999, Fu Kecun et al. gave the estimation mode of the Yellow Sea chlorophyll based on site monitoring data of the sectional radiometer and fluorometer. Chlorophyll concentration is one important parameter reflecting the strength of phytoplankton photosynthesis and algae growth, and the chlorophyll content and water derivative light have certain relevance based on this, in 2002, Ma Yi et al. selected the most relevant band to establish the simple model of correlation between phytoplankton concentration and water derivative light. It is visible that domestic ocean remote sensing application basic research refers mainly to construct some mathematical models. Researches about how to solve the low reflectivity of water body and the scattering impact of atmosphere on blue-violet band spectrum and other problems have not been carried out yet.

9.4.2 Application Research on Resource and Environment Monitoring of Ocean and Coastal Zone

In terms of marine water quality monitoring applications, only the visible spectrum can be used to observe underwater condition, with the most penetrating wavelength range of 0.45–0.6 m (blue light to yellow light), which is called as "ocean window", using an imaging spectrum technique to observe sea conditions including the sedimentary suspended solids, plankton and chlorophyll distribution in the ocean. For example, it can be used for the valuation of marine sediment and chlorophyll content, and the valuation of chlorophyll content can be used to monitor the growth of algae and infer distribution of plankton and fish school location in aquatic product studies. For our country, with the development of industrialization and urbanization in coastal areas, the pollution of the marine environment is getting worse. The marine pollution causes huge economic losses on the country. In 2002, Leng X et al. made use of spectral data obtained by 2001 squid ring red tide barricading experiment to develop the red tide spectral database, which includes statistics query and analysis, spectral data feature extraction, interpolation fitting, curve smoothing and various curve drawing and other functions; in 2000, Zhao Dongzhi et al. also conducted the acquisition and processing work of red tide land class spectral data. Red tide spectral database is a basic work of red tide optical remote sensing and can provide data input for hyperspectral remote sensing information extraction of red tide. Hongbo made use of our country's PHI imaging remote sensing data to successively identify greasy dirt according to the visible and near-infrared band signal strength, and explored the different grey scale in the image based on different thickness of greasy dirt, and made use of the density slicing result to carry out semi-quantitative calculation of the oil content of greasy dirt, thus estimating the coverage area and number, so as to obtain the severity of water pollution; at the same time, made use of specific value data of remote sensing reflectance ratio among different bands combined with satellite data to explore and get the distribution character of water suspension material. In 2003, Fan Xuewei et al. made use of PHI data and the band combination of the maximum covariance and the smallest correlation to synthesize false color image, and proposed the identification method based on reflectance curve and reflectivity differential curve by the comparison of reflectance curves of normal sea and red tide water body and the extraction of reflectance curves in abnormal area as well as the construction of correlation analysis function and other methods, namely, by comparing the reflectance curve and reflectivity differential curve of unknown biological species of red tide and known biological species of red tide in the database, based on correlation analysis to identify the function value of function, exploring the red tide discovery and automated detection, red tide organism dominant species identification and red tide biomass distribution feature extraction technique. In addition, it is planned to carry out the statistical analysis on the reflection characteristics of sea ice and seawater, and get the probability density distribution curve of ice and water albedo in the normal distribution assumption, and distinguish ice and water with the probability density distribution curve, and calculate the sea ice concentration degree. In addition, the researches on spectral characteristics of the land-based source pollution, aquaculture, mud flat and other coastal typical elements are also carried out, and researchers take aerial hyperspectral image as the data source and select land-based source pollution, aquaculture and mud flat as monitored elements for the sensitivity research of spectral bands of the above elements, trying to get the optimal detection band and further developing accurate and rapid identification and detection techniques. The corresponding researches are also carried out in terms of sea surface temperature measurement, sea surface suspended sediment concentration qualitative or semi-quantitative observation and ocean dynamic phenomenon and other aspects.

9.4.3 International Development Trend

At present, the main researches conducted internationally include ① ocean carbon flux research, recognizing its control mechanism and variation; ② relationship research of marine ecosystem and physical properties of the mixed layer; ③ coastal environment monitoring and management. In terms of the ocean carbon flux research: ① mainly making use of long-term sequence data to quantify and analyze the impact of cyclical global climate phenomena (such as El Nino) on the marine environment; ② studying the relationship between sea—gas CO_2 net flux and biological process. This aspect of model emphasizes the important impact of biological and physical processes on the absorption of marine CO_2 and the relationship of global warming; ③ developing global marine primary productivity computing model. Research shows that the model is very sensitive to the input surface chlorophyll concentration field. In terms of biological oceanography and upper ocean process studies: ① making use of ocean water color satellite remote sensing data to verify the authenticity of the distribution of phytoplankton predicted by some numerical models. The chlorophyll field from the satellite data after the assimilation treatment is input into the numerical model to improve the predictive ability of marine simulation; ② making use of ocean water color data to calculate the upper ocean heat balance. This new method gets a good result in the vertical distribution calculation of sea—gas heat flux and upper ocean heat in the Arabian Sea equatorial Pacific Ocean area; ③ making use of ocean water color images to directly observe the impact of the climate and other large-scale phenomena (such as El Nino) on marine chlorophyll distribution.

9.5 Geology

Reflection, transmission, absorption and emission characteristics of land class electromagnetic radiation are the basis of remote sensing technique application and target detection. Different environment, respective components and structures formed by all types of surface rocks result in their different spectral characteristics. Due to the low spectral resolution of multi-spectrum and inadequate spectral characteristics performance of land class, the identification of land class is mainly dependent on the spatial characteristics of land class, including gray scale, color, texture, shape and spatial relationship. Information processing and information extraction mainly apply image enhancement, image conversion and image analysis methods to enhance the differences in tone, color and texture of the image, so as to achieve the purpose of maximizing to distinguish land class. With the successful development of the imaging spectrometer and its industrial development, the remote sensing information extraction has also entered a new era. The recognition of imaging spectrum on land class is mainly dependent on the spectral characteristics of land class to directly carry out the land class identification and quantify land class information. Imaging spectroscopic technique is a leap of multi-spectral technology development, and it can make the spatial characteristics of the target object imaged, at the same time, it can carry out continuous spectral coverage for dozens or even hundreds of narrow spectral bands formed by dispersion or light split of each spatial pixel. The formed remote sensing data can be described with "image cube (three dimension)", with two dimensions representing space and another one dimension characterizing the spectrum. Thus, in the spectral and spatial information integrated three-dimensional space, the land class "continuous" spectrum as well as its diagnostic characteristic spectrum can be obtained arbitrarily, thus directly identifying the target land class based on the land class spectrum knowledge and further obtaining quantitative land class information.

9.5.1 Mineral Identification

The main theories of hyperspectral remote sensing in mineral identification application include based on a single diagnostic absorption characteristic parameters, based on the complete waveform characteristics and based on the spectral knowledge model and so on (Pei and Fu 2007).

1. **based on a single diagnostic absorption characteristic parameters**

Rock and mineral single diagnostic absorption characteristics can use absorption band position, absorption depth, absorption width, absorption area, absorption symmetry, absorption number and sorting parameters as a complete characterization based on a single diagnostic absorption waveform of endmember minerals, the extracted and enhanced parameter information from the imaging spectrometer data can be directly used to identify mineral types, such as the relative absorption depth map (Relative absorption band-depth image) method, continuous interpolation band algorithm (Continuum interpolated band algorithm) and spectral absorption index image (Spectral absorption index image) and so on.

2. **based on the complete waveform characteristics**

Make use of the entire spectrum curve for mineral matching recognition, which can improve the uncertain impact of single waveform to a certain extent (e.g. spectral shift, variation, etc.) and improve the accuracy of recognition. The recognition method based on the entire waveform refers to the two-dimensional space consisting of reference spectrum and pixel spectrum for the rational choice of measurement function metric spectrum or the similarity of actual measurement spectrum and the image spectrum, such as the similarity index method (Similarity index algorithm) and spectral angle recognition method (Spectral angle mapper). In 2000, based on full waveform of minerals, Zhang Zonggui et al. made use of neural networks to automatically identify minerals, and Gan Fuping et al. designed and developed the imaging spectral mineral recognition technology based on full waveform. The above methods are adaptable when there are a number of known land class spectrums, and are more useful for the image land class recognition. However, because the actual land class spectral variation access to data is influenced by the observation angle and the particle size, resulting in the spectral changes. For the land class with few whole spectral characteristic differences, the exact matching is difficult, resulting in the confusion and error of mineral identification and analysis.

3. **based on the spectral knowledge model**

Recognition method based on the spectral model is built based on signal processing technology and method of certain optics, spectroscopy, crystallography and mathematical theory. It can not only overcome the defects of the above methods, but also identify land class types while accurately quantifying the composition and other physical properties of surface materials. For example, the linear mixed spectrum decomposition model (SMA/SUM) based on Hapke bidirectional reflectance spectral theory can construct the linear spectral decomposition model according to the different land class or different pixel spectral reflectance responses. Tompkins proposed the modified spectral mixture analysis (MSMA) model. This model uses a virtual endmember and a damped least squares algorithm to effectively and ultimately choose sub—endmember for spectral decomposition according to some priori knowledge, improving the practicality of SMA. These methods rely on more spectroscopy knowledge and mathematical methods. It is difficult to accurately determine the feature parameters or describe the spectral model in practical applications, which limits the application of such techniques and methods. However, such methods identify land class while quantifying material composition, so in terms of its development trend, with the maturity of a series of techniques, the in-depth development of the spectroscopy and crystallography and other knowledge as well as the improvement of recognition accuracy and quantization ability, its application will be more extensive.

9.5.2 Resource Exploration

1. **Oil exploration**

The use of hyperspectral remote sensing for direct oil detection is to make use of remote sensing image information extraction and other techniques to dig out the hydrocarbon micro-seepage information of remote sensing image and delineate or predict favorable oil and gas exploration target regions. Hyperspectral remote sensing has the advantages of economy, safety and high efficiency, etc., and has great application potential in the oil and gas exploration. With the development of science and technology and the need to find complex hydrocarbon reservoirs, the early ground oil and gas display for the exploration of oil and gas reservoirs method has gradually been replaced by other methods, and seismic, oil and gas exploration and satellite remote sensing methods have been gradually introduced in the oil and gas exploration. Now most remote sensing data applied to the oil and gas exploration is multi-spectral data, and examples of hyperspectral data applications in oil and gas exploration are relatively few, and the development of hyperspectral remote sensing technology can push the oil and gas exploration and application of remote sensing to a higher, more effective and quantitative application level.

United States, Germany, Spain and other countries have successively made use of aerial hyperspectral instrument to detect the alteration anomaly zone of hydrocarbon micro-seepage oil and gas reservoirs. In 1986, Singhroy et al. used the fluorescent linear array imager (spectral range (430–850) nm, 288 channels) to study the condition of vegetation of Michigan Stoney Point oil field area, revealing the changes of spectrum of vegetation caused by oil and gas micro-seepage. Since 2000, foreign countries have been making use of ASD and other wild and Hymap and AVIRIS aerial hyperspectral sensors for the oil and gas exploration, and there have been some successful application cases, establishing the spectral library of oil and gas leakage and oil and gas leakage area soil, and providing basis for other geological researches. Southern California Santa Barbara area made use of mature hyperspectral data processing technique to determine the scope of abnormal areas of vegetation caused by the oil and gas leakage. In 2002, Vander M et al. proposed comprehensive hyperspectral data and relevant geological and geochemical data based on a review of oil and gas micro-leakage method of remote sensing, and made use of relevant decisions to extract possible oil and gas micro-leakage information and carried out the actual verification. In 2003, Noomen M et al. studied the effects of oil and gas micro-seepage on vegetation spectrum through the analysis of laboratory experiments, field spectral measurement and hyperspectral image, with the aim to discover new oil and gas resources by recognizing unusual vegetation. In addition, hyperspectral remote sensing method has also been applied to the detection of oil content in oil sands for assistance in refining of oil in the oil sands, which has been adopted in Alberta of Canada. In 2007, Noomen M et al. extracted abnormal information of surface oil and gas leakage from the hyperspectral imaging through the study of changes of oil and gas leakage on the surface vegetation (wheat and corn) in the high spectral reflectance bands, which has a good effect for oil and gas pipeline monitoring and oil and gas resource exploration. In 2008, the United States Khan S et al. verified that mineral alteration in the rocks and soil was related to hydrocarbon micro-seepage of oil fields. They applied Hyperion sensor to obtain hyperspectral imaging of abnormal regions associated with hydrocarbon micro-seepage in the Patrick Draw region of Wyoming, interpreting the hydrocarbon micro-seepage areas by supervision and classification of images and carrying out verification by the application of the mineral, chemical and carbon isotope method with highly precise interpretation results. X-ray diffraction results show that feldspar component in abnormal area is reducing and contains high clay components.

In 1992, the Chinese Academy of Sciences Institute of Remote Sensing Applications made use of MAIS data to carry out oil and gas exploration and research in Kelpin area of Xinjiang Aksu, distinguishing the strata of the region from the Cambrian, Ordovician, Silurian, Devonian to Permian. Shanghai Institute of Technical Physics made use of MAIS data to carry out oil and gas resource exploration experiments in Shengli Oilfield in 1993 and in Boxing area of Shandong Guangrao in 1994. In 2007, Zhao Xinmei made systematic study and summary of the hydrocarbon micro-leakage phenomenon and the resulting surface alteration to seek remote sensing indication signs from micro-leakage surface soil and rock geochemical anomalies, surface soil absorption hydrocarbon anomalies, surface plant anomalies and geothermal anomalies and other aspects, made full use of the spectrum subdivision feature of the satellite hyperspectral imaging remote sensing data, identified alteration mineral assemblage information associated with hydrocarbon micro-seepage in the known oil fields as the remote sensing interpretation combination sign of gas zone detection, and further identified new oil and gas exploration prospects. In the same year, Xu Daqi et al. proposed macroscopic features of spectral curves of typical gas area measuring points, and gave a reflectance spectrum based on field measurements to determine the surface distribution (namely classification) method of specific alteration. This method was applied to the analysis of reflectance spectrum of filed measurement in an area of Qinghai, and the resulting alteration anomaly area was successively coincide with the existing gas fields in the region, successfully circling three large gas-bearing areas in the measurement area. In the same year, Shen Yuanting et al. analyzed alteration minerals under Qaidam region oil field geology and geographic environment, combined with the satellite hyperspectral remote sensing data Hyperion map to analyze the spectral characteristics of the known gas field area and the background area, determined the favorable wavelength range of oil and gas information identification, and made use of the spectral angle mapping (SAM) technology to extract the spatial distribution information of the oil field and taijnar gas construction prospective area, providing the effective technical way and means for hyperspectral remote sensing oil and gas exploration. In the same year, Tian Shufang et al. took Inner Mongolia Dongsheng as the research area, the oil and gas micro-seepage theory as the basis and the EO-1 satellite Hyperion data with the world's highest spectral resolution among satellite borne sensors as the information source, based on data preprocessing (spectral reconstruction, noise cancellation and band optimization), to make use of diagnostic absorption feature spectral bands of alteration mineral, combined with field measured spectral curve, using the band ratio to separate and extract uranium mineral and stratigraphic spatial distribution information, to determine the spatial location distribution of oil and gas micro-leakage, and to obtain four rich regions of oil and gas micro-leakage from the study area from the perspective of remote sensing, which provided a theoretical basis for oil and gas resource development in Dongsheng area. In 2008, Shen Yuanting et al. made use of Hyperion hyperspectral data to basically realize the classification of natural gas alteration anomaly area in some region in China, and according to geological data in the region, the abnormal display area was coincide with gas reservoir formation conditions of the region. In addition, Shen Yuanting et al. made use of wavelet PCA feature extraction method to effectively extract the surface weak natural gas alteration features of the region, and adopted the unsupervised/supervised classification hybrid training strategy to effectively distinguish disturbance land class, achieving the cluster of natural gas alteration anomaly area. Hyperspectral data are also used in offshore oil and gas source exploration. Subsea oil and gas leakage may form an oil film on the ocean surface, which can be detected with the remote sensing method (including hyperspectral remote sensing). In 2007, Wang Xiangcheng et al. established the spectral band selection and sea surface thin oil film and thick oil film detection mode based on the characteristics of EO-1 satellite hyperspectral remote sensing Hyperion data through Liaodong Bay offshore field spectral experiment and sample collection and analysis and several laboratory simulation experiments and analysis of the oil film spectrum. In 2010, Beijing Research Institute of Uranium Geology made use of high resolution aerial imaging spectrometer CASI/SASI to carry out exploratory study of oil and gas exploration in Qingyang district, and yielded good results, extracting the paradigm for the study of high resolution and hyperspectral imaging in oil and gas exploration. Using remote sensing technology to extract oil and gas micro-seepage information is a non-invasive technique, which has the advantages of economy, safety and efficiency and has great potential. Hyperspectral remote sensing technology has high spectral resolution and uninterrupted spectral coverage, providing a wealth of ground information, optimizing the mineral identification and extraction conditions, enhancing the ability of remote sensing to detect and identify land class and improving the quantitative level of remote sensing technology. The hyperspectral remote sensing technology has an important significance in being used for the extraction of oil and gas micro-seepage information.

2. **Other resources exploration (Pei and Fu** 2007)

In 1982, Goetz et al. applied five bands (the bandwidth of 100 nm) of the shortwave infrared radiometer (SMIRR) on the space shuttle to succeed in identifying kaolinite and carbonate ore in Egypt and succeed in the delineation of the iron oxide ore, clay mineral and alunite ore in Baja California of Mexico. In 1984 and 1985, Fred A of USGS made use of the imaging spectral data of three air strips to carry out the alteration mineral mapping experiment, extracting two types of alteration minerals. In 1994, the imaging spectrometer made by Shanghai Institute of Technical Physics was used to obtain the aerial imaging spectral data in Shandong Changwei region; combined with field ground spectral test data and chemical exploration data, after processing analysis and verification, it was found that the imaging spectrometer can be sensitive to collect hydrocarbons micro-leakage surface abnormal information, with more than 70 % of recombination rate with the known chemical exploration data in survey area. Wang Qinghua et al. made use of the modular aviation 71-band hyperspectral instrument (MAIS) developed by Chinese Academy of Sciences Shanghai Institute of Technical Physics to carry out the mineral identification research in Zhangjiakou region; Gan Fuping et al. took the Qinghai-Tibet Plateau as the experimental area to analyze the foreground of hyperspectral remote sensing technology in geological applications. In addition to the above typical geological applications, imaging spectrum technology has also many demonstration applications in a variety of mineral exploration such as gold, silver, copper, lead, zinc and uranium. The direct application of hyperspectrum onto the uranium ore prospecting has not been published in foreign journals yet. Beijing Research Institute of Uranium Geology applied the hyperspectral technology to identify and delineate the silicified zone associated with uranium ore in granite region of Guangxi Miaoer Mountain. In Dongsheng area of Ordos basin, the satellite hyperspectral (Hyperion) data were used to carry out the alteration mineral mapping, delineating the scope of the kaolinization containing ore deposit. In addition, the systematic uranium mining land class spectrum measurement study was carried out. Hyperspectral remote sensing technology is accompanied by the development of the imaging spectrometer, and with the development of hyperspectral satellite sensors, hyperspectral data will make a breakthrough in larger fields and more respects, with its spatial resolution and spectral resolution continuing to improve and the problem to be solved and the area covered continuing to expand.

9.6 Environment

Applications of hyperspectral remote sensing in environment include atmospheric pollution monitoring, soil erosion monitoring, water environment monitoring and so on. (Qingxi et al. 2006).

9.6.1 Atmospheric Pollution Monitoring

Atmospheric pollution can be reflected in the remote sensing data mainly by aerosol scattering and absorption generated by pollution, so the aerosol can be used as an indicator of atmospheric pollution. In 1994, Kaufman et al. used land—air simultaneous observation data of more than thirty observation stations distributed in the world to obtain empirical distribution of a variety of aerosol scattering phase functions, and gave the empirical formula of aerosol optical thickness and path scattering as well as downstream scattering and upward scattering, which has been successfully applied to aerosol space exploration and remote sensing data atmospheric correction. In 1990, Kaufman et al. also made use of dark background technique to calculate the large-scale distribution of atmospheric pollution, which has been successfully applied to estimate the regional large-scale atmospheric pollution. Domestically, in 2003, Deng Ruru et al. conducted air tests for scattering spectra of typical atmospheric pollution in Beijing. On this basis, they started with the analysis of physical mechanisms constructed by satellite remote sensing pixel electromagnetic wave information to build the synthetic models between the pixel surface composition and the atmosphere spectrum, and then adopted multi-band satellite remote sensing data with the pixel information decomposition method to quantitatively decompose atmospheric pollution information from the ground land class information for the first time, getting the high resolution anthropogenic aerosol turbidity and its spatial distribution results, and making the image map of atmospheric pollution cumulative concentration distribution field in the research area.

9.6.2 Soil Erosion Monitoring

Soil properties such as texture, organic matter, iron oxide content and nutrient are the key variables that affect vegetation growth and can be reflected by its specific spectral response. In 1998, Palacios O et al. showed the great advantages of hyperspectral remote sensing in the estimation of some soil properties. In 2001, Chang et al. evaluated the ability of using near-infrared reflectance spectra to predict different soil properties. The results showed that near-infrared reflectance spectra can be used to quickly estimate the soil properties with the acceptable prediction accuracy. In 2002, Liu Weidong carried out the relevant analysis of soil spectral reflectance and soil Munsell color properties based on a number of soil laboratory spectra, and established the inversion model of soil characteristic parameters and the surprised spatial distribution diagram of soil parameters.

9.6.3 Water Environment Monitoring

Using remote sensing to monitor water quality is also called ocean color remote sensing, which develops from ocean remote sensing. The water quality parameters, which could be inverted after using hyperspectral remote sensing data, include chlorophyll, suspended solids, yellow substance, transparency, turbidity etc. The inversion of water quality can be divided into 2 types, which are semi-empirical method on the basis of spectral analysis and analytical method on the basis of the bio-optical model. Semi-empirical method refers to using the statistical relation between remote sensing data or band combination and synchronous ground monitoring to establish water quality algorithm. It is the most commonly used water quality inversion algorithm. In 2002, Thiemann S and others used AVIRIS to measure multi-temporal hyperspectral data and synchronous monitoring ground data to invert the chlorophyll concentration of lake water. In 2000, Shu Xiaozhou and others used the imaging spectrometer which has been developed by our country to conduct surface water quality remote sensing experiment in the Taihu Lake region, estimated the chlorophyll concentration distribution in the survey region, and made an comparison of remote sensing estimated value and ground sampling data. The foundation of the analytical method based on the bio-optical model is using theoretical model of light field under the water to confirm the relation between the ratio of absorption coefficient and backscattering coefficient and surface albedo. After determining the relationship, the ratio of the actual absorption coefficient and backscattering coefficient could be calculated with the reflectance value measured by the remote sensing. Combining with the characteristic absorption coefficient and backscattering coefficient of the water components, the content of water component shall be obtained. In 1993, after measuring and studying the optical properties of 19 different types inland water body, Dekker A made use of analitical method based on bio-optical model to establish inverse algorithm of inverting chlorophyll and phycobilin concentration inland water body for the first time. In 1998, Hoogenboom H put forward the matrix inversion algorithm which could use the underwater irradiance obtained by the aerial hyperspectral remote sensing to invert the content of the water components in the inland water body.

9.7 Military Affairs

The application of hyperspectral remote sensing technology mainly includes the detailed investigation on the battlefield, recognition of camouflaged target, detection and calculating of the actual temperature of target and emissivity and so on (Chaoyang et al. 2008).

Target camouflage technology has already appeared in the modern hi-tech warfare. It is especially important for the military target detection to recognize the camouflage equipment and camouflage target to refer to the surface substance of the target recognized by the object spectrum database. It is increasingly urgent to reveal the technical features of hyperspectral imaging investigation and explore hyperspectral camouflage.

The most essential feature of hyperspectral data is the integration of traditional image dimension and spectrum dimension information, which could obtain the continuous spectrum information of each object unit when obtaining the surface spatial image. The outstanding character of hyperspectral imaging investigation technology is finding camouflage equipment made of artificial materials and camouflage materials from the natural background, revealing the military target in disguise, determining the nature of military target, and realizing the object component information inversion and object recognition on the basis of object spectrum features. Thus, it could play an important role in the military object recognition, camouflage object detection, camouflage effect detection, beachhead selection, battlefield environment assessment, battlefield poison gas detection, minefield detection and other aspects.

Hyperspectrum remote sensing use the characteristic spectrum formed by the interaction of object and electromagnetic wave to study the object on the surface of the earth; and the wavelength band mainly includes visible spectrum, near-infrared, thermal infrared band. In the visible spectrum and near-infrared band, ground objects mainly reflect the sun's energy. The reflectance characteristics spectrum could be analyzed to recognize the target. In the thermal infrared band, ground object radiation energy is mainly thermal radiation. The target emissivity and radiation temperature are the main recognition information. As new type investigation technology, hyperspectral remote sensing has its unique feature which is different from the traditional investigation technology. Firstly, it has the features of more working waveband s, narrow breadth and strong recognition capacity. The peak half width of spectral characteristics of the ground object is generally 20–40 nm, which could not be discernible for traditional multi-spectral and other remote sensing technology. But hyperspectral remote sensing could better recognize them. The past qualitative and semi-quantitative remote sensing has entered into the quantitative remote sensing era. Secondly, when the hyperspectrum is imaging, the spectral curve of the ground object could be obtained, which integrates the advantages of the imaging analysis and spectral analysis. Thirdly, due to the various wave bands, there are huge hyperspectral remote sensing data volume and superfluous data, which are difficult to handle. On account of the advantage of hyperspectral remote sensing on the ground object identification, it has already been applied in the military field and gradually replaced the multi-spectral remote sensing as the main investigation method.

1. Detailed investigation of the battlefield. The hyperspectral remote sensing instruments could detect the targets on the continuous working waveband at the same time, which could directly reflect the fine spectral characteristics of the object tested and identify the component and state of the surface of the object. Thus, the existing accurate congruent relationship between the space exploration information and the actual ground target could be obtained. Israeli scientists used CASI hyperspectral imaging spectrometer to conduct a research in Tel Aviv. The typical ground objects were selected from the CASI images as the metadata, the identification effects on the rivers, sandy soils, road surface and plants are well. Hyperspectral imager designed by the Navy of the United States could provide spectral data of 210 wave band within the spectrum scope of 0.4–2.5 μm, and obtain the dynamic performance of offshore environment targets, such as transparency of seawater, ocean depth, ocean current, seabed features, underwater hazardous substance, oil leak and other imaging data, which could provide reference for the inshore battle of American Navy. The United States puts forward the digital earth observation, whose target is to establish the database on each square meters of the earth including many parameters such as targets and background features. After establishing the database, the military targets anywhere in the world shall be under monitoring and hit.
2. Camouflage target identification. In the aspect of military target investigation and camouflage identification, the hyperspectral remote sensing could find out military equipment according to the different spectrum characters of background and camouflage targets. The constituents of the target could be inverted through the spectrum characteristic curve to reveal the target and its disguise which is different from the background environment. One of the important green camouflage materials detection methods is using the red edge effect of the vegetation. Within the scope of 680–720 nm, the reflectivity of vegetation rises sharply. The species and state of the vegetation could be identified through testing its characteristics of location and slope. The current spectrum curve of green camouflage materials could match up with the vegetation. Thus, under the condition of the multi-spectrual investigation, it could satisfy the camouflage requirements. However, under the tiny resolution capability of hyperspectrum, the camouflage target could be found out. Regarding the red edge of vegetation as the basic identification characteristic by Liu Kailong, the identification accuracy could reach more than 99 %.

3. Detection and calculation of the actual temperature and emissivity of the targets. In the current thermal infrared detection, use Planck law to combine emissivity and temperature as one parameter. It is called assumed temperature or radiation temperature in radiance thermometry. Assumed temperature is the coupling temperature of actual temperature and spectral emissivity, which could not reflect the actual temperature of the object tested. The thermal infrared camouflage of military targets mainly uses low emissivity defilade to reduce the radiation energy, which could make the target temperature close to the coupling temperature. Thus, it is hard for thermal infrared detectors to find out and identify the camouflage targets. However, if adopting hyperspectrum to detect, use linear hypothesis structure equation in the thermal infrared wave band, the actual temperature and emissivity on the surface of the target could be calculated. Hyperspectrum has broken the limitation of assumed temperature measurement, which makes the temperature measurement solution much more approximate to the actual temperature on the surface of the object. Camouflage target and background could be better identified. From the development of the hyperspectral remote sensing instrument, different kinds of AVIRIS have developed well. The foreign satellite-borne instruments have been put into use. Hyperspectrum has segmented spectrum characteristics of the target tested on the basis of multi-spectrum. It could identify different artificial targets and camouflage targets. Using hyperspectrum to detect the actual temperature and emissivity on the surface of the object tested in the thermal infrared wave band could improve the disadvantages of thermal imaging detection. It could better identify the camouflage targets which use traditional thermal infrared camouflage technology.

References

Lou QS, Chen L et al (2008) The application and prospecting of the hyperspectral remote sensing technology in marine research. Trans Oceanol Limnol 3:168–173

Pei CK, Fu J (2007) Status and prospect of hyperspectral remote sensing technique in rock and mineral identification. World Nucl Geosci 24(1):32–38

Tan BX, Li ZY et al (2008) Research advance in forest information extraction from hyperspectral remote sensing Data. For Res 21(supplement):105–111

Tan BX, Li ZY, Chen E et al (2006) Estimating forest crown closure using hyperion hyperspectral data. J Beijing For Univ 28(3):95–101

Teng AG, Gao F et al (2009) Progress of hyperspectral technology application research in agriculture. Jiangsu Agric Sci 3:8–11

Tong QX, Zhang B, Lanfen Z (2006) Hyperspectral remote sensing: principle, technology and application. Higher Education Press, Beijing

Zhang CY, Cheng HF et al (2008) The development of hyperspectral remote sensing and its threatening to military equipments. Electro-optic Technol Appl 23(1):10–12

Zhou L, Xin XP et al (2009) Application of hyperspectral remote sensing in the monitoring of the grassland. Sci Grass Ind 26(4):20–27

Appendix

Related Articles and Patents of the Authors

1. Liguo Wang, Siyuan Hao, Qunming Wang, Ying Wang. Semi-supervised Classification for Hyperspectral Imagery Based On Spatial-Spectral Label Propagation. ISPRS Journal of Photogrammetry and Remote Sensing. 2014, 97 (1): 123–137.
2. Liguo Wang, Siyuan Hao, Qunming Wang. Spatial-spectral Information-Based Semi-supervised Classification Algorithm For Hyperspectral Imagery. IEEE Journal of Volume: 7, Issue: 8. DOI: 10.1109/JSTARS.2014.2333233.
3. Liguo Wang, Siyuan Hao, Qunming Wang. Composite Kernel for Hyperspectral Imagery Classification. Remote Sensing Letters. To be published.
4. Qunming Wang, Wenzhong Shi, and Liguo Wang. Allocating classes for soft-then-hard sub-pixel mapping algorithms in units of class. IEEE Transactions on Geoscience and Remote Sensing. 2014, 52(5): 2940–2959.
5. Wang Qunming, Shi Wenzhong, Wang Liguo. Indicator Cokriging-Based Sub pixel Land Cover Mapping with Shifted Images. IEEE Journal of Selected Topics in Applied Earth Observation and Remote Sensing. 2014, 7(1): 327–339.
6. J.H. Yang, Liguo Wang, J.X. Qian. Hyperspectral Imagery Classification Based on Spatial-Spectral Features and Sparse Representation. Applied Geophysics. 2014, 11(4): 489–499.
7. Jinghui Yang, Liguo Wang, Jinxi Qian. Research on the Optimal Classification Method for Remote Sensing Image based on the Gabor-PCA Analysis. WCCAIT 2014, Advanced Materials Research. 2014, 989–994: 3617–3620.
8. Liguo Wang, Danfeng Liu, Qunming Wang. Geometric Method of Fully Constrained Least Squares Linear Spectral Mixture Analysis. IEEE Transactions on Geoscience and Remote Sensing. 2013, 51(6): 3558–3566.
9. Liguo Wang, Qunming Wang. Spectral Unmixing Model based on Least Squares Support Vector Machine with Unmixing Residue Constraints. IEEE Geoscience and Remote Sensing Letters. 2013, 10(6): 1592–1596.

L. Wang and C. Zhao, *Hyperspectral Image Processing*,
DOI 10.1007/978-3-662-47456-3

10. Liguo Wang, Qunming Wang. Sub-pixel Mapping Using Markov Random Field with Multiple Spectral Constraints from Sub-pixel Shifted Remote Sensing Images. IEEE Geoscience and Remote Sensing Letters. 2013, 5(3), 598–602.
11. Liguo Wang, Fangjie Wei, Danfeng Liu, Qunming Wang. Fast Implementation of Maximum Simplex Volume-based Endmember Selection in Original Hyperspectral Data Space. IEEE Journal of Selected Topics in Applied Earth Observations and Remote Sensing. 2013, 6(2): 516–521.
12. Liguo Wang, Danfeng Liu, Liang Zhao. A Color Visualization Method based on Sparse Representation of Hyperspectral Imagery. Applied Geophysics. 2013, 10(2): 210–221.
13. Liguo Wang, Qunming Wang, Danfeng Liu. Geometric estimation method of spectral unmixing. Journal of Infrared and Millimeter Waves. 2013, 32(1), 56–61.
14. Qunming Wang, Liguo Wang. Integration of spatial attractions between and within pixels for sub-pixel mapping. Journal of Systems Engineering and Electronics. 2012, 23(2): 293–303.
15. Qunming Wang, Liguo Wang. Particle Swarm Optimization-based Sub-pixel Mapping for Remote Sensing Imagery. International Journal of Remote Sensing. 2012, 33(20): 6480–6496.
16. Liguo Wang, Luqun Deng, Jing Zhang. Endmember Selection Algorithm based on Linear Least Square Support Vector Machines. Spectroscopy and Spectral Analysis. 2010, 30(3), 748–747.
17. Liguo Wang, Yan Zhao. MAP-based Super-Resolution Method for Hyperspectral Imagery. Spectroscopy and Spectral Analysis. 2010, 30(4), 1044–1048.
18. Liguo Wang, Jing Zhang, Danfeng Liu. Distance Measurement-based Methods from Endmember Selection to Spectral Unmixing. Journal of Infrared and Millimeter Waves. 2010(6): 471–475.
19. Liguo Wang and Xiuping Jia. Integration of Soft and Hard Classification using Extended Support Vector Machines. IEEE transaction on Geoscience and Remote Sensing Letters. 2009, 6(3), 548–547.
20. Liguo Wang, Ye Zhang, Junping Zhang. A new weighted least squares support vector machines and its sequential minimal optimization algorithm. Chinese Journal of Electronics. 2008, 17(2) 285–288.
21. Liguo Wang, Chunhui Zhao, Wanhai Chen, Yulong Qiao. Research on All-around Weighting Methods of Hyperspectral Imagery Classification. Journal of Infrared and Millimeter Waves. 2008, 27(6), 442–446.
22. Liguo Wang, Xiuping Jia, Ye Zhang. A novel geometry-based feature-selection technique for hyperspectral imagery. IEEE Geoscience and Remote Sensing Letters. 2007, 4(1): 171–175.
23. Liguo Wang, Y. Zhang, C.H. Zhao. Combination of Linear Support Vector Machines and Linear Spectral Mixed Model for Spectral Unmixing. LNCIS. 2006 (0345), 767–772.
24. Liguo Wang, Ye Zhang, Jiao Li. BP Neural Network-based Sub-Pixel Mapping Method. LNCIS. 2006 (0345): 755–760.

25. Liguo Wang, Xiuping Jia, Ye Zhang. Construction of Fast and Robust N-FINDR Algorithm. LNCIS. 2006 (0345): 791–796.
26. Liguo Wang, Fangjie Wei. Band selection for hyperspectral imagery based on combination of genetic algorithm and ant colony algorithm. Journal of Image and Graphics. 2013, 18(2): 235–242.
27. Liguo Wang, Fangjie Wei. Artificial physics optimization algorithm combined band selection for hyperspectral imagery. Journal of Harbin Institute of Technology. 2013, 45(9): 100–106.
28. Danfeng Liu, Liguo Wang. Color display of hyperspectral data in three levels. Infrared and Laser Engineering. 2012, 41(9): 2527–2533.
29. Liguo Wang, Danfeng Liu. Exploring Support Vector Machine in Spectral Unmixing. Workshop on Hyperspctral Image and Signal Processing: Evolution in Remote Sensing (WHISPERS) 2012. 2012: 1–4.
30. Danfeng Liu, Wang, Liguo, Visual Attention based Hyperspectral Imagery Visualization. IEEE Conf. 2012 Symposium on Photonics and Optoelectronics (SOPO). 2012: 1–4.
31. Xiaofeng Lia, Liguo Wang and Xiuping Jia. Spectral Unmixing based on Improved extended Support Vector Machines. IEEE International Geoscience and Remote Sensing Symposium (IGARSS). 2012: 4118–4121.
32. Qunming Wang, Liguo Wang, Danfeng Liu, Wang Zhengyan. A novel Super-resolution mapping method for hyperspectral imagery. Journal of Harbin Institute of Technology. 2012, 44(7): 92–96.
33. Qunming Wang, Liguo Wang, Danfeng Liu, Wang Zhengyan. Sub-pixel Mapping for Land Class with Linear Features Using Least Square Support Vector Machine. Infrared and Laser Engineering. 2012, 41(6): 1669–1675.
34. Wang, L. & Jia, X., 2011. Fuzzy Accuracy Assessment of Subpixel Analysis of Multi/Hyperspectral Image Data, International CiSE 2011: Conference on Computational Intelligence and Software. 2011: 9–11.
35. Liguo Wang, Qunming Wan. Sub-pixel mapping based on sub-pixel to sub-pixel spatial attraction model. IEEE International Geoscience and Remote Sensing Symposium (IGARSS), 2011: 593–596.
36. Liguo Wang, Jing Zhang, Chengyuan Liu, Chaozhu Zhang. Construction and solution of an new spectral unmixing model. Journal of Optoelectronics Laser. 2011, 22(11): 1731–1734.
37. Liguo Wang, Jing Zhang. An improved spectral unmixing modeling based on linear spectral mixing modeling. Journal of Optoelectronics Laser. 2010, 21(8), 1222–1226.
38. Jiao Li, Liguo wang, Ye Zhang, and Yanfeng Gu. Sub-Pixel Mapping Method based on BP Neural Network. Journal of Harbin Institute of Technology (New Series). 2009, 16(2): 99–103.
39. Liguo Wang, Jing Zhang, and Luqun Deng. Spectral Unmixing Technique based on Flexibly Selected End members. 2009 World Congress on Computer Science and Information Engineering (CSIE). 2009: 148–151.

40. Liguo Wang, Deng Luqun, and Lei Ming. Hyperspectral Imagery Classification Aiming at Protecting Classes of Interest. 2009 World Congress on Computer Science and Information Engineering (CSIE). 2009: 144–147.
41. Chandrama Dey, Xiuping Jia, D. Fraser, L. Wang. Mixed Pixel Analysis for Flood Mapping Using Extended Support Vector Machine Proceedings of the 2009 Digital Image Computing: Techniques and Applications (DICTA). 2009: 291–295.
42. Liguo Wang, Ye Zhang. Speed-up for N-FINDR Algorithm. Journal of Harbin Institute of Technology(New Series), 2008. 15(1) 141–144.
43. Liguo Wang, Ye Zhang, and Hao Chen. Spectral unmixing based on robust support vector machine. Journal of Jilin University Engineering and Technology Edition. 2007, 37(1) 155–159.
44. Liguo Wang, Chunhui Zhao, and Xiaojun Bi. Application of endmember extraction method to band selection. Journal of Jilin University Engineering and Technology Edition. 2007, 37(4) 915–919.
45. Liguo Wang, Chunhui Zhao, and Ye Zhang. Subpixel Mapping of Raw Hyperspectral Imagery. DCDIS-B. 2007, 14(S2): 1770–1773.
46. Liguo Wang, Lei Yuan. A differential optical flow algorithm based on second-order gradient constraint equation. DCDIS-B. 2007, 14(S2): 1774–1778.
47. Liguo Wang, Chunhui Zhao, and Ye Zhang. Double Weighted Least Square Support Vector Machines. DCDIS-B. 2007, 14(S2): 1765–1769.
48. Liguo Wang, Ye Zhang, Chunhui Zhao. Base Vector Selection Method based on Iterative Weighted Eigenvector Fitting. Geoinformatics 2006, Proc. SPIE 6420, 64201N.
49. Liguo Wang, Ye Zhang, Yanfeng Gu. Unsupervised band selection method based on improved N-FINDR algorithm for spectral unmixing. ISSCAA 2006, 1018–1021.
50. Liguo Wang, Yanfeng Gu, Ye Zhang. Band selection method based on combination of support vector machines and subspatial partition. Systems Engineering and Electronics. 2005, 27(6):974–977.
51. Liguo Wang, Yanfeng Gu, Ye Zhang. Image interpolation based on adaptive edge-preserving algorithm. Journal of Harbin Institute of Technology. 2005, 37 (1): 18–21.
52. Liguo Wang, Ye Zhang, Yanfeng Gu. The Research of Simplification of Structure of Multi-class Classifier of Support Vector Machine. Journal of Image and Graphics. 2005, 10(5):571–574
53. Chunhui Zhao, Jie Li, Feng Mei. A kernel weighted RX algorithm for anomaly detection in hyperspectral imagery. Journal of Infrared and Millimeter Waves. 2010, 29(5), 372–377.
54. Chunhui Zhao, Yaxin Ji. Fusion of hyperspectral images based on second generation curvelet transform and pulse-coupled neural networks. Journal of Harbin Engineering University. 2008, 29(7): 729–734.
55. Chunhui Zhao, Zhiqiu Zhu. Fusion classification of hyperspectral remote sensing images by dyadic ridgelet transform. Journal of Harbin Engineering University. 2008, 29(11): 1222–1226.

56. Chunhui Zhao, Yulei Wang, Feng Mei. Kernel ICA Feature Extraction for Anomaly Detection in Hyperspectral Imagery. Chinese Journal of Electronics. 2012, 21(2): 265–269.
57. Chunhui Zhao, Wanhai Chen, Lingyan Zhang. A compression algorithm of hyperspectral remote sensing image based on vector quantization. Journal of Harbin Engineering University. 2006, 27(3): 447–452.
58. Chunhui Zhao, Wanhai Chen, Lingyan Zhang. A compression algorithm of hyperspectral remote sensing image based on lifting scheme. Journal of Harbin Engineering University. 2006, 27(4): 588–592.
59. Baozhi Cheng, Chunhui Zhao, Yulei Wang. Anomaly detection of hyperspectral image for band subsets based on fourth order cumulant. Journal of Optoelectronics Laser. 2012, 23(8): 1582–1588.
60. Chunhui Zhao, Chunhong Liu, Kecheng Wang. Research on Fusion of Hyperspectral Remote Sensing Image based on Second Generation Wavelet. Acta Optica Sinica. 2005, 25(7): 891–896.
61. Chunhui Zhao, Chunhong Liu. Research and Analysis of Hyperspectral Remote Sensing Image Dimensional Reduction. Chinese Space Science and Technology. 2004, 24(5): 28–36.
62. Chunhui Zhao, Bin Qi, Yi Zhang. Hyperspectral image classification based on variational relevance vector machine. Acta Optica Sinica. 2012, 32(8): 0828004-1-0828004-6.
63. Jie Li, Chunhui Zhao, Feng Mei. Detecting hyperspectral anomaly by using background residual error data. Journal of Infrared and Millimeter Waves. 2010, 29(2), 150–155.
64. Feng Mei, Chunhui Zhao. Spatial filter-based anomaly detection algorithm for hyperspectral imagery kernel RX detectors. Journal of Harbin Engineering University. 2009, 30(6), 697–702.
65. Bin Qi, Chunhui Zhao, Eunseog Youn, and Christian Nansen. Use of weighting algorithms to improve traditional support vector machine-based classifications of reflectance data. Optics Express. 2011, 19(27): 26816–26826.
66. Chunhui Zhao, Yi Zhang, Yulei Wang. Relevant Vector Machine Classification of Hyperspectral Image based on Wavelet Kernel Principal Component Analysis. Journal of Electronics & Information Technology, 2012, 34(8): 1905–1910.
67. Chunhui Zhao, Baozhi Cheng, Weichao Yang. Algorithm for hyperspectral unmixing using constrained nonnegative matrix factorization. Journal of Harbin Engineering University. 2012, 33(3): 377–382.
68. Mei Feng, Zhao Chunhui, Liguo Wang, Huo, Hanjun. Anomaly Detection in Hyperspectral Imagery based on Kernel ICA Feature Extraction. 2nd International Symposium on Intelligent Information Technology Application (IITA). 2008, 869–873.
69. Mei Feng, Zhao Chunhui, Huo Hanjun; Sun Yan. An adaptive kernel method for anomaly detection in hyperspectral imagery. 2nd International Symposium on Intelligent Information Technology Application (IITA). 2008, 874–878.